Subcellular Biochemistry

Volume 70

Series editor

Robin Harris, Northumberland, UK

For further volumes:
http://www.springer.com/series/6515

Mary Ann Asson-Batres · Cécile Rochette-Egly
Editors

The Biochemistry of Retinoic Acid Receptors I: Structure, Activation, and Function at the Molecular Level

Springer

Editors
Mary Ann Asson-Batres
Department of Biological Sciences
Tennessee State University
Nashville, TN
USA

Cécile Rochette-Egly
IGBMC
Illkirch Cedex
France

ISSN 0306-0225
ISBN 978-94-024-0340-4 ISBN 978-94-017-9050-5 (eBook)
DOI 10.1007/978-94-017-9050-5
Springer Dordrecht Heidelberg New York London

Printed on acid-free paper

Springer is part of Springer Science+Business Media (www.springer.com)

Foreword

The nutritional importance of liposoluble compounds, such as vitamin A, for good health has been known for a long time, but vitamin A itself, was only discovered 100 years ago. Since that time, vitamin A has been the focus of many scientific investigations aimed at understanding its mechanism of action. A number of epidemiologic studies performed with vitamin A-deficient populations of the Third World and with vitamin A-deficient animals have indicated that vitamin A is essential for vision, reproduction, growth, and development. The overall conclusion was that vitamin A maintains good health from birth to death by controlling the development and well-being of all tissues in a time and concentration dependent manner.

Early research in the field focused on vision, and in the 1940s, it was established that a metabolite of vitamin A, *11-cis*-retinaldehyde, is the visual chromophore. However, the mechanism of action of vitamin A in the regulation of cell proliferation and differentiation throughout life was not uncovered until 1987, when nuclear receptors for retinoic acid, another active derivative of vitamin A, were cloned. Since that time, there has been an explosion of new techniques and concepts that have revealed how retinoic acid and its receptors regulate gene expression at the molecular level and impact development and homeostasis. Some of these new findings have been summarized in recent specialized reviews, but it has been over 20 years since a comprehensive account of vitamin A function and metabolism has been published (*The Retinoids Biology, Chemistry and Medicine* edited by Sporn, Roberts and Goodman, Raven Press, New York 1993; *Vitamin A in Health and Disease*, edited by Rune Blomhoff, Dekker, New York, 1994).

In view of the recent explosion in this field, it is timely to publish a contemporary, comprehensive, book series recapitulating the most exciting developments in the field and covering fundamental research in molecular mechanisms of vitamin A action, its role in physiology, development, and continued well-being, and the potential of vitamin A derivatives and synthetic mimetics to serve as therapeutic treatments for cancers and other debilitating human diseases.

Here, we present the first volume of a multivolume series on Retinoic Acid Signaling that will cover all aspects of this broad and diverse field. One aim of Volume I is to present a compilation of topics related to the biochemistry of nuclear retinoic acid receptors, from their architecture when bound to DNA and associated with their coregulators to their ability to regulate target gene transcription. A second aim is to provide insight into recent advances that have been made in identifying novel targets and nongenomic effects of retinoic acid.

Volume I is divided into ten chapters contributed by prominent experts in their respective fields. Each chapter starts with the history of the area of research. Then, the key findings that contributed to development of the field are described, followed by a detailed look at key findings and progress that is being made in current, ongoing research. Each chapter is concluded with a discussion of the relevance of the research and a perspective on missing pieces and lingering gaps that the authors recommend will be important in defining future directions in vitamin A research.

The volume begins with a retrospective of the vitamin A story and of the cloning of the nuclear retinoic acid receptors. Then, it is organized into three broad areas. The first area focuses on fundamental research covering the architecture of DNA bound RARs, the structural basis for coregulator interaction and exchange, the evolution of the receptors, and the role of the RXR heterodimerization partner. The second area addresses the complexity of the RAR-mediated transcriptional regulatory programs, focusing on the epigenetic changes at the gene promoters and on recent integrative genomics. The third area presents new mechanisms of action of RARs, including nongenomic effects, novel targets, and microRNAs.

We thank all of the authors for their efforts in preparing this volume. They comprehensively reviewed the literature and provided stimulating ideas that will serve to guide continued development of the retinoid field. They pointed out many questions that remain unsolved and noted that these answers will require new state-of-the-art techniques.

We also thank and acknowledge Meran Owen for his invitation to put together this "Retinoic Acid Signaling" book series, his assistant, Tanja van Gaans for her constant help, and Springer Publishing for its support of this project.

It is our hope that this book will serve as an illuminating introduction to the fascinating field of vitamin A biology for those who are not familiar with the amazing molecular intricacies of retinoic acid signaling, as well as a frequent reference for the current and next generations of scientists working in the field of retinoids and nuclear receptors.

Mary Ann Asson-Batres
Cécile Rochette-Egly

Contents

Chapter 1
History of Retinoic Acid Receptors

Doris M. Benbrook, Pierre Chambon, Cécile Rochette-Egly and Mary Ann Asson-Batres

Abstract The discovery of retinoic acid receptors arose from research into how vitamins are essential for life. Early studies indicated that Vitamin A was metabolized into an active factor, retinoic acid (RA), which regulates RNA and protein expression in cells. Each step forward in our understanding of retinoic acid in human health was accomplished by the development and application of new technologies. Development cDNA cloning techniques and discovery of nuclear receptors for steroid hormones provided the basis for identification of two classes of retinoic acid receptors, RARs and RXRs, each of which has three isoforms, α, β and ɣ. DNA manipulation and crystallographic studies revealed that the receptors contain discrete functional domains responsible for binding to DNA, ligands and cofactors. Ligand binding was shown to induce conformational changes in the receptors that cause release of corepressors and recruitment of coactivators to create functional complexes that are bound to consensus promoter DNA sequences called retinoic acid response elements (RAREs) and that cause

D. M. Benbrook (✉)
Department of Obstetrics and Gynecology, University of Oklahoma Health Sciences Center, 975 NE 10th St BRC1374, Oklahoma City 73165, USA
e-mail: doris-benbrook@ouhsc.edu

D. M. Benbrook
Department of Pharmaceutical Sciences, University of Oklahoma Health Sciences Center, 975 NE 10th St BRC1374, Oklahoma City 73165, USA

P. Chambon · C. Rochette-Egly
IGBMC (Institut de Génétique et de Biologie Moléculaire et Cellulaire), INSERM, U964, CNRS, UMR7104, Université de Strasbourg, 1 rue Laurent Fries, BP 10142, 67404 Illkirch Cedex, France
e-mail: chambon@igbmc.fr

C. Rochette-Egly
e-mail: cegly@igbmc.fr

M. A. Asson-Batres
Department of Biological Sciences, Tennessee State University, 3500 John A Merritt Blvd, Nashville 37209, USA
e-mail: massonbatres@tnstate.edu

M. A. Asson-Batres and C. Rochette-Egly (eds.), *The Biochemistry of Retinoic Acid Receptors I: Structure, Activation, and Function at the Molecular Level*, Subcellular Biochemistry 70, DOI: 10.1007/978-94-017-9050-5_1, © Springer Science+Business Media Dordrecht 2014

opening of chromatin and transcription of adjacent genes. Homologous recombination technology allowed the development of mice lacking expression of retinoic acid receptors, individually or in various combinations, which demonstrated that the receptors exhibit vital, but redundant, functions in fetal development and in vision, reproduction, and other functions required for maintenance of adult life. More recent advancements in sequencing and proteomic technologies reveal the complexity of retinoic acid receptor involvement in cellular function through regulation of gene expression and kinase activity. Future directions will require systems biology approaches to decipher how these integrated networks affect human stem cells, health, and disease.

Abbreviations

ChIP	Chromatin immunoprecipitation
ChIP-seq	Chromatin immunoprecipitation coupled with deep sequencing
cDNA	Complementary DNA
CRABP	Cellular retinoic acid binding protein
CRBP	Cellular retinol binding protein
DNA	Deoxyribonucleic acid
DBD	DNA binding domain
LBD	Ligand binding domain
NMR	Nuclear magnetic resonance
RA	Retinoic acid
RAR	Retinoic acid receptor
RARE	Retinoic acid response element
RNA	Ribonucleic acid
RNA-seq	High throughput RNA sequencing
RXR	Retinoic X receptor
VAD	Vitamin A deficiency

Introduction: In Quest of a Mechanism of Action for Vitamin A

The idea that essential factors other than proteins, fat, starch, sugar, or minerals are present in food was a novel concept before the late 1880s. Ultimately this notion was verified by a series of human dietary supplementation studies and controlled experiments in animal models conducted between 1880 and 1920, which demonstrated that removal of these factors from the diet caused debilitating illnesses and death. The first discoveries in the field were made by Christiaan Eijkman and Frederick Gowland Hopkins who found that rice polishings contain substances preventing beriberi. These investigators received the Nobel prize for their work in 1929 [17]. In 1912, Casimer Funk identified the active fraction, which was named water-soluble factor b, later described as thiamine [16]. Since this substance, which was vital for life belonged to a class of organic compounds called amines, he named it "vitamine" (vital amines) [98].

During this time, two groups, one led by Elmer McCollum and the other by Thomas Osborne and Lafayette Mendel, independently provided evidence for another essential substance that was named fat-soluble factor a [113]. By 1920, several other low abundance dietary factors were also being described, and Jack Drummond argued to the American Chemical Society that since there was no evidence for the presence of an amine in all of these "vitamines", it would be easiest for classification purposes to drop the *-e* on Funk's general reference to vital factors and refer to them as Vitamin A, Vitamin B, Vitamin C, etc. [29].

The structure of vitamin A, also known as retinol, was first reported by Paul Karrer and his collaborators in 1931 [55, 111], confirmed by the group of Heilbron the following year [44], and crystallized in 1937 [47]. The molecule is composed of 20 carbon atoms arranged as a beta-ionone ring with a conjugated isoprene tail that terminates with an alcohol functional group (Fig. 1.1).

Ongoing studies revealed that vitamin A serves as a precursor for active derivatives that impart two very different physiologic effects: (1) an aldehyde derivative (11-*cis* retinal), which is the active chromophore of vision [2, 7, 48, 128], and (2): an acid derivative (all-*trans* retinoic acid), which has the ability to reverse developmental defects in vitamin A deficient (VAD) animals [4, 5, 123] (Fig. 1.1). Further experiments by Arens and van Dorp suggested that retinoic acid (RA) could not be converted into vitamin A in vivo and thus, they concluded that RA was itself a hormone involved in cell growth and in development. Since that time, several other active vitamin A metabolites have been identified, and active compounds have been synthesized. In 1976, all of them were grouped as *retinoids* [117].

Clues as to how retinol and retinoic acid work inside cells came from studies performed with other fat soluble (lipophilic) hormones such as estrogens and glucocorticoids. In the 1960s, advancements in methods to synthesize radiolabelled hormones with sufficient specific activity for in vivo use and in techniques to count tritium in animal tissues [51] allowed the identification of binding proteins for these hormones [52], and also suggested that there was a link between their physiological action and transcription in the nucleus. This was the first indication that lipophilic hormones regulate gene transcription through nuclear receptors functioning as transcription factors.

More support for this concept came from Pierre Chambon's finding that administration of estradiol to immature chickens elicited an increase in liver *aggregate* polymerase that preceded an induction of protein synthesis [19, 130]. Subsequent studies found that estradiol induces translocation of an estrogen binding protein from the cytoplasm to the nucleus [53] and that steroid hormones induce transcription of specific subsets of genes [82, 96]. The generality of this phenomena in the animal kingdom was shown by studies in flies demonstrating that insect ecdysteroids induce alterations in chromatin dynamics, observed as *puffing* of chromosomes [6]. The final pieces of the puzzle came together in the 1980s with the discovery that glucocorticoid receptor proteins could be proteolytically cleaved into independently-functional, ligand-binding (LBD) and DNA-binding (DBD)

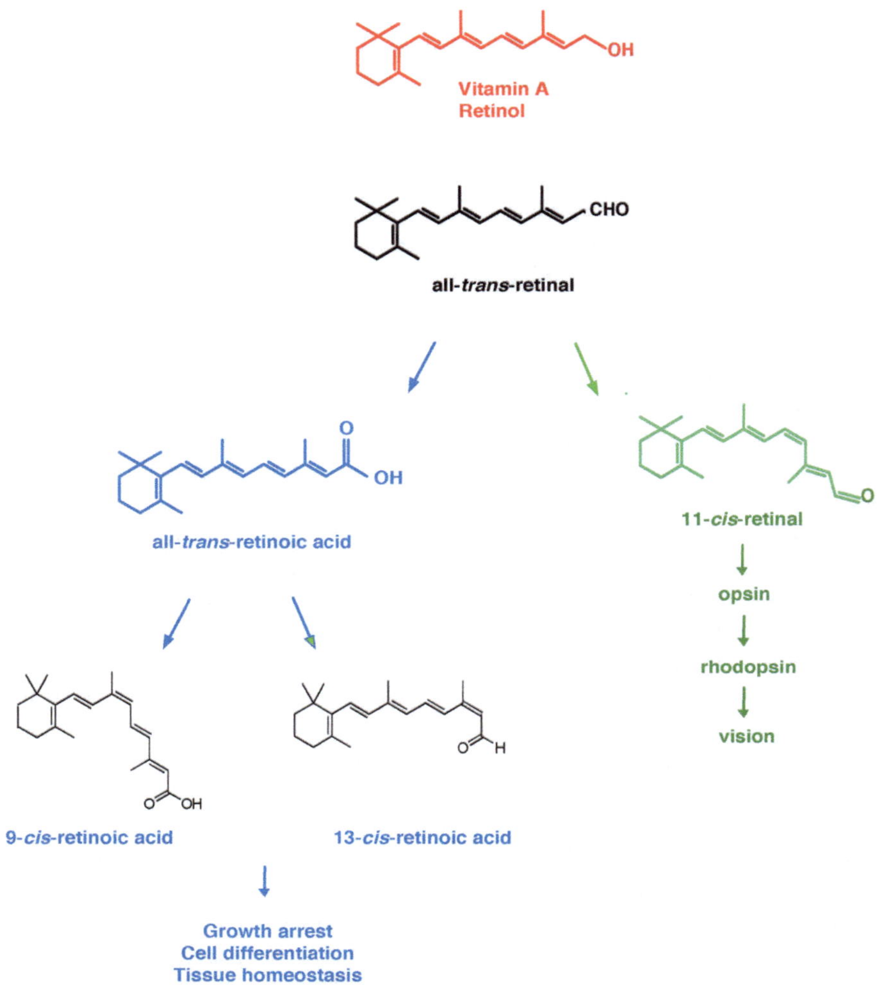

Fig. 1.1 Retinol and its main metabolites

domains [132], and that the DBD could bind to specific DNA sequence elements conferring glucocorticoid regulation of adjacent genes [20, 109].

At the same time, experimental evidence was accumulating that vitamin A and its derivatives also influence RNA and protein synthesis [24, 54, 134] and can bind intracellular proteins [8, 90]. In 1978, the group of Frank Chytil purified and characterized cytosolic proteins that bound retinol (CRBPI and CRBPII) or retinoic acid (CRABPI and CRABPII) [23, 91, 92]. However, subsequent studies performed by the same group revealed that CRBPs and CRABPs are present essentially in the cytosol and that these proteins merely serve as vehicles or *shuttles* transferring the ligand into the nucleus to specific binding sites on the chromatin [70, 71, 120, 121]. Also at that time, there appeared to be clear evidence that other retinoid receptors were present in the nucleus that could modulate the transcription of specific genes [43].

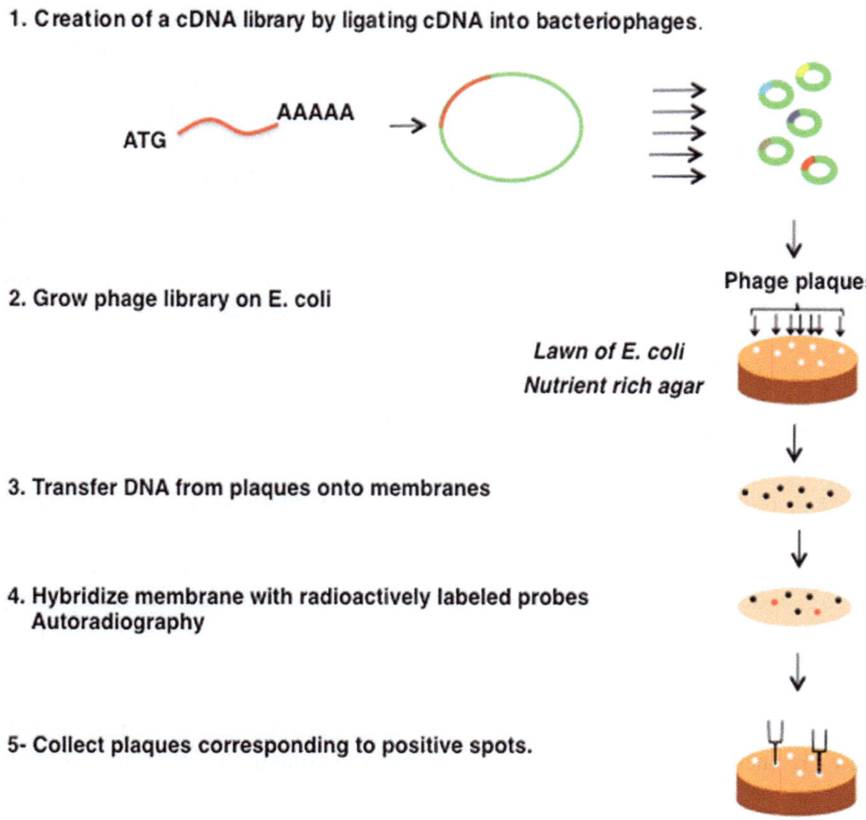

Fig. 1.2 Gene cloning using phage cDNA libraries

History: Cloning of the Nuclear Retinoic Acid Receptors

In the 1970s, progress in genetic technologies increased rapidly with promising possibilities for gene mapping. New procedures for DNA hybridization [27] and gene transfer [131] made possible gene localization to specific regions of chromosomes and gene cloning into DNA plasmids. A huge advance in molecular genetic studies was also provided by the purification and characterization of a microbial reverse transcriptase enzyme that could be engineered to generate DNA complementary (cDNA) to RNA in animals [116, 122]. This ability to generate DNA sequences representing processed RNA circumvented the problems of non-coding introns and exon splicing, and allowed mRNA isolated from cells to be copied into cDNA sequences that could then be individually cloned into plasmids [112] or bacteriophages to create libraries of all the DNA translation products expressed in a particular cell type.

In 1985, these technologies were used by the group of Pierre Chambon to clone the human estrogen [129] and glucocorticoid receptors [41] (Fig. 1.2). Phage libraries were infected into bacteria growing on a petri dish at densities that allowed individual phage to be separately identified by the clear areas they produced in the bacterial lawn, called plaques. The cDNA in the plaques was transferred to a nitrocellulose membrane, which was then screened with a radioactively labeled query probe (for example, a nucleotide sequence with homology to putative hormone receptors). When exposed to X-RAY film, plaques hybridizing with the radioactive probe produced a spot that allowed their identification and isolation. Then, the isolated cDNA's could be used to produce large quantities of protein and could be genetically manipulated to mutate or excise coding regions for specific parts of the proteins, which allowed precise identification of functional areas.

Cloning of RARs

The first retinoic acid receptor, RARα, was independently identified in 1987 by the laboratories of Pierre Chambon and Ronald Evans. Chambon's group cloned RARα by using a consensus oligonucleotide probe corresponding to a highly conserved sequence in the DBD of several members of the nuclear receptor family (human and chicken estrogen receptor, human and rat glucocorticoid receptor, human progesterone receptor and viral oncogene erbA) [97]. This probe was radiolabeled and hybridized with a phage cDNA library created from the human breast cancer cell lines MCF-7 and T47D. Several phage plaques giving positive signals were obtained and rescreened with probes corresponding to the DBD of the human estrogen, progesterone and glucocorticoids receptors in order to eliminate clones of these receptors. The cDNA inserts of the remaining positive clones were subcloned into a plasmid vector and sequenced. One of the cDNA inserts contained an open reading frame encoding a 432 amino-acid protein with a predicted molecular mass of 47,682 Da. This sequence was referred to as hRAR.

Analysis of the cDNA-deduced hRAR protein sequence revealed that the regions corresponding to the DBD and the LBD possess a high degree of similarity with that of all other nuclear receptors. Several experiments were designed to define whether the cloned hRAR protein binds RA and whether, by analogy with the other members of the nuclear receptor family, it could act as a ligand-inducible transcription factor. A cDNA fragment encoding the LBD was subcloned into an expression vector and introduced in HeLa cells. Incubation of the extracts with radiolabelled ligands confirmed that the hRAR protein binds RA selectively and with high affinity. A chimeric receptor in which the DBD of hRAR is replaced by the DBD of the estrogen receptor was constructed and cotransfected into HeLa cells with a vit-tk-CAT reporter gene under the control of an estrogen response element. Addition of RA resulted in an increase in CAT reporter activity indicating that hRAR is a ligand-inducible trans-activator of transcription. Thus the first human nuclear retinoic acid receptor was cloned and is now named RARα.

A few month later, Evans' group published that they had independently cloned the same receptor using a different probe corresponding to a novel sequence with striking similarity to the DBD of the steroid hormone receptors that they fortuitously identified in a human hepatocellular carcinoma [36]. Close on the heels of these groundbreaking publications, two independent groups led by Magnus Phfahl and Chambon used the same probe [9, 13, 25] and cloned a second human nuclear retinoic acid receptor, which depicted high homology with RARα. Both groups created chimeric proteins by swapping the DBD for that of the estrogen receptor and verified that RA could activate the chimeric protein to transactivate a reporter gene under the control of an estrogen responsive DNA sequence element. One group called the new receptor, RARε, based on its high expression in epithelial tissues [9], but the name proposed by the other group, RARβ, became the commonly known name for this second RAR [13]. Finally, efforts to clone the murine counterparts of the human RARα and RARβ receptors revealed the existence of a third RA receptor, RARγ [137]. DNA sequences of the mouse RARγ gene were then used to clone the human RARγ gene [58].

Subsequent research indicated that the three RARs are encoded by three distinct genes located in different chromosomes [50, 81]. Several isoforms of each RAR subtype were identified that differ in the N-terminal region as a result of alternative splicing or of the use of different promoters upstream of the gene. Two major isoforms were identified for RARα (RARα1 and RARα2) [67], four isoforms for RARβ (RARβ1, RARβ2, RARβ3, and RARβ4) [89, 138], and two isoforms for RARγ (RARγ1 and RARγ2) [37, 57]. Official classification of the nuclear receptors identified RARα as NR1B1, RARβ as NR1B2 and RARγ as NR1B3 [35].

Cloning of a Second Family of Nuclear Retinoid Receptors: The RXRs

In the late 1980s a novel strategy was used to isolate cDNA clones encoding DNA binding domains. The technique involves absorbing the proteins produced by a phage cDNA library onto nitrocellulose filters and probing the filters with radioactive, double-stranded DNA [114, 125]. In 1989, the Keiko Ozato group used this technique to isolate from mouse liver a cDNA clone encoding a protein capable of binding the conserved MHC class I regulatory element (CRE). Interestingly, this protein, H-2RIIBP (H-2 region II binding protein), had modular domains characteristic of the nuclear hormone receptors and could also bind estrogen response elements [42]. More research needed to be done however, to determine whether the H-2RIIBP protein was a nuclear hormone receptor. Then Mangelsdorf et al. performed a low stringency screen of human liver and kidney cDNA libraries using a probe corresponding to the DBD of RARα and isolated a novel nuclear receptor, which was substantially different from RARα and was referred to as hRXRα [79]. A few month later, the Michael Rosenfeld group screened a cDNA phage library from a thyroid tumor using a RAR response element and isolated an additional nuclear receptor, which exhibited remarkable homology to RXRα and which differed by only 2 amino acids from the H-2RIIBP protein. This protein was finally named RXRβ [133]. Subsequently, three murine RXRs (RXRα, RXRβ and RXRγ) encoded by different genes were isolated [78]. Several studies attempted to clone the human RXRγ gene, but without results. Nevertheless, a human genomic locus encoding RXRγ has been mapped in chromosome 1 using florescence in situ hybridization (FISH) [3]. According to the official classification of nuclear receptors, RXRs are now identified as NR2B1 (RXRα), NR2B2 (RXRβ), and NR2B3 (RXRγ) [35].

The interesting feature of the RXR proteins is that they can activate transcription in response to RA, but are unable to bind all-*trans* RA due to the fact that they do not share significant homology with the LBD of RARs. Instead, *9-cis* retinoic acid was identified as a high affinity ligand for RXRs [45, 68]. However, today it is clear that *9-cis* RA cannot be detected in most tissues, and the existence of a physiological RXR ligand is still being investigated [34]. Nevertheless, several synthetic compounds that bind RXRs and not RARs, rexinoids, have been designed and have provided useful tools [94].

Establishment of the Basis of RARs and RXRs Mechanism of Action (1990–1995)

Once they were cloned, the sequences of the different RAR and RXR cDNAs were analyzed, aligned and compared to that of the other nuclear receptors [61]. This analysis revealed that all nuclear hormone receptors including RARs and RXRs exhibit a modular structure composed of 6 conserved regions designated A–F (Fig. 1.3a) with different degrees of conservation [65]. Region C, which corresponds to the DBD is the most conserved region with 94–97 % identity between RARs and with 60 % identity between RARs and RXRs. Region E encompasses the LBD and is also well conserved between RARs (84–90 % identity), but differs considerably between RARs and RXRs (27 % identity). Finally the A and F regions differ markedly among the 3 RARs, and the F region is lacking in RXRs.

During the same time, the promoter regions of endogenous genes that are controlled by RA were characterized, leading to the identification of specific *cis*-acting DNA elements that bind RARs (Fig. 1.3c). The first natural RA response element (RARE) identified was that in the RARβ2 promoter [26, 46, 119]. It consists of a direct repetition of 2 motifs (G/AGTTCA) separated by 5 base pairs and was called a DR5. During ensuing years, other RAREs that have different spacings between the direct repeats (DR1 and DR2) were discovered in the promoters of several other RA-responsive genes [31, 115].

RARs from crude cell extracts were able to bind RAREs, however several groups observed that high affinity binding required interactions with other nuclear factors [38, 66]. Chambon's group purified the nuclear accessory factor that enhanced the binding of RARs to RAREs in vitro and discovered that this protein was RXRβ [66]. This group also showed that RXRs form heterodimers with RARs and that interaction regions overlap with the LBDs and the DBDs of both partners [66]. Concomitantly, Rosenfeld's group also identified RXRβ as the factor that stimulates the binding of RARs to RAREs [133].

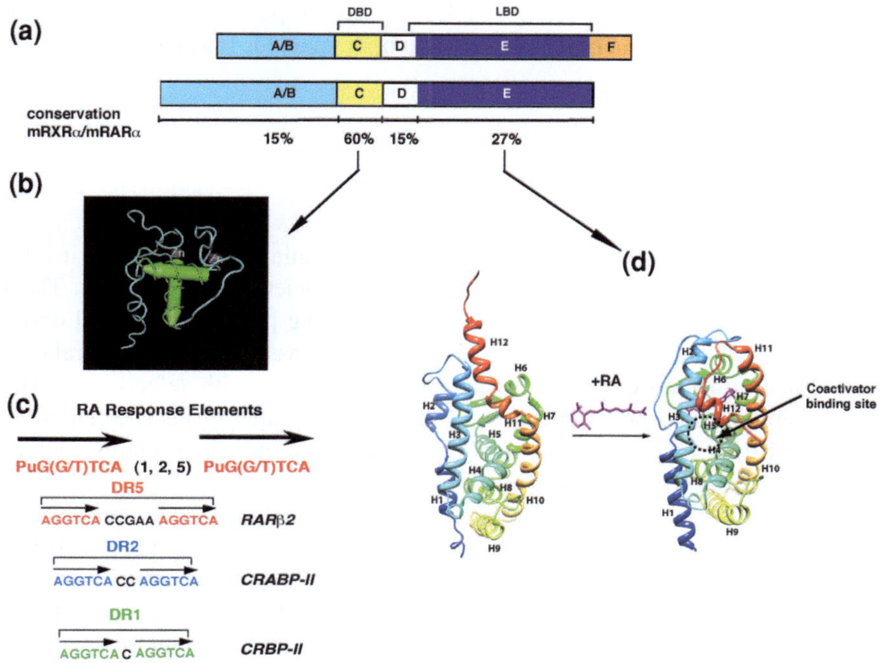

Fig. 1.3 Structure of RARs and of their DNA binding sites. **a** Schematic representation of the modular organization of RARs and RXRs with the functional domains. The conservation between RARα and RXRα is shown. **b** High resolution structure of the RXRα DNA binding domain NMR (mmdbId: 8588). **c** Description of the classical retinoic acid response elements (RAREs). **d** Structural changes induced upon RA binding. The crystal structures of the unliganded RXRα and liganded RARγ LBDs are shown with the binding domain for corepressors. Helices are represented as ribbons and labeled from H1 to H12. Adapted from Protein Data Bank 1lbd and 2lbd

In 1994–1995, knowledge of the mechanism(s) of action of RAR/RXR heterodimers and most nuclear receptors increased tremendously due to the discovery of co-activators that bind specific sequences of the LBD in response to RA [30, 62, 126, 127] and characterization of the three-dimensional structures of the DBD [64] (Fig. 1.3b) and the LBD [12, 100] (Fig. 1.3d). Most notably, the description of the crystallographic structures of the LBD in the absence and presence of RA revealed that the *on* switch for gene transcription by liganded RARs relies on structural rearrangements that create binding surfaces for co-regulators [18, 87] (Fig. 1.3d).

During this same time, peptides corresponding to amino acid sequences specific for each RAR and RXR subtypes were synthesized, allowing generation of antibodies recognizing not only the RAR and RXR proteins produced in vitro (recombinant proteins), but also endogenous RARs and RXRs [32, 105]. In addition to their utility in localizing endogenous receptor expression in tissues, the antibodies had an additional interesting benefit of revealing that endogenous RARs and RXRs migrated to different positions on polyacrylamide gels when compared with their recombinant counterparts. This later finding suggested the native proteins had a higher relative molecular weight, possibly due to post translational modifications, such as phosphorylation, in vivo [33, 104, 106].

Genetic Evidence that RARs Transduce the Retinoid Signals in Vivo

Until the 1990s, the physiological functions of vitamin A and retinoids in vivo were mainly inferred from studies on vitamin A-deficient (VAD) animals. These studies demonstrated that vitamin A is required during pre- and post-natal development as well as in adult life. The VAD syndrome is associated with several congenital malformations during embryonic development and with defects in growth, vision, reproduction and maintenance of several tissues after birth. The cloning of RARs and RXRs raised the question of whether these nuclear receptors mediate the physiological effects of vitamin A and retinoids in vivo.

The development of the homologous recombination technology for targeted disruption of specific genes allowed the generation of mice lacking CRABPs, RARs or RXRs. Single or double knock out of CRABP I and II did not generate overt phenotypic abnormalities corroborating that these retinoic acid binding proteins are not required for the effects of RA during development [40, 60]. In contrast, knock out of RARα or RARγ exhibited congenital malformations and displayed some of the defects of the postnatal VAD syndrome with reduced postnatal viability [72, 74, 75]. Double mutants were also generated that recapitulated all of the fetal VAD syndrome malformations and that exhibited a dramatically reduced viability [73, 83]. All these results confirmed that the effects of vitamin A in development are indeed mediated by RARs. Subsequently, RXR knock out mice were also engineered, but the interpretation of the phenotypes was more complicated since RXR heterodimerize not only with RARs, but also with multiple other types of nuclear hormone receptors [56, 118]. Nevertheless, compound mutants in both RXR and RAR greatly exacerbated the VAD phenotypes [56] suggesting that the RAR/RXR heterodimers are the functional units that mediate vitamin A signaling in vivo.

Development of the Field: A Huge Explosion Has Occurred in the Field of RARs During the Last Two Decades

Between 1987 and 1995, RARs and RXRs were cloned and the basis for their mechanism of action was established (Fig. 1.4). During the next twenty years, up to now, knowledge in the field of RARs and RXRs has increased tremendously due to the development of novel genetic, biophysical and high throughput molecular technologies. Integration of these approaches has provided an in-depth view of the mechanism of action of RARs and of all nuclear receptors. These findings have been the subject of several recent comprehensive reviews, some of which are recapitulated in this book volume. In addition, several databases have been developed and are now publically

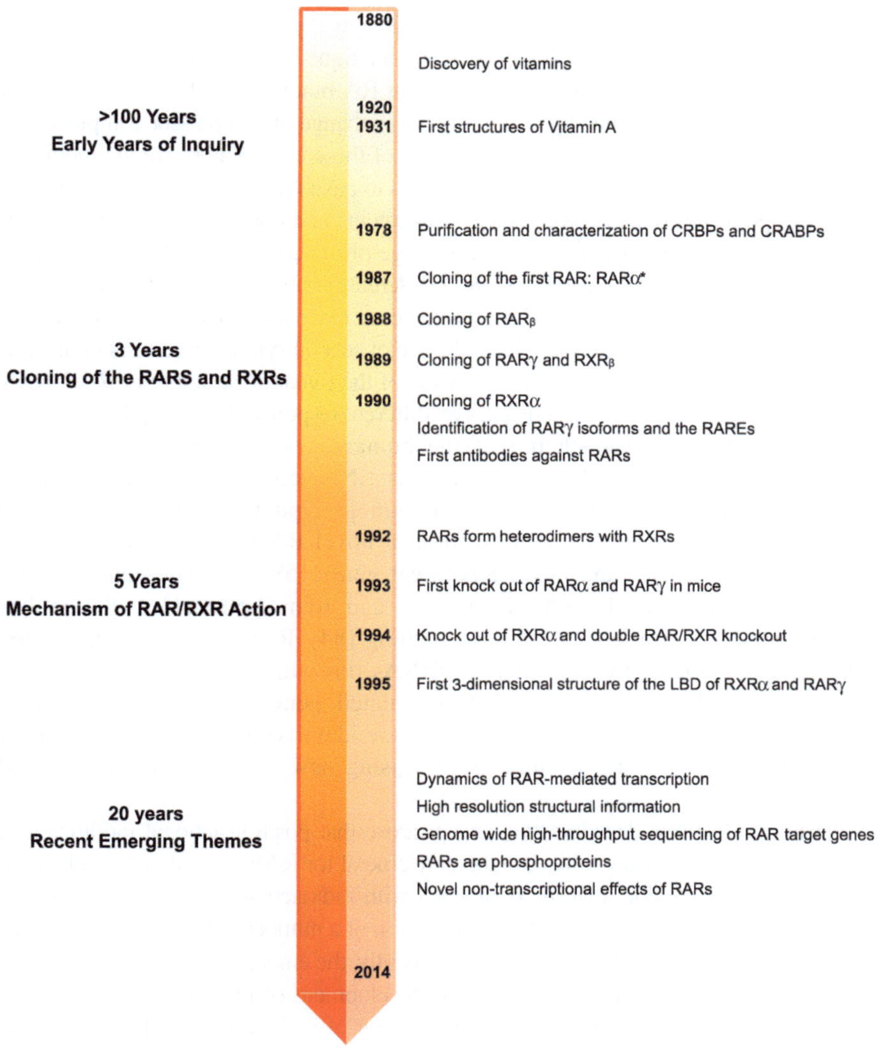

Fig. 1.4 Chronology of the main events in the field of vitamin A and RARs. *Petkovich et al. 1987; Giguere et al. 1987

available including [NURSA (http://www.nursa.org), Transcriptomine (http://www.nursa.org/transcriptomine) and IUPHAR (http://www.iuphar-db.org)].

Briefly, the novel findings discovered during the two last decades can be summarized as follows:

A large number of coactivator and corepressors have been identified and found to be components of multisubunit coregulator complexes exhibiting an ever-expanding diversity of enzymatic and epigenetic activities, exemplified by ATP-dependent nucleosome remodeling complexes, histone acetyl/deacetyl transferases (HATs and HDACs), histone methyl transferases (HMTs) and histone demethylases [39]. The development of the chromatin immunoprecipitation (ChIP) technique revealed that transcription of RAR-target genes is a dynamic process and that RA binding switches RARs from an inactive state to an active state by promoting the exchange of corepressor complexes for coactivators [95, 102]. The consummate effects of these coregulators are the modification, remodeling and decompaction of chromatin to pave the way for the recruitment of the transcription machinery [28]. Today new proteins and complexes are still continuously being discovered increasing our knowledge of the complexity of RAR-mediated transcription. Moreover, the combination of different biophysical methods [X-ray crystallography, small angle X-ray (SAXs), fluorescence resonance energy transfer (FRET), nuclear magnetic resonance (NMR)] and the use of several synthetic agonists or antagonists for RARs and RXRs is providing a view of the dynamic structure of RAR-RXR heterodimers associated with coactivators on different response elements [101].

Subsequent to the sequencing of the human genome, the development of genome-wide profiling technologies such as RNA-seq (high throughput qPCR sequencing) and ChIP-seq (chromatin immunoprecipitation coupled with deep sequencing) has allowed the identification of novel RA response elements with different spacings [88] and novel RA-target genes [59, 77]. The genome-wide integrative analysis of RAR/RXR binding and transcriptional regulation has also provided a dynamic view of RA signaling [84]. Recently, RARs have been found to target the expression of microRNAs, introducing yet another level of complexity in the regulation of RAR-regulated gene transcription [93, 135, 136]. These recent approaches and findings are identifying novel mechanisms of action for RARs and are opening up promising new avenues for research and development.

During the last two decades, the concept arose that posttranslational modifications such as phosphorylation and ubiquitination are crucial for RARs activity [2]. Early studies using phosphopeptide analysis performed with radioactivity and large amounts of recombinant receptors resulted in the identification of a number of phosphorylation sites on RARs and RXRs [1, 63, 103, 107]. Subsequently, the emergence of new methods for enrichment of phosphopeptide samples and development of phosphospecific antibodies provided the ability to analyze phosphorylation of endogenous RARs in response to their cognate ligand or signaling pathways [14]. Now it is clear that RARs, as well as several other proteins, are rapidly phosphorylated in response to RA, subsequent to RA-induced activation of kinase cascades via a pool of RARs that are present in membrane lipid rafts [80, 99]. A concept that is gaining support is that phosphorylations induce subtle changes in the conformation of the receptors that modulate the association/dissociation of new coregulators [21, 108]. Another developing concept is that phosphorylation is a signal for the degradation of RARs by the ubiquitin proteasome system, a process that signals the end of transcription [11].

In situ analysis of endogenous RAR protein expression profiles became possible with the generation of purified highly-specific antibodies [15, 124]. Moreover, the development of novel conditional gene targeting strategies based on the use of the Cre recombinase allowed the generation of somatic mutations in individual genes in a specific cell type and at a given time in the life of a transgenic mouse [85, 86]. This novel strategy has yielded remarkable advances in understanding the roles played by RARs and RXRs because it circumvents the limitations of previous transgenic approaches that led to early lethality and that were compromised by redundancies in receptor isoform expression in tissues under investigation [49, 69].

Future Directions

Today, 25 years after the cloning of RARs (Fig. 1.4), knowledge continues to evolve in the field of retinoid biology. The structures of RAR/RXR heterodimers bound to DNA have been solved, but they are still lacking the N-terminal domain which exhibits a quasi absence of defined secondary structures but confers considerable flexibility to RARs. The integration of data from several sources and from high-resolution biophysical approaches should provide the structure of the RAR/RXR heterodimers as full-length proteins bound to DNA with their coregulators intact.

A driving goal for future studies will be the discovery of the rules for cell fate specification integrated into a systems biology view of RAR/RXR actions and RA signaling. Application of computational models and programs to reconstruct differentiation-related gene networks obtained from different cell types should allow the prediction of RA-regulated gene network intricacies and the identification of key factors that direct cells towards a particular differentiation phenotype. Stem cells, which are pluripotent cells capable of generating all the differentiated cell types present in the body and which are responsive to RA, are currently, a most promising tool for such cell fate studies. Extrapolation of data generated from stem cell differentiation models should have applicability to a deeper understanding RAR-dysfunctional diseases that interfere with normal cell homeostasis and redirect normal cells to a more primitive, mitosis-driven state.

Recent findings are stretching the boundaries of our understanding of vitamin A action. These newer studies are indicating that the effects of vitamin A retinol and RA are not mediated only by RAR/RXR heterodimers and transcriptional processes, and this is opening up new avenues in the field. As an example, it has been found that RA can activate other nuclear receptors such as the peroxisome proliferator-activated receptor β/δ [110], providing a rationale for the long-noted, but poorly understood function of vitamin A in regulating energy balance. Moreover, recent findings are hinting that RA, as well as vitamin A itself, can have extranuclear, non-transcriptional effects and can activate kinase-signaling pathways [2, 10]. Consequently, one can speculate that, in addition to affecting the transcriptome, RA and retinol could also affect the phospho-proteome. Next generation, dual linear ion trap mass spectrometers coupled with Orbitrap technology should allow the identification of new panels of proteins that are phosphorylated in response to retinol or RA. The future objectives should be to integrate the RA-induced variations in the phospho-proteome with the transcriptome. Such an integrative study should pave the way to breakthroughs in disease-related research. The recent observation that RARs are present in the cytosol of specific cell types [22, 76] continues to open new areas in the mechanisms of action of vitamin A and RA.

Great progress has been made in deciphering how specific molecules and signaling pathways interact to mediate vitamin A/RA action. But much is left to be done to fully understand the complexities of their action at the cellular and sub-cellular levels and of their regulation in time and space throughout the life of an organism.

Acknowledgments We thank Gabriel R Batres for assistance with the design and preparation of Fig. 1.4.

References

1. Adam-Stitah S, Penna L, Chambon P, Rochette-Egly C (1999) Hyperphosphorylation of the retinoid X receptor alpha by activated c-Jun NH2-terminal kinases. J Biol Chem 274:18932–18941
2. Al Tanoury Z, Piskunov A, Rochette-Egly C (2013) Vitamin A and retinoid signaling: genomic and nongenomic effects: thematic review Series: Fat-Soluble Vitamins: Vitamin A. J Lipid Res 54:1761–1775
3. Almasan A, Mangelsdorf DJ, Ong ES, Wahl GM, Evans RM (1994) Chromosomal localization of the human retinoid X receptors. Genomics 20:397–403
4. Arens JF, Van Dorp DA (1946) Activity of vitamin A-acid in the rat. Nature 158:622
5. Arens JF, Van Dorp DA (1946) Synthesis of some compounds possessing vitamin A activity. Nature 157:190
6. Ashburner M, Chihara C, Meltzer P, Richards G (1974) Temporal control of puffing activity in polytene chromosomes. Cold Spring Harb Symp Quant Biol 38:655–662
7. Ball S, Goodwin TW, Morton RA (1948) Studies on vitamin A; the preparation of retinene1, vitamin A aldehyde. Biochem J 42:516–523
8. Bashor MM, Toft DO, Chytil F (1973) In vitro binding of retinol to rat-tissue components. Proc Natl Acad Sci USA 70:3483–3487
9. Benbrook D, Lernhardt E, Pfahl M (1988) A new retinoic acid receptor identified from a hepatocellular carcinoma. Nature 333:669–672
10. Berry DC, Noy N (2012) Signaling by vitamin A and retinol-binding protein in regulation of insulin responses and lipid homeostasis. Biochim Biophys Acta 1821:168–176
11. Bour G, Lalevee S, Rochette-Egly C (2007) Protein kinases and the proteasome join in the combinatorial control of transcription by nuclear retinoic acid receptors. Trends Cell Biol 17:302–309
12. Bourguet W, Ruff M, Chambon P, Gronemeyer H, Moras D (1995) Crystal structure of the ligand-binding domain of the human nuclear receptor RXR-alpha. Nature 375:377–382
13. Brand N, Petkovich M, Krust A, Chambon P, de The H, Marchio A, Tiollais P, Dejean A (1988) Identification of a second human retinoic acid receptor. Nature 332:850–853
14. Bruck N, Vitoux D, Ferry C, Duong V, Bauer A, de The H, Rochette-Egly C (2009) A coordinated phosphorylation cascade initiated by p38MAPK/MSK1 directs RARalpha to target promoters. EMBO J 28:34–47
15. Buchanan FQ, Rochette-Egly C, Asson-Batres MA (2011) Detection of variable levels of RARalpha and RARgamma proteins in pluripotent and differentiating mouse embryonal carcinoma and mouse embryonic stem cells. Cell Tissue Res 346:43–51
16. Carpenter KJ (2012) The discovery of thiamin. Ann Nutr Metab 61:219–223
17. Carpenter KJ, Sutherland B (1995) Eijkman's contribution to the discovery of vitamins. J Nutr 125:155–163
18. Chambon P (1996) A decade of molecular biology of retinoic acid receptors. Faseb J 10:940–954
19. Chambon P (2004) How I became one of the fathers of a superfamily. Nat Med 10:1027–1031
20. Chandler VL, Maler BA, Yamamoto KR (1983) DNA sequences bound specifically by glucocorticoid receptor in vitro render a heterologous promoter hormone responsive in vivo. Cell 33:489–499
21. Chebaro Y, Amal I, Rochel N, Rochette-Egly C, Stote R, Dejaegere A (2013) Phosphorylation of the Retinoic Acid Receptor alpha induces a mechanical allosteric regulation and changes in internal dynamics. Plos Comp Biol 9:e1003012

22. Chen N, Napoli JL (2008) All-trans-retinoic acid stimulates translation and induces spine formation in hippocampal neurons through a membrane-associated RARalpha. Faseb J 22:236–245
23. Chytil F, Ong DE (1978) Cellular vitamin A binding proteins. Vitam Horm 36:1–32
24. De Luca L, Little EP, Wolf G (1969) Vitamin A and protein synthesis by rat intestinal mucosa. J Biol Chem 244:701–708
25. de The H, Marchio A, Tiollais P, Dejean A (1987) A novel steroid thyroid hormone receptor-related gene inappropriately expressed in human hepatocellular carcinoma. Nature 330:667–670
26. de The H, Vivanco-Ruiz MM, Tiollais P, Stunnenberg H, Dejean A (1990) Identification of a retinoic acid responsive element in the retinoic acid receptor beta gene. Nature 343:177–180
27. Deisseroth A, Nienhuis A, Turner P, Velez R, Anderson WF, Ruddle F, Lawrence J, Creagan R, Kucherlapati R (1977) Localization of the human alpha-globin structural gene to chromosome 16 in somatic cell hybrids by molecular hybridization assay. Cell 12:205–218
28. Dilworth FJ, Chambon P (2001) Nuclear receptors coordinate the activities of chromatin remodeling complexes and coactivators to facilitate initiation of transcription. Oncogene 20:3047–3054
29. Drummond JC (1920) The nomenclature of the so-called accessory food factors (Vitamins). Biochem J 14:660
30. Durand B, Saunders M, Gaudon C, Roy B, Losson R, Chambon P (1994) Activation function 2 (AF-2) of retinoic acid receptor and 9-cis retinoic acid receptor: presence of a conserved autonomous constitutive activating domain and influence of the nature of the response element on AF-2 activity. EMBO J 13:5370–5382
31. Durand B, Saunders M, Leroy P, Leid M, Chambon P (1992) All-trans and 9-cis retinoic acid induction of CRABPII transcription is mediated by RAR-RXR heterodimers bound to DR1 and DR2 repeated motifs. Cell 71:73–85
32. Gaub MP, Lutz Y, Ruberte E, Petkovich M, Brand N, Chambon P (1989) Antibodies specific to the retinoic acid human nuclear receptors alpha and beta. Proc Natl Acad Sci USA 86:3089–3093
33. Gaub MP, Rochette-Egly C, Lutz Y, Ali S, Matthes H, Scheuer I, Chambon P (1992) Immunodetection of multiple species of retinoic acid receptor alpha: evidence for phosphorylation. Exp Cell Res 201:335–346
34. Germain P, Chambon P, Eichele G, Evans RM, Lazar MA, Leid M, De Lera AR, Lotan R, Mangelsdorf DJ, Gronemeyer H (2006) International union of pharmacology. LXIII. Retinoid X receptors. Pharmacol Rev 58:760–772
35. Germain P, Staels B, Dacquet C, Spedding M, Laudet V (2006) Overview of nomenclature of nuclear receptors. Pharmacol Rev 58:685–704
36. Giguere V, Ong ES, Segui P, Evans RM (1987) Identification of a receptor for the morphogen retinoic acid. Nature 330:624–629
37. Giguere V, Shago M, Zirngibl R, Tate P, Rossant J, Varmuza S (1990) Identification of a novel isoform of the retinoic acid receptor gamma expressed in the mouse embryo. Mol Cell Biol 10:2335–2340
38. Glass CK, Devary OV, Rosenfeld MG (1990) Multiple cell type-specific proteins differentially regulate target sequence recognition by the alpha retinoic acid receptor. Cell 63:729–738
39. Glass CK, Rosenfeld MG (2000) The coregulator exchange in transcriptional functions of nuclear receptors. Genes Dev 14:121–141
40. Gorry P, Lufkin T, Dierich A, Rochette-Egly C, Decimo D, Dolle P, Mark M, Durand B, Chambon P (1994) The cellular retinoic acid binding protein I is dispensable. Proc Natl Acad Sci USA 91:9032–9036
41. Govindan MV, Devic M, Green S, Gronemeyer H, Chambon P (1985) Cloning of the human glucocorticoid receptor cDNA. Nucleic Acids Res 13:8293–8304
42. Hamada K, Gleason SL, Levi BZ, Hirschfeld S, Appella E, Ozato K (1989) H-2RIIBP, a member of the nuclear hormone receptor superfamily that binds to both the regulatory element of major histocompatibility class I genes and the estrogen response element. Proc Natl Acad Sci USA 86:8289–8293

43. Haq R, Chytil F (1988) Retinoic acid rapidly induces lung cellular retinol-binding protein mRNA levels in retinol deficient rats. Biochem Biophys Res Commun 156:712–716
44. Heilbron IM, Morton RA, Webster ET (1932) The structure of vitamin A. Biochem J 26:1194–1196
45. Heyman RA, Mangelsdorf DJ, Dyck JA, Stein RB, Eichele G, Evans RM, Thaller C (1992) 9-cis retinoic acid is a high affinity ligand for the retinoid X receptor. Cell 68:397–406
46. Hoffmann B, Lehmann JM, Zhang XK, Hermann T, Husmann M, Graupner G, Pfahl M (1990) A retinoic acid receptor-specific element controls the retinoic acid receptor-beta promoter. Mol Endocrinol 4:1727–1736
47. Holmes HN, Corbet RE (1937) A crystalline Vitamin a concentrate. Science 85:103
48. Hubbard R, Wald G (1951) The mechanism of rhodopsin synthesis. Proc Natl Acad Sci USA 37:69–79
49. Imai T, Jiang M, Kastner P, Chambon P, Metzger D (2001) Selective ablation of retinoid X receptor alpha in hepatocytes impairs their lifespan and regenerative capacity. Proc Natl Acad Sci USA 98:4581–4586
50. Ishikawa T, Umesono K, Mangelsdorf DJ, Aburatani H, Stanger BZ, Shibasaki Y, Imawari M, Evans RM, Takaku F (1990) A functional retinoic acid receptor encoded by the gene on human chromosome 12. Mol Endocrinol 4:837–844
51. Jacobson HI, Gupta GN, Fernandez C, Hennix S, Jensen EV (1960) Determination of tritium in biological material. Arch Biochem Biophys 86:89–93
52. Jensen EV (1962) On the mechanism of estrogen action. Perspect Biol Med 6:47–59
53. Jensen EV, Suzuki T, Kawashima T, Stumpf WE, Jungblut PW, DeSombre ER (1968) A two-step mechanism for the interaction of estradiol with rat uterus. Proc Natl Acad Sci USA 59:632–638
54. Johnson BC, Kennedy M, Chiba N (1969) Vitamin A and nuclear RNA synthesis. Am J Clin Nutr 22:1048–1058
55. Karrer P, Morf R, Schöpp K (1931) Zur Kenntnis des Vitamins-A aus Fischtranen. Helv Chim Acta 14:1036–1040
56. Kastner P, Grondona JM, Mark M, Gansmuller A, LeMeur M, Decimo D, Vonesch JL, Dolle P, Chambon P (1994) Genetic analysis of RXR alpha developmental function: convergence of RXR and RAR signaling pathways in heart and eye morphogenesis. Cell 78:987–1003
57. Kastner P, Krust A, Mendelsohn C, Garnier JM, Zelent A, Leroy P, Staub A, Chambon P (1990) Murine isoforms of retinoic acid receptor gamma with specific patterns of expression. Proc Natl Acad Sci USA 87:2700–2704
58. Krust A, Kastner P, Petkovich M, Zelent A, Chambon P (1989) A third human retinoic acid receptor, hRAR-gamma. Proc Natl Acad Sci USA 86:5310–5314
59. Lalevee S, Anno YN, Chatagnon A, Samarut E, Poch O, Laudet V, Benoit G, Lecompte O, Rochette-Egly C (2011) Genome-wide in silico identification of new conserved and functional retinoic acid receptor response elements (direct repeats separated by 5 bp). J Biol Chem 286:33322–33334
60. Lampron C, Rochette-Egly C, Gorry P, Dolle P, Mark M, Lufkin T, LeMeur M, Chambon P (1995) Mice deficient in cellular retinoic acid binding protein II (CRABPII) or in both CRABPI and CRABPII are essentially normal. Development 121:539–548
61. Laudet V, Gronemeyer H (2001) Nuclear receptor factsbook. Acedemic, London
62. Le Douarin B, Zechel C, Garnier JM, Lutz Y, Tora L, Pierrat P, Heery D, Gronemeyer H, Chambon P, Losson R (1995) The N-terminal part of TIF1, a putative mediator of the ligand-dependent activation function (AF-2) of nuclear receptors, is fused to B-raf in the oncogenic protein T18. EMBO J 14:2020–2033
63. Lee HY, Suh YA, Robinson MJ, Clifford JL, Hong WK, Woodgett JR, Cobb MH, Mangelsdorf DJ, Kurie JM (2000) Stress pathway activation induces phosphorylation of retinoid X receptor. J Biol Chem 275:32193–32199
64. Lee MS, Kliewer SA, Provencal J, Wright PE, Evans RM (1993) Structure of the retinoid X receptor alpha DNA binding domain: a helix required for homodimeric DNA binding. Science 260:1117–1121

65. Leid M, Kastner P, Chambon P (1992) Multiplicity generates diversity in the retinoic acid signalling pathways. Trends Biochem Sci 17:427–433
66. Leid M, Kastner P, Lyons R, Nakshatri H, Saunders M, Zacharewski T, Chen JY, Staub A, Garnier JM, Mader S et al (1992) Purification, cloning, and RXR identity of the HeLa cell factor with which RAR or TR heterodimerizes to bind target sequences efficiently. Cell 68:377–395
67. Leroy P, Krust A, Zelent A, Mendelsohn C, Garnier JM, Kastner P, Dierich A, Chambon P (1991) Multiple isoforms of the mouse retinoic acid receptor alpha are generated by alternative splicing and differential induction by retinoic acid. EMBO J 10:59–69
68. Levin AA, Sturzenbecker LJ, Kazmer S, Bosakowski T, Huselton C, Allenby G, Speck J, Kratzeisen C, Rosenberger M, Lovey A et al (1992) 9-cis retinoic acid stereoisomer binds and activates the nuclear receptor RXR alpha. Nature 355:359–361
69. Li M, Indra AK, Warot X, Brocard J, Messaddeq N, Kato S, Metzger D, Chambon P (2000) Skin abnormalities generated by temporally controlled RXRalpha mutations in mouse epidermis. Nature 407:633–636
70. Liau G, Ong DE, Chytil F (1981) Interaction of the retinol/cellular retinol-binding protein complex with isolated nuclei and nuclear components. J Cell Biol 91:63–68
71. Liau G, Ong DE, Chytil F (1985) Partial characterization of nuclear binding sites for retinol delivered by cellular retinol binding protein. Arch Biochem Biophys 237:354–360
72. Lohnes D, Kastner P, Dierich A, Mark M, LeMeur M, Chambon P (1993) Function of retinoic acid receptor gamma in the mouse. Cell 73:643–658
73. Lohnes D, Mark M, Mendelsohn C, Dolle P, Dierich A, Gorry P, Gansmuller A, Chambon P (1994) Function of the retinoic acid receptors (RARs) during development (I). Craniofacial and skeletal abnormalities in RAR double mutants. Development 120:2723–2748
74. Lufkin T, Lohnes D, Mark M, Dierich A, Gorry P, Gaub MP, LeMeur M, Chambon P (1993) High postnatal lethality and testis degeneration in retinoic acid receptor alpha mutant mice. Proc Natl Acad Sci USA 90:7225–7229
75. Luo J, Pasceri P, Conlon RA, Rossant J, Giguere V (1995) Mice lacking all isoforms of retinoic acid receptor beta develop normally and are susceptible to the teratogenic effects of retinoic acid. Mech Dev 53:61–71
76. Maghsoodi B, Poon MM, Nam CI, Aoto J, Ting P, Chen L (2008) Retinoic acid regulates RARalpha-mediated control of translation in dendritic RNA granules during homeostatic synaptic plasticity. Proc Natl Acad Sci USA 105:16015–16020
77. Mahony S, Mazzoni EO, McCuine S, Young RA, Wichterle H, Gifford DK (2011) Ligand-dependent dynamics of retinoic acid receptor binding during early neurogenesis. Genome Biol 12:R2
78. Mangelsdorf DJ, Borgmeyer U, Heyman RA, Zhou JY, Ong ES, Oro AE, Kakizuka A, Evans RM (1992) Characterization of three RXR genes that mediate the action of 9-cis retinoic acid. Genes Dev 6:329–344
79. Mangelsdorf DJ, Ong ES, Dyck JA, Evans RM (1990) Nuclear receptor that identifies a novel retinoic acid response pathway. Nature 345:224–229
80. Masia S, Alvarez S, de Lera AR, Barettino D (2007) Rapid, nongenomic actions of retinoic acid on phosphatidylinositol-3-kinase signaling pathway mediated by the retinoic acid receptor. Mol Endocrinol 21:2391–2402
81. Mattei MG, Riviere M, Krust A, Ingvarsson S, Vennstrom B, Islam MQ, Levan G, Kautner P, Zelent A, Chambon P et al (1991) Chromosomal assignment of retinoic acid receptor (RAR) genes in the human, mouse, and rat genomes. Genomics 10:1061–1069
82. McKnight GS, Palmiter RD (1979) Transcriptional regulation of the ovalbumin and conalbumin genes by steroid hormones in chick oviduct. J Biol Chem 254:9050–9058
83. Mendelsohn C, Lohnes D, Decimo D, Lufkin T, LeMeur M, Chambon P, Mark M (1994) Function of the retinoic acid receptors (RARs) during development (II). Multiple abnormalities at various stages of organogenesis in RAR double mutants. Development 120:2749–2771

84. Mendoza-Parra MA, Walia M, Sankar M, Gronemeyer H (2011) Dissecting the retinoid-induced differentiation of F9 embryonal stem cells by integrative genomics. Mol Syst Biol 7:538
85. Metzger D, Chambon P (2001) Site- and time-specific gene targeting in the mouse. Methods 24:71–80
86. Metzger D, Clifford J, Chiba H, Chambon P (1995) Conditional site-specific recombination in mammalian cells using a ligand-dependent chimeric Cre recombinase. Proc Natl Acad Sci USA 92:6991–6995
87. Moras D, Gronemeyer H (1998) The nuclear receptor ligand-binding domain: structure and function. Curr Opin Cell Biol 10:384–391
88. Moutier E, Ye T, Choukrallah MA, Urban S, Osz J, Chatagnon A, Delacroix L, Langer D, Rochel N, Moras D, Benoit G, Davidson I (2012) Retinoic acid receptors recognise the mouse genome through binding elements with diverse spacing and topology. J Biol Chem 287:26328–26341
89. Nagpal S, Zelent A, Chambon P (1992) RAR-beta 4, a retinoic acid receptor isoform is generated from RAR-beta 2 by alternative splicing and usage of a CUG initiator codon. Proc Natl Acad Sci USA 89:2718–2722
90. Ong DE, Chytil F (1974) Multiple retinol binding proteins in rabbit lung. Biochem Biophys Res Commun 59:221–229
91. Ong DE, Chytil F (1978) Cellular retinoic acid-binding protein from rat testis. Purification and characterization. J Biol Chem 253:4551–4554
92. Ong DE, Chytil F (1978) Cellular retinol-binding protein from rat liver. Purification and characterization. J Biol Chem 253:828–832
93. Pelosi A, Careccia S, Lulli V, Romania P, Marziali G, Testa U, Lavorgna S, Lo-Coco F, Petti MC, Calabretta B, Levrero M, Piaggio G, Rizzo MG (2013) miRNA let-7c promotes granulocytic differentiation in acute myeloid leukemia. Oncogene 32:3648–3654
94. Perez E, Bourguet W, Gronemeyer H, de Lera AR (2012) Modulation of RXR function through ligand design. Biochim Biophys Acta 1821:57–69
95. Perissi V, Rosenfeld MG (2005) Controlling nuclear receptors: the circular logic of cofactor cycles. Nat Rev Mol Cell Biol 6:542–554
96. Peterkofsky B, Tomkins GM (1968) Evidence for the steroid-induced accumulation of tyrosine-aminotransferase messenger RNA in the absence of protein synthesis. Proc Natl Acad Sci USA 60:222–228
97. Petkovich M, Brand NJ, Krust A, Chambon P (1987) A human retinoic acid receptor which belongs to the family of nuclear receptors. Nature 330:444–450
98. Piro A, Tagarelli G, Lagonia P, Tagarelli A, Quattrone A (2010) Casimir funk: his discovery of the vitamins and their deficiency disorders. Ann Nutr Metab 57:85–88
99. Piskunov A, Rochette-Egly C (2012) A retinoic acid receptor RARalpha pool present in membrane lipid rafts forms complexes with G protein alphaQ to activate p38MAPK. Oncogene 31:3333–3345
100. Renaud JP, Rochel N, Ruff M, Vivat V, Chambon P, Gronemeyer H, Moras D (1995) Crystal structure of the RAR-gamma ligand-binding domain bound to all-trans retinoic acid. Nature 378:681–689
101. Rochel N, Ciesielski F, Godet J, Moman E, Roessle M, Peluso-Iltis C, Moulin M, Haertlein M, Callow P, Mely Y, Svergun DI, Moras D (2011) Common architecture of nuclear receptor heterodimers on DNA direct repeat elements with different spacings. Nat Struct Mol Biol 18:564–570
102. Rochette-Egly C (2005) Dynamic combinatorial networks in nuclear receptor-mediated transcription. J Biol Chem 280:32565–32568
103. Rochette-Egly C, Adam S, Rossignol M, Egly JM, Chambon P (1997) Stimulation of RAR alpha activation function AF-1 through binding to the general transcription factor TFIIH and phosphorylation by CDK7. Cell 90:97–107
104. Rochette-Egly C, Gaub MP, Lutz Y, Ali S, Scheuer I, Chambon P (1992) Retinoic acid receptor-beta: immunodetection and phosphorylation on tyrosine residues. Mol Endocrinol 6:2197–2209

105. Rochette-Egly C, Lutz Y, Pfister V, Heyberger S, Scheuer I, Chambon P, Gaub MP (1994) Detection of retinoid X receptors using specific monoclonal and polyclonal antibodies. Biochem Biophys Res Commun 204:525–536
106. Rochette-Egly C, Lutz Y, Saunders M, Scheuer I, Gaub MP, Chambon P (1991) Retinoic acid receptor gamma: specific immunodetection and phosphorylation. J Cell Biol 115:535–545
107. Rochette-Egly C, Oulad-Abdelghani M, Staub A, Pfister V, Scheuer I, Chambon P, Gaub MP (1995) Phosphorylation of the retinoic acid receptor-alpha by protein kinase A. Mol Endocrinol 9:860–871
108. Samarut E, Amal I, Markov G, Stote R, Dejaegere A, Laudet V, Rochette-Egly C (2011) Evolution of nuclear retinoic acid receptors alpha (RARa) phosphorylation sites. Serine gain provides fine-tuned regulation. Mol Biol Evol 28:2125–2137
109. Scheidereit C, Geisse S, Westphal HM, Beato M (1983) The glucocorticoid receptor binds to defined nucleotide sequences near the promoter of mouse mammary tumour virus. Nature 304:749–752
110. Schug TT, Berry DC, Shaw NS, Travis SN, Noy N (2007) Opposing effects of retinoic acid on cell growth result from alternate activation of two different nuclear receptors. Cell 129:723–733
111. Semba RD (2012) On the 'discovery' of vitamin A. Ann Nutr Metab 61:192–198
112. Sim GK, Kafatos FC, Jones CW, Koehler MD, Efstratiadis A, Maniatis T (1979) Use of a cDNA library for studies on evolution and developmental expression of the chorion multigene families. Cell 18:1303–1316
113. Simoni RD, Hill RH, Vaughan M (2002) Nutritional biochemistry and the discovery of Vitamins: the work of Elmer Verner McCollum. J Biol Chem 277:e8
114. Singh H, LeBowitz JH, Baldwin AS Jr, Sharp PA (1988) Molecular cloning of an enhancer binding protein: isolation by screening of an expression library with a recognition site DNA. Cell 52:415–423
115. Smith WC, Nakshatri H, Leroy P, Rees J, Chambon P (1991) A retinoic acid response element is present in the mouse cellular retinol binding protein I (mCRBPI) promoter. EMBO J 10:2223–2230
116. Spiegelman S, Watson KF, Kacian DL (1971) Synthesis of DNA complements of natural RNAs: a general approach. Proc Natl Acad Sci USA 68:2843–2845
117. Sporn MB, Dunlop NM, Newton DL, Smith JM (1976) Prevention of chemical carcinogenesis by vitamin A and its synthetic analogs (retinoids). Fed Proc 35:1332–1338
118. Sucov HM, Dyson E, Gumeringer CL, Price J, Chien KR, Evans RM (1994) RXR alpha mutant mice establish a genetic basis for vitamin A signaling in heart morphogenesis. Genes Dev 8:1007–1018
119. Sucov HM, Murakami KK, Evans RM (1990) Characterization of an autoregulated response element in the mouse retinoic acid receptor type beta gene. Proc Natl Acad Sci USA 87:5392–5396
120. Takase S, Ong DE, Chytil F (1979) Cellular retinol-binding protein allows specific interaction of retinol with the nucleus in vitro. Proc Natl Acad Sci USA 76:2204–2208
121. Takase S, Ong DE, Chytil F (1986) Transfer of retinoic acid from its complex with cellular retinoic acid-binding protein to the nucleus. Arch Biochem Biophys 247:328–334
122. Temin HM, Baltimore D (1972) RNA-directed DNA synthesis and RNA tumor viruses. Adv Virus Res 17:129–186
123. Van Dorp DA, Arens JF (1946) Biological activity of vitamin A acid. Nature 158:60
124. Vernet N, Dennefeld C, Rochette-Egly C, Oulad-Abdelghani M, Chambon P, Ghyselinck NB, Mark M (2006) Retinoic acid metabolism and signaling pathways in the adult and developing mouse testis. Endocrinology 147:96–110
125. Vinson CR, LaMarco KL, Johnson PF, Landschulz WH, McKnight SL (1988) In situ detection of sequence-specific DNA binding activity specified by a recombinant bacteriophage. Genes Dev 2:801–806

126. Voegel JJ, Heine MJ, Zechel C, Chambon P, Gronemeyer H (1996) TIF2, a 160 kDa transcriptional mediator for the ligand-dependent activation function AF-2 of nuclear receptors. EMBO J 15:3667–3675
127. vom Baur E, Zechel C, Heery D, Heine MJ, Garnier JM, Vivat V, Le Douarin B, Gronemeyer H, Chambon P, Losson R (1996) Differential ligand-dependent interactions between the AF-2 activating domain of nuclear receptors and the putative transcriptional intermediary factors mSUG1 and TIF1. EMBO J 15:110–124
128. Wald G (1948) The synthesis from vitamin A1 of retinene 1 and of a new 545 m-mu chromogen yielding light-sensitive products. J Gen Physiol 31:489–504
129. Walter P, Green S, Greene G, Krust A, Bornert JM, Jeltsch JM, Staub A, Jensen E, Scrace G, Waterfield M et al (1985) Cloning of the human estrogen receptor cDNA. Proc Natl Acad Sci USA 82:7889–7893
130. Weill JD, Busch S, Chambon P, Mandel P (1963) The effect of estradiol injections upon chicken liver nuclei ribonucleic acid polymerase. Biochem Biophys Res Commun 10:122–126
131. Willecke K, Ruddle FH (1975) Transfer of the human gene for hypoxanthine-guanine phosphoribosyltransferase via isolated human metaphase chromosomes into mouse L-cells. Proc Natl Acad Sci USA 72:1792–1796
132. Wrange O, Gustafsson JA (1978) Separation of the hormone- and DNA-binding sites of the hepatic glucocorticoid receptor by means of proteolysis. J Biol Chem 253:856–865
133. Yu VC, Delsert C, Andersen B, Holloway JM, Devary OV, Naar AM, Kim SY, Boutin JM, Glass CK, Rosenfeld MG (1991) RXR beta: a coregulator that enhances binding of retinoic acid, thyroid hormone, and vitamin D receptors to their cognate response elements. Cell 67:1251–1266
134. Zachman RD (1967) The stimulation of RNA synthesis in vivo and in vitro by retinol (vitamin A) in the intestine of vitamin A deficient rats. Life Sci 6:2207–2213
135. Zardo G, Ciolfi A, Vian L, Billi M, Racanicchi S, Grignani F, Nervi C (2012) Transcriptional targeting by microRNA-polycomb complexes: a novel route in cell fate determination. Cell Cycle 11:3543–3549
136. Zardo G, Ciolfi A, Vian L, Starnes LM, Billi M, Racanicchi S, Maresca C, Fazi F, Travaglini L, Noguera N, Mancini M, Nanni M, Cimino G, Lo-Coco F, Grignani F, Nervi C (2012) Polycombs and microRNA-223 regulate human granulopoiesis by transcriptional control of target gene expression. Blood 119:4034–4046
137. Zelent A, Krust A, Petkovich M, Kastner P, Chambon P (1989) Cloning of murine alpha and beta retinoic acid receptors and a novel receptor gamma predominantly expressed in skin. Nature 339:714–717
138. Zelent A, Mendelsohn C, Kastner P, Krust A, Garnier JM, Ruffenach F, Leroy P, Chambon P (1991) Differentially expressed isoforms of the mouse retinoic acid receptor beta generated by usage of two promoters and alternative splicing. EMBO J 10:71–81

Chapter 2
Architecture of DNA Bound RAR Heterodimers

Natacha Rochel and Dino Moras

Abstract Nuclear Retinoic Acid receptors (RARs) consist of three subtypes, α, β, and γ, encoded by separate genes. They function as ligand-dependent transcriptional regulators, forming heterodimers with Retinoid X receptors (RXRs). RARs mediate the effects of retinoic acid (RA), the active metabolite of Vitamin A, and regulate many biological functions such as embryonic development, organogenesis, homeostasis, vision, immune functions, and reproduction. During the two last decades, a number of in-depth structure–function relationship studies have been performed, in particular with drug design perspectives in the therapeutics for cancer, dermatology, metabolic disease, and other human diseases. Recent structural results concerning integral receptors in diverse functional states, obtained using a combination of different methods, allow a better understanding of the mechanisms involved in molecular regulation. The structural data highlight the importance of DNA sequences for binding selectivity and the role of promoter response elements in the spatial organization of the protein domains into functional complexes.

Abbreviations

AF-1	Activation function 1
AF-2	Activation function 2
ChIP	Chromatin immuno-precipitation

N. Rochel (✉) · D. Moras (✉)
Department of Integrative Structural Biology, Institut de Génétique et de Biologie Moléculaire et Cellulaire (IGBMC), Institut National de Santé et de Recherche Médicale (INSERM), U964, Centre National de Recherche Scientifique (CNRS), UMR7104, Université de Strasbourg, 67404 Illkirch, France
e-mail: rochel@igbmc.fr

D. Moras
e-mail: moras@igbmc.fr

M. A. Asson-Batres and C. Rochette-Egly (eds.), *The Biochemistry of Retinoic Acid Receptors I: Structure, Activation, and Function at the Molecular Level*, Subcellular Biochemistry 70,
DOI: 10.1007/978-94-017-9050-5_2,

DBD	DNA binding domain
DR	Direct repeat
Cryo-EM	Cryo-electron microscopy
FRET	Fluorescence resonance energy transfer
GR	Glucocorticoid receptor
HDX	Hydrogen deuterium exchange
IR	Inverted repeat
LBD	Ligand binding domain
LBP	Ligand binding pocket
NR	Nuclear receptor
NTD	N-terminal domain
PPAR	Peroxisome proliferator-activated receptor
RA	*All-trans* retinoic acid
RAR	Retinoic acid nuclear receptor
RARE	Retinoic acid nuclear receptor response element
RXR	Retinoid X nuclear receptor
SANS	Small angle neutron scattering
SAXS	Small angle X-ray scattering
VDR	Vitamin D nuclear receptor

Introduction

Following the pioneering work of Max Perutz and the British school of biocrystallography in the 1950s, protein crystallography has become one of the most powerful techniques for the analysis of the structure of macromolecules. Milestones have included resolution of the crystal structures of tRNA-Phe in 1974 [21, 47] and nucleic acid protein complexes (aminoacyl-tRNA synthetases/tRNAs, ribosome, RNA polymerase) that have enlightened the structure-function relationships of key steps of the translation of genetic information [9, 49, 50, 60]. The first structure of a membrane protein was resolved in 1984. This accomplishment paved the way for an incursion into an essential domain of biology [10]. Most of our knowledge of macromolecular interactions at the atomic level originates from these pioneering studies.

The field of structural investigation has expanded in the last decade to allow more ambitious questions, such as the study of transient complexes, to be undertaken. A three-dimensional view of molecular interactions, conformational changes, and dynamics of association can now be reconstructed at atomic or near atomic resolution using a variety of approaches and technologies that provide information at different time scales. Combination of these complementary data on the same molecular complex is called integrative structural biology [3].

The structural investigation of nuclear receptors (NR) started in the early nineties. There has been an impressive increase in knowledge about the structure and mechanism of action of NRs from the first reported atomic structure of a DNA binding domain [54] to our current appreciation of the structure of full length

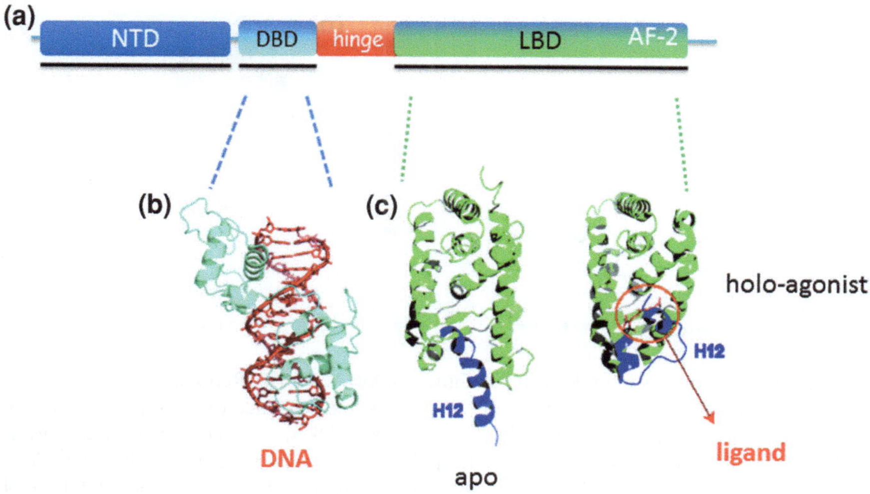

Fig. 2.1 Structural organization of RAR together with the atomic resolution structures of the isolated domain. **a** RAR, like other nuclear receptors, has a modular structure with an unstructured N-terminal domain (NTD) and two well-structured domains, the DNA binding domain (DBD) and the ligand binding domain (LBD). The LBD contains the ligand dependent activation function, AF-2. **b** Crystal structure of the RXR DBD homodimer bound to its DR1 DNA response element (PDB ID: 1BY4). Helices are represented as ribbon in *cyan*. **c** Structural changes in the ligand binding domain induced upon agonist ligand binding. Crystal structures of unliganded (apo) RXRα (PDB ID: 1LBD) and liganded (holo) RARγ bound to *all-trans* retinoic acid (PDB ID: 2LBD) are shown. Helices are represented as ribbon in green with the C-terminal helix H12 labeled and shown in *blue*

receptors bound to DNA and coactivator proteins (review in [17]). Crystallography has provided the bulk of the available information at atomic resolution with some interesting data contributed by researchers using Nuclear Magnetic Resonance (NMR) and Mass Spectrometry (review in [19]).

History: Structural Analysis of Nuclear Receptor Isolated Domains

NRs control a large number of physiological events through their interactions with DNA sequence elements and downstream actions that are set in motion to regulate gene transcription. NR activation is controlled by ligands and cofactors that include repressors, activators, and bridging proteins [25, 39]. Additional fine-tuning is provided by post-translational modifications of NRs that result from cross-talk between different signaling pathways.

NRs share a common structural organization that is comprised of a variable N-terminal domain (NTD) harboring a ligand-independent activation function (AF-1), a conserved DNA binding domain (DBD) and a C-terminal ligand binding domain

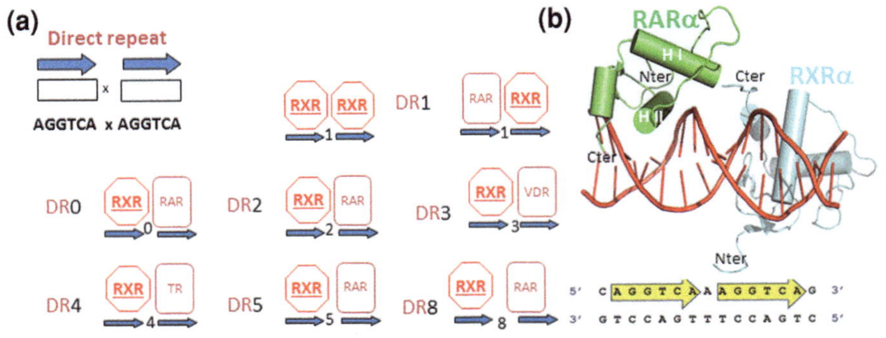

Fig. 2.2 DNA binding of RAR-RXR heterodimers to DNA. **a** DNA Retinoid response elements are composed of direct repeats (DR) of the hexanucleotide sequence (5′-(A/G)G(G/T)TCA-3′) separated by separated by 0 (DR0), 1 (DR1), 2 (DR2), 5 (DR5) or 8 (DR8) nucleotides. RAR-RXR binds to these elements with a specific polarity. **b** Crystal structure of the heterodimer RARα (*green*)-RXRα (*cyan*) DBD in complex with the retinoic response element DR1. (PDB ID: 1DSZ). The DBD core is composed of approximately 66 amino acid residues which form a tertiary structure composed of an N-terminal β-hairpin and two α-helices (H I and H II) followed by a short C-terminal helix and an extension. The response elements are indicated with *yellow arrows*

(LBD). The LBD contains the ligand dependent activation function, AF-2 (Fig. 2.1). The LBD also harbors several interaction surfaces for homodimerization or heterodimerization and for the binding of coregulators. Structural studies of individual DBD and LBD have shed light on the molecular basis of transcription regulation by nuclear receptors. Unfortunately, structural information is still not available for the NTDs containing the ligand-independent function AF-1 or the hinge region connecting the DNA and ligand binding domains. These two domains are highly variable in size and sequence and show poorly defined secondary structures [30].

Studies of DNA Binding Domains

Retinoic acid receptors (RAR) are members of the NR family. The pleiotropic effects of retinoic acid (RA) are mediated through binding interactions with RARs. The RARs, as most of the NRs, function as dimers. RARs partner with retinoid receptors (RXR) to form heterodimers. A general model proposes that RARs bind to cognate RA-response elements (RAREs) both with and without RA bound to the receptor. Recent chromosome-protein precipitation analyses coupled to massive parallel sequencing and bioinformatics analyses carried out using different cell types have led to the identification of thousands of genomic RAR binding sites and RA-regulated gene networks [11, 18, 28, 29, 33]. Analyses of RAR bound loci have confirmed the presence of direct repeat consensus sequences composed of the hexanucleotide motif (5'-(A/G)G(G/T)TCA-3') separated by 1 (DR1), 2 (DR2) or 5 (DR5) nucleotides [2] (Fig. 2.2a). In vitro studies have also shown that a significant number of RAR-RXR heterodimer-occupied sites in embryoid bodies or F9 embryonal carcinoma cells have divergent, non-canonical half-site spacings, including DR0, DR8 and inverted repeat 0 (IR0) elements [33]. RAR-RXR heterodimers bind to the asymmetric DRs with specific polarities [27, 41]. RAR DBDs bind to the half-site at the 3' end of DR5 or DR2 elements, while RXR binds to the 5' half-site of these RAREs. The polarity is reversed in the case of DR1, with RXR bound to the 3'half-site and RAR or other binding partners bound at the 5' end (Fig. 2.2a) [27, 41].

NMR provided the first 3D structure of estrogen NR DBDs [54], and crystallography unraveled the atomic details of DNA-DBD interactions in the glucocorticoid receptor (GR) [26]. It was determined that the DBD core is composed of approximately 66 amino acid residues which form a tertiary structure composed of an N-terminal β-hairpin and two α-helices followed by a short C-terminal helix and an extension (Fig. 2.1). The N-terminal α-helix (helix I) fits into the major groove of the DNA and makes direct and water mediated hydrogen bonds with the nucleotide sequence. In addition, there are a number of interactions between amino acid side chains and the phosphate backbone of the DNA. Helix II (Fig. 2.1) is perpendicular to the N-terminal helix I and stabilizes the core of the DBD.

Current understanding of DNA recognition by RAR-RXR at the atomic level is limited to the crystal structure of the RAR-RXR DBDs bound to a consensus DR1 with identical half-sites, as illustrated in Fig. 2.2b [45]. Unfortunately this structure does not clarify how non-canonical elements are recognized nor how flanking and spacer nucleotides influence the interactions. Consensus motifs could induce potential artifacts [26]. The use of natural DNA sequences from target genes for crystallization may be required to reveal the selectivity process. Such an approach has been used to study GR by the group of Yamamoto who demonstrated that the sequence of the GR binding sites differentially affects receptor conformation and transcriptional activity [32]. Although only minor structural changes could be determined by comparing the numerous GR crystal structures, the study showed that DNA can act as an allosteric effector to modulate GR activity [32]. A recent NMR study of GR DBD-DNA complexes confirms this allosteric mechanism [57]. The molecular structure of several other NR DBDs, such as the thyroid nuclear receptor (TR) or the Vitamin D nuclear receptor (VDR), either in their free states or bound to target DNA, have indicated that DNA sequences specify specific recognition and facilitate allosteric regulation [43, 51, 61].

Studies of Ligand Binding Domains

The first structures of LBDs to be determined were those of unliganded RXRα [5] and liganded RARγ bound to all-trans RA [46]. These structures revealed a novel fold comprising 12 α-helices (H1 to H12) and a short β-turn, arranged in three layers to form an anti-parallel α-helical sandwich (Fig. 2.1). The overall fold has proven to be a prototype for the NR family [58].

The LBD is a key regulatory domain containing the ligand binding pocket (LBP) and multiple interaction surfaces for homo- or hetero- dimerization and for interactions with corepressors, coactivators, and other cofactors that participate in sending signals to the basal transcriptional machinery [40]. Available structural data of NR dimers suggest a conserved interface in the LBD, with helices H7, H9 and H10 of each NR contributing a contact surface of between 1,000 and 1,500 square angstroms. In the absence of ligand, the RAR-RXR heterodimers are associated with corepressor complexes with histone-deacetylase activity that modify chromatin to establish and maintain a repressed transcriptional state [15, 34]. The binding of RA induces a structural transition in the LBD leading to release of the corepressors and the formation of a novel interaction surface for coactivators, including histone acetyltransferases and methyltransferases, as well as chromatin remodeling complexes or components of the basal transcription machinery [1].

LBD crystal structures for most NRs in different oligomeric states have now been determined, and the changes brought about by the binding of a large number of natural and synthetic RAR and RXR selective ligands, including agonists, antagonists, and inverse agonists, have been characterized (review in [19]). The RAR and RXR ligands can be classified by their actions on coregulator recruitment and dissociation. Ligand binding induces allosteric conformational changes that promote or repress receptor-coregulator interactions. Structural data analyses have clearly linked (1) coactivator recruitment with receptor binding of agonists (2) coactivator dissociation with receptor binding of antagonists and (3) corepressor stabilization with receptor binding of inverse agonists.

Agonist ligands induce a unique closed conformation of RAR or RXR LBD with the LBP sealed by helix H12 allowing coactivator to interact. This conformation is referred to as the "holo" or "active" conformation. Comparison of the unliganded RXRα LBD and the "holo" RARγ LBD suggest a ligand-triggered activation mechanism that is accompanied by a repositioning of the C-terminal helix (Fig. 2.1) [46]. Helix H12 (which contains the residues of the AF-2 domain) of Apo-RXRα extends outwards to the solvent, whereas this helix in RA-bound RAR LBDs folds back over the ligand binding pocket (LBP) such that the ligand is entirely buried in a predominantly hydrophobic pocket. Structures of liganded RXR confirm the proposed mouse trap mechanism [12].

Some synthetic ligands bind to RAR or RXR with high affinity, but in contrast to natural ligands that act as agonist, they fail to stabilize the receptors's active conformation and prevent coactivator recruitment using two molecular mechanisms. (1) Inverse agonists induce a conformational change of the receptor protein that stabilizes its interactions with the corepressor. The compound BMS493 that strengthens corepressor interaction with RARa is an example of this type of ligand [24]. (2) Antagonists prevent helix H12 from adopting the active conformation, which disrupts the interaction surface with coactivators. The structural basis of antagonism was provided by the structures of RARα LBD in complex with the synthetic antagonist BMS614 [6] and those of RXRα LBD in complex with LG100754 [52].

The first crystal structure of a functional heterodimer, RAR-RXR showed that hydrophobic interactions play an important role in the relative positioning and stabilization of the dimers [6]. RAR-RXR is a non-permissive heterodimer meaning

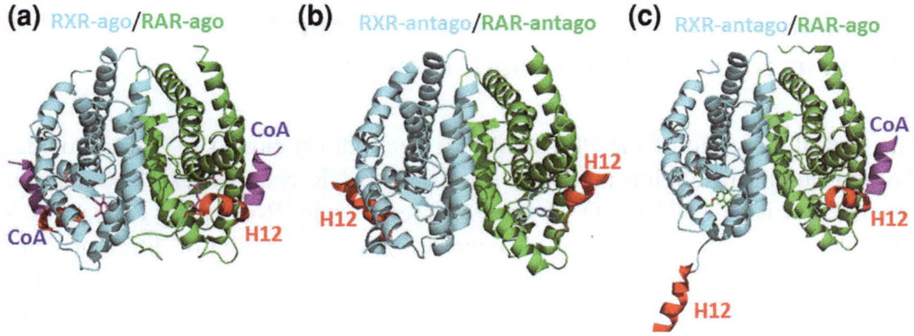

Fig. 2.3 Dimers of RAR-RXR LBDs. The crystal structures of RAR-RXR heterodimers with (**a**) both LBDs bound to agonists (ago) (RAR-*9-cis*-RA-RXR-*9-cis*-RA) (PDB ID: 1XDK), or with (**b**) both LBDs bound to antagonists (antago) (RAR-BMS614-RXR-oleic acid) (PDB ID: 1DKF), or with (**c**) one LBD bound to an agonist (RAR-*all-trans*-RA) and one LBD bound to an antagonist (RXR-LG100754) (PDB ID: 3A9E) have been reported. RXR and RAR are shown as ribbon in *cyan* and *green*, respectively with the C-terminal helices H12 shown in *red*. The coactivator peptide (CoA) that binds to the agonist-bound NR is shown in *pink*

that RAR agonists can activate transcription upon binding to the RAR LBD even if a ligand is not bound to the RXR LBD. In contrast, full responses to RXR ligands only occur if the RAR LBD is occupied by an agonist [13]. Although transactivation will not occur, RXR ligands are able to bind to the heterodimer even in the absence of RAR ligand [20].

The crystal structures of RAR-RXR heterodimers with both LBDs bound to agonists (RAR-9c RA-RXR-9c RA) or of RAR-RXR with both LBDs bound to antagonists (RAR-BMS614-RXR-oleic acid), or with one LBD bound to an agonist (RAR-at RA) and one LBD bound to an antagonist (RXR-LG100754) have been reported (Fig. 2.3) [6, 42, 52]. The observation that the RARα antagonist, BMS614, prevented adoption of the open, active, RARα conformation in full antagonist-occupied heterodimers indicated that helix H12 in RARα occludes the coactivator binding site when the receptor is not activated [6]. Previous studies showed that binding of LG100754 to RXRα led to transactivation mediated by RAR, an effect referred to as the "phantom effect" [23]. Examination of the RXR-LG100754 interactions indicated that this ligand acts as a true RXR antagonist; that is, it prevents helix H12 of the RXRα from folding into an agonist position and instead, causes helix H12 to flip out to the solvent. The antagonism of LG100754 on the RXR LBD does not affect dimerization of RXR with RAR nor does it have any effect on RAR adopting an active conformation upon binding RA.

Taken together, structural studies have indicated that RAR can bind RA and activate transcription, but only if it is interacting with RXR; RXR does not need to have a ligand. On the other hand, RXR can bind a ligand even when RA is not bound to RAR, but the complex will be transcriptionally inactive. Thus, the 'phantom ligand effect' of LG100754 is explained by the fact that direct binding of RA to RAR induces coactivator binding and transcriptional activation independent of RXR LBD antagonism [52].

Development of the Field: Towards Structural Characterization of Full-Length Proteins

The crystal structure of the PPARγ-RXRα heterodimer bound to DNA provided the first atomic resolution model of a full-length NR complex [7]. The solution structures of RARα -RXRα-DNA were also reported by Rochel and colleagues in 2011. These last structures were obtained using a combination of different structural methods, including small angle X-ray (SAXS), neutron scattering, and electron microscopy methods. The cryo-EM structure of VDR-RXRα-DNA provided another high resolution view of the complex that fitted the SAXs data [36]. The work has produced a clear picture of receptor architecture and receptor interaction dynamics with DNA and coregulators [48].

NRs have well-defined domains separated by unstructured linkers that make them inherently flexible. Their N-terminal domains (NTD) are highly variable in both length and sequence and characteristically, unfolded. The hinge domains connecting the DBD to the LBD are also flexible (a requirement for the NRs to recognize and adapt their conformations to DNA response elements of various topologies). Additionally, their conformations are adaptable. These features make it a challenge to determine the full-length structure of NR proteins using crystallographic methods and, even more so, to trap a meaningful functional conformer in the process [31]. As well, the quaternary structure of macromolecules can be affected by crystal packing forces that create artifacts in the crystal structure [55]. Alternate approaches have been developed to address these challenges. Small-angle scattering of X-rays (SAXS) is a method that is specifically tailored for the structural analysis of multi-domain proteins with flexible linkers [44, 56]. SAXS can determine the low-resolution, three-dimensional structure of a macromolecule in close-to-native conditions in a time-resolved manner that also provides information about the kinetics and dynamics of interactive elements and biological processes. Data from NMR [16], mutagenesis [14], fluorescence resonance energy transfer (FRET), or small angle neutron scattering (SANS) are often used to complement and validate SAXS models [44]. The SANS method has the unique capability of measuring diffraction data from samples where part of the multi-component complex can be masked. Another promising method that provides macromolecular solution structures at near atomic resolution is cryo-electron microscopy (Cryo-EM). This method can attain high resolution while avoiding the pitfalls of crystal packing artifacts.

Dynamics of functional complexes have been illustrated using proton exchange methods [22]. In the case of the VDR-RXR-DNA complex, Zhang and colleagues observed that binding of ligand to VDR or RXR causes changes within both the cognate receptor LBD and the receptor partner LBD. A number of these changes map to dimerization regions as well as more distant regions in the complex [61]. These studies suggest that crosstalk between the DBD and LBD promotes allosteric regulation of receptor binding with DNA and cofactors that ultimately tune gene expression.

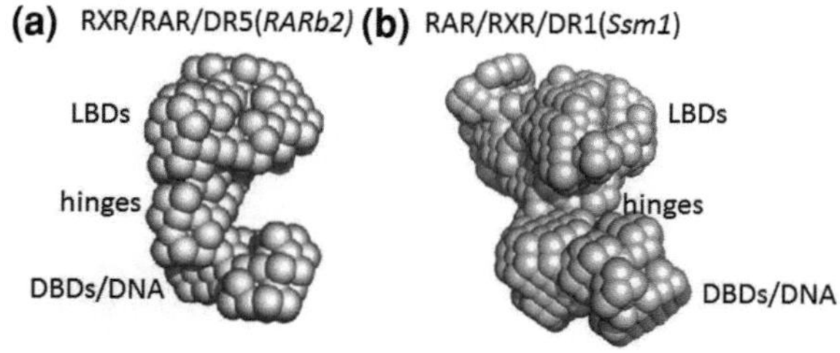

Fig. 2.4 Molecular envelopes calculated from Small Angle X-ray Scattering data of full-length complexes. **a** RAR-RXR-DR5. **b** RAR-RXR-DR1. These envelopes correspond to the low resolution architecture of the complexes showing the positioning of the two main regions of the dimer—the ligand-binding domains (LBDs) and the DNA-binding domains (DBDs) bound to the corresponding DNA

Current State of the Field: Architecture of Full-Length RAR/RXR/DNA Complexes

The atomic model of full-length PPARγ-RXRα bound to a canonical DR1 response element [7] confirmed previous structural information obtained from studies of isolated DBDs bound to DR1 [45] and PPARγ-RXR LBD heterodimer complexes [35, 59]. The PPAR monomer adopted a 'closed' conformation with extensive interactions between the PPAR LBD and DBD and the RXR LBD, hinge, and DBD [7]. In contrast, the RXR monomer had an 'open' conformation with the hinge region extended to create a surface for PPAR binding. The hinge regions adopt different conformations when different ligands are bound, whereas the PPAR LBD exhibits an agonist conformation even when antagonists serve as ligands. In all cases, the NTD was not visible in the electron density map and is probably unfolded. Variable conformations of the hinge regions were observed when comparing the three crystal structures of PPARγ-RXRα in complex with different ligands. The functional correlation of the novel interdomain interactions is limited to a single point mutation having an observable effect on transactivation.

The 'closed' conformation of the PPAR-RXR complex was not observed in solution structure studies carried out using SAXS [48, 38]. Analysis of a number of RXR-NR heterodimers bound to different response elements using SAXS, SANS, and FRET clearly established the existence of a single, or largely dominant, conformer in solution. In contrast with the crystal structure, the solution structures of all heterodimers exhibited an extended asymmetric shape without additional interdomain contacts between the DBDs and LBDs beyond the connection through the hinge regions. The calculated molecular envelopes of RAR-RXR complexes illustrate this important result (Fig. 2.4). For both complexes with DR5 (Fig. 2.4a) and DR1 (Fig. 2.4b), the LBD dimers are positioned at the 5′ end of the target response element with an orientation orthogonal to the DNA axis. The hinge regions are in extended conformations permitting the ordering of the LBDs over the 5′ half-site of the DNA element. The pseudo-atomic models of RAR-RXR-DNA complexes have

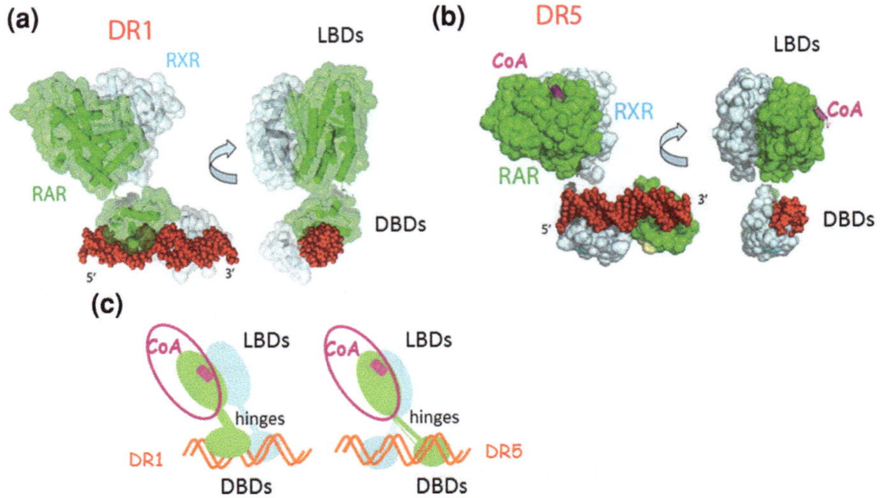

Fig. 2.5 Solution structures of full-length RAR-RXR-DNA complexes. (**a**, **b**) Two views of the solution structure of RAR-RXR (NTDs are truncated from the NRs) in complex with DR1 (**a**) and DR5 (**b**) DNA response elements. RAR is shown in *green*, RXR in *cyan*, DNA in *red*, and the coactivator peptide (CoA) interacting with RAR is shown in *pink*. The molecular models were obtained by docking the crystal structures of the LBD heterodimer and the DBDs bound to the DNA into the ab initio envelopes. The structures were refined as two rigid bodies using experimental diffraction data. Note that in the DR1 complex, the RAR and RXR DBDs are positioned on opposite side on the DNA, whereas in the DR5 complex, the RAR and RXR DBDs are positioned side by side on the DNA. **c** Schematic view of the two complexes that summarizes the main features of the structures: the absence of contacts between the DBDs and LBDs, the same orientation of the LBD heterodimer despite the different polarities of the DBD-DNA, and relative positions of the two DRs. In both cases, the relative position of the coactivator, represented by a *pink circle*, is similar

been determined, combining the available crystal structures of the domains, the SAXS analysis, and biophysical data (Fig. 2.5). Regardless of the different polarity of the bound RAR/RXR heterodimers on DR5 or DR1, the full-length RAR-RXR complexes exhibit a similar extended asymmetric conformation with the LBD dimer positioned on top of the 5' hexanucleotide (Fig. 2.5a for the RAR-RXR-DR1 and Fig. 2.5b for the RAR-RXR-DR5). In both cases, the two domains form an L-shaped structure with the dimer LBD's pseudo two-fold symmetry axis nearly orthogonal to the DNA-bound DBDs. The LBDs can rotate around the two-fold axis with the position controlled by the hinge domains that link to the DBDs. The DBDs are anchored to the DNA response elements and their location is dictated by the relative position of the binding motifs. Addition of one base pair to the spacer sequence induces a shift of approximately 3.5 A and a rotation of 36° of the second hexameric target. Different orientations of heterodimer LBDs on RAREs is a consequence of the geometry dictated by the DNA sequence.

Cryo-EM provided a model of the VDR-RXR-DNA complex (frozen solution) that fits perfectly with the SAXS solution data and gives near-atomic details for the full-length heterodimer complex [36]. The results from these analyses confirmed the flexibility of the full-length receptors, pointed to a lack of secondary structure in the connecting hinge domains of RXR, and underscored the importance of the hinge region in the positioning of the LBDs. Altogether the structural studies suggest that the discrepancy between the crystal and the solution structures are due to crystal packing artifacts. The latter could also explain the structural similarity between agonist and antagonist bound PPARs, a minor conformer present in the crystallization solution being trapped and stabilized by crystal packing forces. The recently published crystal structure of an HNF-4a homodimer bound to a consensus DR1 [8] revealed an asymmetric conformation that is very similar to the one observed in the solution structure of the RAR-RXR-DR1 complex, with a similar positioning of the LBDs and extended conformations of the hinges. The result confirms the concept of common architecture for DNA bound NRs and its extension to homodimers.

The ability of NRs to modulate the expression of target genes results from a combinatorial, coordinated, and sequentially orchestrated exchange between NRs and their coregulators [40]. Several structural models have been proposed for the binding of coactivators to a conserved anchoring cleft within the AF-2 in the LBD. Based on the finding that the primary sequence of the cofactor binding domain usually exhibits two or three LXXLL binding motifs, it was postulated that either two cofactors could bind to one heterodimer (RAR-RXR) or only one to both receptors using two motifs (the "hat model"). In numerous crystal structure models of LBD dimers in complex with short cofactor peptides, the stoichiometry is always 2:2, supporting both models. This observation may be an artifact arising from the addition of excess peptide during crystallization and the low binding affinity of the complex for the peptide compared to that of a larger coactivator domain. The first unambiguous structural evidence for a 2:1 stoichiometry for the receptor-coactivator complex was provided by solution studies of full-length receptors (RAR-RXR and VDR-RXR bound to DNA) bound to DNA and large coactivator protein fragments [37, 48]. Each heterodimer was shown to bind only one coactivator protein via the RXR partner. This preferential binding was controlled by affinity rather than by steric exclusion. Indeed, RAR antagonists prevent coactivator binding, whereas mutation of residues in the RXR coactivator binding cleft of RXR have no effect on the stoichiometry. The molecular model resulting from the experimental diffraction data indicate that the coactivator interacting domain is on one side of the DNA opposite to the RXR LBD and DBDs (Fig. 2.5c).

Functional Relevance

The solution structures of NRs bound to RARE reveal two key features (1) the position of LBDs at the 5′ end of the target DNAs is conserved regardless of the polarity of the response elements (2) the binding of only one coactivator molecule per heterodimer through the RXR partner [48]. The combination of these two features explains the key role of DNA in NR dependent transcription regulation. The response elements direct the relative position of the LBDs and the DNA helix, which in turn fixes the binding site of the cofactors (Fig. 2.6).

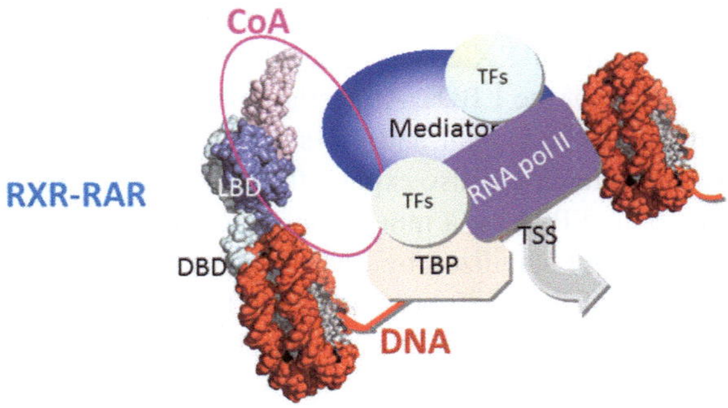

Fig. 2.6 Functional implication of the conserved relative positions of RAR and the bound coactivator. The model of RAR-RXR bound to the RARβ2 promoter illustrates the importance of the DNA sequence in orienting the protein complex and its association with coactivator to the transcriptional initiation complex. To begin transcription, eukaryotic RNA polymerase II (RNA Pol II) requires the general transcription factors (TFs) to be associated at the promoter. The promoter contains a DNA sequence called the TATA box, located 25 nucleotides away from the site where transcription is initiated (TSS), that is recognized by the TATA box binding protein (TBP). The rest of the general transcription factors (TFs) as well as the Mediator assemble at the promoter. Docking of the complex RAR-RXR-DNA on a nucleosome of the RARβ2 promoter using experimental DNA protection data shows that the position of the RAR LBD orients the bound coactivator (CoA) allowing its association with the transcription machinery

In summary the molecular structures show that DNA binding controls the architecture of the complexes. The polarity of the binding motifs and the number of dinucleotide spacers modulates rotation of the LBDs and the relative positions of the receptors. As a result, the environment of the accessible surface in the active complex is different for each receptor. In addition to tethering the NR near the transcription start site of target genes (Fig. 2.6), the architecture of the DNA response element can also serve as an allosteric regulator of receptor function and receptor association with coregulators [4, 32, 43, 51, 61].

The solution structures make it possible to address another functionally related question: is there a sequential order for complex formation? The structural data provide several snapshots of different functional states that suggest that the heterodimer forms first and then binds to DNA. Such a process, which combines two DBDs for a simultaneous recognition of the response element, is an efficient way to overcome the specificity problem with low affinity constants for each single DBD. The structural data also show that the extended conformation is recognized and maintained during the subsequent step, namely coactivator binding.

Future Directions

A combination of structural methods has elucidated the architecture of full-length RAR-RXR complexes bound to DR5 or DR1 RARE. However, the precise structural organization of RAR-RXR-coregulator complexes on consensus and non-consensus DR and IR elements has not been determined. The polarity of the DNA molecule raises several questions in relation to the specific sequence of the target elements and their relative position on DR or IR. A complete understanding of the role and more specifically, the allosteric effects of DNA binding elements will require further data regarding the structure and organization of individual domains and full length NRs bound to different elements in different functional states. The capacity of NRs to specifically interact with numerous partners, such as DNA and protein cofactors, has functional consequences which are driven by mechanisms that are yet to be revealed. NRs and their coregulators are subject to post-translational modifications, including phosphorylation, acetylation, methylation and sumoylation that allosterically influence their functions. The physico-chemical details underlying the assembly and coordination of these large, transient, dynamic macromolecular complexes and the impact of post-translational modifications are yet unknown. Future studies will utilize multiple structural approaches to assemble information on complexes in multiple functional states, a first step towards "cellular structural biology".

References

1. Aoyagi S, Archer TK (2008) Dynamics of coactivator recruitment and chromatin modifications during nuclear receptor mediated transcription. Mol Cell Endo 280:1–5
2. Balmer JE, Blomhoff R (2005) A robust characterization of retinoic acid response elements based on a comparison of sites in three species. J Steroid Biochem Mol Biol 96:347–354
3. Berger I, Blanco AG, Boelens R, Cavarelli J, Coll M, Folkers GE, Nie Y, Pogenberg V, Schultz P, Wilmanns M, Moras D, Poterszman A (2011) Structural insights into transcription complexes. J Struct Biol 175:135–146
4. Billas I, Moras D (2013) Allosteric controls of nuclear receptor function in the regulation of transcription. J Mol Biol 425:2317–2329
5. Bourguet W, Ruff M, Chambon P, Gronemeyer H, Moras D (1995) Crystal structure of the ligand-binding domain of the human nuclear receptor RXR-alpha. Nature 375:377–382
6. Bourguet W, Vivat V, Wurtz JM, Chambon P, Gronemeyer H, Moras D (2000) Crystal structure of a heterodimeric complex of RAR and RXR ligand-binding domains. Mol Cell 5:289–298
7. Chandra V, Huang P, Hamuro Y, Raghuram S, Wang Y, Burris TP, Rastinejad F (2008) Structure of the intact PPAR-gamma-RXR- nuclear receptor complex on DNA. Nature 456:350–356
8. Chandra V, Huang P, Potluri N, Wu D, Kim Y, Rastinejad F (2013) Multidomain integration in the structure of the HNF-4alpha nuclear receptor complex. Nature 495:394–398
9. Cramer P, Bushnell DA, Kornberg RD (2001) Structural basis of transcription: RNA polymerase II at 2.8 angstrom resolution. Science 292:1863–1876
10. Deisenhofer J, Epp O, Miki K, Huber R, Michel H (1984) X-ray structure analysis of a membrane protein complex. Electron density map at 3 A resolution and a model of the chromophores of the photosynthetic reaction center from Rhodopseudomonas viridis. J Mol Biol 180:385–398
11. Delacroix L, Moutier E, Altobelli G, Legras S, Poch O, Choukrallah MA, Bertin I, Jost B, Davidson I (2010) Cell-specific interaction of retinoic acid receptors with target genes in mouse embryonic fibroblasts and embryonic stem cells. Mol Cell Biol 30:231–244
12. Egea PF, Mitschler A, Rochel N, Ruff M, Chambon P, Moras D (2000) Crystal structure of the human RXRalpha ligand-binding domain bound to its natural ligand: 9-cis retinoic acid. EMBO J 19:2592–2601
13. Germain P, Iyer J, Zechel C, Gronemeyer H (2002) Co-regulator recruitment and the mechanism of retinoic acid receptor synergy. Nature 415:187–192

14. Gherardi E, Sandin S, Petoukhov MV, Finch J, Youles ME, Ofverstedt LG, Miguel RN, Blundell TL, Vande Woude GF, Skoglund U, Svergun DI (2006) Structural basis of hepatocyte growth factor/scatter factor and MET signalling. Proc Natl Acad Sci USA 103:4046–4051
15. Glass CK, Rosenfeld MG (2000) The coregulator exchange in transcriptional functions of nuclear receptors. Genes Dev 14:121–141
16. Hammel M (2012) Validation of macromolecular flexibility in solution by small-angle X-ray scattering (SAXS). Eur Biophys J 41:789–799
17. Helsen C, Claessens F (2013) Looking at nuclear receptors from a new angle. Mol Cell Endocrinol 382:97–106
18. Hua S, Kittler R, White KP (2009) Genomic antagonism between retinoic acid and estrogen signaling in breast cancer. Cell 137:1259–1271
19. Jin L, Li Y (2010) Structural and functional insights into nuclear receptor signaling. Adv Drug Deliv Rev 62:1218–1226
20. Kersten S, Dawson MI, Lewis BA, Noy N (1996) Individual subunits of heterodimers comprised of retinoic acid and retinoid X receptors interact with their ligands independently. Biochemistry 35:3816–3824
21. Kim SH, Suddath FL, Quigley GJ, McPherson A, Sussman JL, Wang AH, Seeman NC, Rich A (1974) Three-dimensional structure of yeast phenylalanine transfer RNA. Science 185:435–440
22. Konermann L, Pan J, Liu YH (2011) Hydrogen exchange mass spectrometry for studying protein structure and dynamics. Chem Soc Rev 40:1224–1234
23. Lala DS, Mukherjee R, Schulman IG, Koch SS, Dardashti LJ, Nadzan AM, Croston GE, Evans RM, Heyman RA (1996) Activation of specific RXR heterodimers by an antagonist of RXR homodimers. Nature 383:450–453
24. le Maire A, Teyssier C, Erb C, Grimaldi M, Alvarez S, de Lera AR, Balaguer P, Gronemeyer H, Royer CA, Germain P, Bourguet W (2010) A unique secondary-structure switch controls constitutive gene repression by retinoic acid receptor. Nat Struct Mol Biol 17:801–807
25. Lonard DM, O'Malley BW (2012) Nuclear receptor coregulators: modulators of pathology and therapeutic targets. Nat Rev Endocrinol 8:598–604
26. Luisi BF, Xu WX, Otwinowski Z, Freedman LP, Yamamoto KR, Sigler PB (1991) Crystallographic analysis of the interaction of the glucocorticoid receptor with DNA. Nature 352:497–505
27. Mader S, Chen JY, Chen Z, White J, Chambon P, Gronemeyer H (1993) The patterns of binding of RAR, RXR and TR homo- and heterodimers to direct repeats are dictated by the binding specificites of the DNA binding domains. EMBO J 12:5029–5041
28. Martens JH, Brinkman AB, Simmer F, Francoijs KJ, Nebbioso A, Ferrara F, Altucci L, Stunnenberg HG (2010) PML-RARalpha/RXR Alters the Epigenetic Landscape in Acute Promyelocytic Leukemia. Cancer Cell 17:173–185
29. Martens JH, Rao NA, Stunnenberg HG (2011) Genome-wide interplay of nuclear receptors with the epigenome. Biochim Biophys Acta 1812:818–823
30. McEwan IJ (2012) Nuclear hormone receptors: allosteric switches. Mol Cell Endocrinol 348:345–347
31. McEwan IJ, Lavery D, Fischer K, Watt K (2007) Natural disordered sequences in the amino terminal domain of nuclear receptors: lessons from the androgen and glucocorticoid receptors. Nucl Recept Signal 5:e001
32. Meijsing SH, Pufall MA, So AY, Bates DL, Chen L, Yamamoto KR (2009) DNA binding site sequence directs glucocorticoid receptor structure and activity. Science 324:407–410
33. Moutier E, Ye T, Choukrallah MA, Urban S, Osz J, Chatagnon A, Delacroix L, Langer D, Rochel N, Moras D, Benoit G, Davidson I (2012) Retinoic acid receptors recognize the mouse genome through binding elements with diverse spacing and topology. J Biol Chem 287:26328–26341
34. Nagy L, Kao HY, Love JD, Li C, Banayo E, Gooch JT, Krishna V, Chatterjee K, Evans RM, Schwabe JW (1999) Mechanism of corepressor binding and release from nuclear hormone receptors. Genes Dev 13:3209–3216

35. Nolte RT, Wisely GB, Westin S, Cobb JE, Lambert MH, Kurokawa R, Rosenfeld MG, Willson TM, Glass CK, Milburn MV (1998) Ligand binding and co-activator assembly of the peroxisome proliferator-activated receptor-gamma. Nature 95:137–143
36. Orlov I, Rochel N, Moras D, Klaholz BP (2012) Structure of the full human RXR/VDR nuclear receptor heterodimer complex with its DR3 target DNA. EMBO J 31:291–300
37. Osz J, Pethoukhov MV, Sirigu S, Svergun DI, Moras D, Rochel N (2012) Solution structures of PPARgamma2/RXRalpha complexes. PPAR Res 701412
38. Osz J, Brélivet Y, Peluso-Iltis C, Cura V, Eiler S, Ruff M, Bourguet W, Rochel N, Moras D (2012) Structural basis for a molecular allosteric control mechanism of cofactor binding to nuclear receptors. Proc Natl Acad Sci USA 109:E588–E594
39. Pawlak M, Lefebvre P, Staels B (2012) General molecular biology and architecture of nuclear receptors. Curr Top Med Chem 12:486–504
40. Perissi V, Rosenfeld MG (2005) Controlling nuclear receptors: the circular logic of cofactor cycles. Nat Rev Mol Cell Biol 6:542–554
41. Perlmann T, Rangarajan PN, Umesono K, Evans RM (1993) Determinants for selective RAR and TR recognition of direct repeat HREs. Genes Dev 7:1411–1422
42. Pogenberg V, Guichou JF, Vivat-Hannah V, Kammerer S, Pérez E, Germain P, de Lera AR, Gronemeyer H, Royer CA, Bourguet W (2005) Characterization of the interaction between retinoic acid receptor/retinoid X receptor (RAR/RXR) heterodimers and transcriptional coactivators through structural and fluorescence anisotropy studies. J Biol Chem 280:1625–1633
43. Putcha BD, Fernandez EJ (2009) Direct interdomain interactions can mediate allosterism in the thyroid receptor. J Biol Chem 284:22517–22524
44. Rambo RP, Tainer JA (2013) Super-resolution in solution x-ray scattering and its applications to structural systems biology. Annu Rev Biophys. 42:415–441
45. Rastinejad F, Wagner T, Zhao Q, Khorasanizadeh S (2000) Structure of the RXR-RAR DNA-binding complex on the retinoic acid response element DR1. EMBO J 19:1045–1054
46. Renaud JP, Rochel N, Ruff M, Vivat V, Chambon P, Gronemeyer H, Moras D (1995) Crystal structure of the RAR-gamma ligand-binding domain bound to all-trans retinoic acid. Nature 378:681–689
47. Robertus JD, Ladner JE, Finch JT, Rhodes D, Brown RS, Clark BF, Klug A (1974) Structure of yeast phenylalanine tRNA at 3 A resolution. Nature 250:546–551
48. Rochel N, Ciesielski F, Godet J, Moman E, Roessle M, Peluso-Iltis C, Moulin M, Haertlein M, Callow P, Mély Y, Svergun DI, Moras D (2011) Common architecture of nuclear receptor heterodimers on DNA direct repeat elements with different spacings. Nat Struct Mol Biol 18:564–570
49. Rould MA, Perona JJ, Söll D, Steitz TA (1989) Structure of E. coli glutaminyl-tRNA synthetase complexed with tRNA(Gln) and ATP at 2.8 A resolution. Science 246:1135–1142
50. Ruff M, Krishnaswamy S, Boeglin M, Poterszman A, Mitschler A, Podjarny A, Rees B, Thierry JC, Moras D (1991) Class II aminoacyl transfer RNA synthetases: crystal structure of yeast aspartyl-tRNA synthetase complexed with tRNA(Asp). Science 252:1682–1689
51. Santos GM, Fairall L, Schwabe JW (2011) Negative regulation by nuclear receptors: a plethora of mechanisms. Trends Endocrinol Metab 22:87–93
52. Sato Y, Ramalanjaona N, Huet T, Potier N, Osz J, Antony P, Peluso-Iltis C, Poussin-Courmontagne P, Ennifar E, Mély Y, Dejaegere A, Moras D, Rochel N (2010) The phantom effect of the Rexinoid LG100754: structural and functional insights. PLoS ONE 5:e15119
53. Schulman IG, Li C, Schwabe JW, Evans RM (1997) The phantom ligand effect: allosteric control of transcription by the retinoid X receptor. Genes Dev 11:299–308
54. Schwabe JW, Neuhaus D, Rhodes D (1990) Solution structure of the DNA-binding domain of the oestrogen receptor. Nature 348(6300):458–461
55. Svergun DI, Petoukhov MV, Koch MH, König S (2000) Crystal versus solution structures of thiamine diphosphate-dependent enzymes. J Biol Chem 275:297–302
56. Svergun DI, Petoukhov MV, Koch MH (2001) Determination of domain structure of proteins from X-ray solution scattering. Biophys J 80:2946–2953
57. Watson LC, Kuchenbecker KM, Schiller BJ, Gross JD, Pufall MA, Yamamoto KR (2013) The glucocorticoid receptor dimer interface allosterically transmits sequence-specific DNA signals. Nat Struct Mol Biol 20(7):876–883

58. Wurtz JM, Bourguet W, Renaud JP, Vivat V, Chambon P, Moras D, Gronemeyer H (1996) A canonical structure for the ligand-binding domain of nuclear receptors. Nat Struct Biol 3:87–94
59. Xu HE, Lambert MH, Montana VG, Plunket KD, Moore LB, Collins JL, Oplinger JA, Kliewer SA, Gampe RT Jr, McKee DD, Moore JT, Willson TM (2001) Structural determinants of ligand binding selectivity between the peroxisome proliferator-activated receptors. Proc Natl Acad Sci U S A 98:13919–13924
60. Yusupov MM, Yusupova GZ, Baucom A, Lieberman K, Earnest TN, Cate JH, Noller HF (2001) Crystal structure of the ribosome at 5.5 A resolution. Science 292:883–896
61. Zhang J, Chalmers MJ, Stayrook KR, Burris LL, Wang Y, Busby SA, Pascal BD, Garcia-Ordo nez RD, Bruning JB, Istrate MA, Kojetin DJ, Dodge JA, Burris TP, Griffin PR (2011) DNA binding alters coactivator interaction surfaces of the intact VDR-RXR complex. Nat Struct Mol Biol 18:556–563

Chapter 3
Retinoic Acid Receptors: Structural Basis for Coregulator Interaction and Exchange

Albane le Maire and William Bourguet

Abstract In the form of heterodimers with retinoid X receptors (RXRs), retinoic acid receptors (RARs) are master regulators of gene expression in humans and important drug targets. They act as ligand-dependent transcription factors that regulate a large variety of gene networks controlling cell growth, differentiation, survival and death. The biological functions of RARs rely on a dynamic series of coregulator exchanges controlled by ligand binding. Unliganded RARs exert a repressor activity by interacting with transcriptional corepressors which themselves serve as docking platforms for the recruitment of histone deacetylases that impose a higher order structure on chromatin which is not permissive to gene transcription. Upon ligand binding, the receptor undergoes conformational changes inducing corepressor release and the recruitment of coactivators with histone acetylase activities allowing chromatin decompaction and gene transcription. In the following, we review the structural determinants of the interaction between RAR and either type of coregulators both at the level of the individual receptor and in the context of the RAR-RXR heterodimers. We also discuss the molecular details of the fine tuning of these associations by the various pharmacological classes of ligands.

A. le Maire (✉) · W. Bourguet
Inserm U1054, Centre de Biochimie Structurale, 29 rue de Navacelles, 34090 Montpellier, France
e-mail: lemaire@cbs.cnrs.fr

W. Bourguet
e-mail: bourguet@cbs.cnrs.fr

A. le Maire · W. Bourguet
CNRS UMR 5048, Universités Montpellier 1 & 2, 34090 Montpellier, France

M. A. Asson-Batres and C. Rochette-Egly (eds.), *The Biochemistry of Retinoic Acid Receptors I: Structure, Activation, and Function at the Molecular Level*, Subcellular Biochemistry 70,
DOI: 10.1007/978-94-017-9050-5_3, © Springer Science+Business Media Dordrecht 2014

Abbreviations

AF	Activation function
CoRNR	Corepressor/nuclear receptor
DBD	DNA-binding domain
HAT	Histone acetyltransferase
HDAC	Histone deacetylase
LBD	Ligand-binding domain
LBP	Ligand-binding pocket
N-CoR	Nuclear receptor corepressor
NR	Nuclear receptor
NTD	N-terminal domain
PPAR	Peroxisome proliferator-activated receptor
RAR	Retinoic acid receptor
RXR	Retinoid X receptor
SMRT	Silencing mediator for retinoid and thyroid hormone receptors

A Brief History of Nuclear Receptor Coregulators

Squelching experiments led to the prediction that additional factors were assisting NRs in transmitting the hormone-induced signal to the transcription machinery. Squelching occurs if a receptor inhibits the activity of the same (autosquelching) or a different (heterosquelching) receptor that is competing to bind to a common and limiting interacting protein [5, 50, 72]. Experiments in yeast confirmed that core polymerase proteins and additional helper proteins facilitate communication between transcription factors and the polymerase II complex [4, 63]. Biochemical and expression cloning approaches showed that some factors interact with NRs in either a ligand-independent or ligand-dependent manner, and cotransfection assays demonstrated that many of these factors could potentiate NR activity. Numerous cell culture and animal model studies carried out with each class of coregulatory protein have elucidated their biological roles [26, 59].

The discovery of coactivators (Fig. 3.1a) and corepressors (Fig. 3.1b) revealed that NR-mediated transcription is subject to both positive and negative regulation (Fig. 3.1c). Factors such as the steroid receptor coactivator (SRC-1 [56]) and the transcriptional intermediary factor 2 (TIF-2 [77]) have been shown to serve as coactivators. The observation that unliganded thyroid hormone receptor could function as a transcriptional repressor [2, 3] led to the characterization and cloning of corepressor factors, including the Nuclear receptor CoRepressor (N-CoR/NCoR1/RIP13 [30]) and the Silencing Mediator of Retinoic acid receptor and Thyroid hormone receptor (SMRT/NCoR2/TRAC [11]).

When NR coregulatory proteins were first described, it was predicted that only a few coactivators and corepressors were interacting with NRs. However, contrary to prediction, approximately 350 coregulators have been reported to date (http://www.

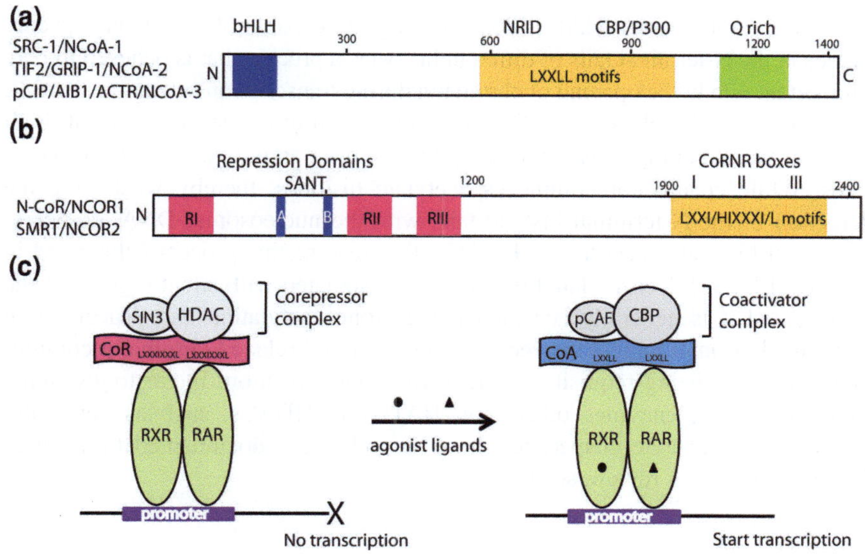

Fig. 3.1 Functional organization of coactivators and corepressors. **a** The functional domains of the members of the SRC-1/p160 family of coactivators illustrated are: LXXLL motifs belonging to the NR and CBP-interaction domains; a basic helix-loop-helix domain (bHLH) at the N-terminus and a glutamine rich region at the C-terminus. **b** The corepressors (N-CoR and SMRT) functional domains illustrated are: C-terminal CoRNR boxes I, II and III including the LXXI/HIXXXI/L motifs mediating NR interactions; RI, RII and RIII representing domains that harbor intrinsic repression activity when tethered to the DNA-binding domain of GAL4; the two SANT domains (**a** and **b**). **c** Model for the exchange of a corepressor (CoR) complex with a coactivator (CoA) complex. In the absence of ligand, the RAR/RXR heterodimer interacts with a corepressor complex via ligand-independent interactions with LXXIXXL motifs of CoR, inducing gene repression. In the presence of agonist ligands, RXR/RAR recruits the coactivator complex through interaction with the LXXLL helices of coactivators, resulting in transcriptional activation

NURSA.org). Interaction of NRs with this vast array of coregulators provides a complex regulatory framework for controlling the transcription of target genes. High-throughput mass spectrometric analyses of coregulator complexes have revealed that coregulators do not work alone, but instead function as multicomponent protein complexes [34, 46]. The primary amino acid sequences of these proteins indicate a wide range of enzymatic functions that control transcription [48, 49]. Thus, coregulators are not merely "bridging" agents, but are a vast enzymatic tool chest necessary to execute the many steps of transcription [44, 68].

Coactivators can be subdivided into two groups. Secondary coactivators represent a subgroup of molecules that are constituents of multisubunit coactivator complexes and that also contribute to the enhancement of NR-mediated transcription, but that do not directly contact the NRs. Primary coactivators, such as those of the TIF-2/SRC-1/RAC3 (p160) family, mediate the interaction of coactivator complexes with NRs. CBP, p300, P/CAF, and some p160 coactivators themselves are reported to act as histone acetyltransferases (HATs) [26, 43]. They are capable of acetylating specific residues in the N-terminal tails of different histones, a process that is believed to play an important role in the opening of chromatin during transcription activation [10, 33].

Corepressors N-CoR and SMRT have been shown to reside in, or recruit, high molecular weight complexes that display the opposite activity of coactivator complexes. While coactivator complexes acetylate histones, thereby weakening the interaction of the N-terminal histone tails with the nucleosomal DNA, corepressors recruit histone deacetylases (HDACs) that reverse this process (illustrated in Fig. 3.1c) [29, 53]. Deacetylated histones are associated with silent regions of the genome, and it is generally accepted that histone acetylation and deacetylation shuffle nucleosomal targets between a condensed and relaxed chromatin configuration, the latter being requisite for transcriptional activation. Interestingly, some histone-modifying enzymes other than HATs and HDACs, such as chromatin remodelers and histone chaperones, have recently been documented to serve as NR coregulators (for review, see [35]).

Discovery of the Basic Principles of NR-Coregulator Interaction

Cloning of NR coregulators was rapidly followed by the characterization of coregulator/NR consensus sequence signature boxes. The amino acid sequence motif, LxxLL, where L represents leucine and x is any amino acid, was found to be embedded in a short α-helical peptide of coactivator proteins such as SRC-1 and TIF-2 (Fig. 3.2a) [28, 40, 73, 76]. Early functional and biochemical studies established that these motifs are important for coregulator interaction with NRs and inferred a putative interaction site [18, 20]. It is now well established that NR boxes are necessary and sufficient for ligand-dependent direct interaction of coactivators with a cognate surface in the NR LBD that constitutes the transcriptional activation function, AF-2. Different coactivators, including closely related ones from the TIF-2/SRC-1/RAC3 family, are observed to display some degree of NR selectivity [13, 18, 47]. The fact that some coactivators contain multiple LxxLL motifs (four in SRC-1 and three in TIF-2, Fig. 3.2a), most of which appear to be functional in terms of NR binding in vitro, brings up the still unsolved question of whether this multiplicity reflects redundancy or confers some specificity to the interface.

Three conserved corepressor/NR signature box motifs (CoRNR box 1–3) LxxI/HIxxxI/L have been identified in SMRT and N-CoR (Fig. 3.2b) [31, 54, 60]. They were predicted to adopt a longer amphipathic helical conformation than the LxxLL motif and bind to a receptor surface that overlaps the coactivator recognition surface. It was also discovered that CoRNR boxes are not equivalent, as for example, RAR interacts strongly with CoRNR1 but very weakly with CoRNR2. Further biochemical studies revealed that residues flanking the core helical sequence determine NR specificity [32, 60, 82].

```
(a)
                 LXXLL
SRC-1 NR1   SQTSHKLVQLLTTTAE
SRC-1 NR2   TARHKILHRLLQEGSP
SRC-1 NR3   SKDHQLLRYLLDKDEK
SRC-1 NR4   AQQKSLLQQLLTE*
TIF-2 NR1   SKGQTKLLQLLTTKSD
TIF-2 NR2   KEKHKILHRLLQDSSS
TIF-2 NR3   KKENALLRYLLDKDDT
Sec. Str.

(b)
                 LXXXIXXIIXXXF
S-CoRNR1    RVVTLAQHISEVITQDYTRHHPQ
N-CoRNR1    RLITLADHICQIITQDFARNQVS
Sec. Str.

                     LXXIIXXXL
S-CoRNR2    EHASTNMGLEAIIRKALMGKYDQ
N-CoRNR2    ADPASNLGLEDIIRKALMGSFDD
Sec. Str.

                     IXXIIXXXI
S-CoRNR3    TTLTAATFIDAIITRQIAHERGP
N-CoRNR3    TTITAANFIDVIITRQIASDKDA
```

Fig. 3.2 Corepressor and coactivator receptor interaction motifs. **a** Amino acid sequence alignment of the NR boxes present in two coactivators, SRC-1 and TIF-2. **b** Amino acid sequence alignment of the three CoRNR motifs of the two corepressors, SMRT and N-CoR. Residues of the consensus motifs are highlighted. Secondary-structure (Sec. str.) elements as observed in the crystal structures are indicated

Current Structural View of RAR-Coregulator Interactions

Overall Structure of the RAR Ligand-Binding Domain

Similar to all other NRs, the LBD of RARs is organized as an antiparallel α-helical sandwich with 12 conserved helices arranged in three layers with a β-turn (S1 and S2) situated between helices H5 and H6 [25, 36, 37, 41, 64]. Helices H4, H5, H8, H9 and H11 are sandwiched between H1, H2 and H3 on one side and H6, H7 and H10 on the other side (Fig. 3.3). In contrast, H12 is flexible and can adopt various conformations depending on the ligation state. This LBD architecture generates a ligand-binding pocket (LBP) primarily made up of hydrophobic residues that are present in helices H3, H5, H11 and the β-sheet segment (Fig. 3.3). Crystallographic studies have revealed that bound ligands, such as all-*trans*-retinoic acid (RA), are stabilized in the RAR LBP through extensive van der Waals contacts and a network of ionic and hydrogen bonds between the carboxylate moiety of RA, a conserved arginine in H5, and water molecules [7, 42, 64]. The shape of the LBP matches the volume of the ligand, maximizing the hydrophobic contacts and contributing to the selectivity of ligand binding [22, 36, 37].

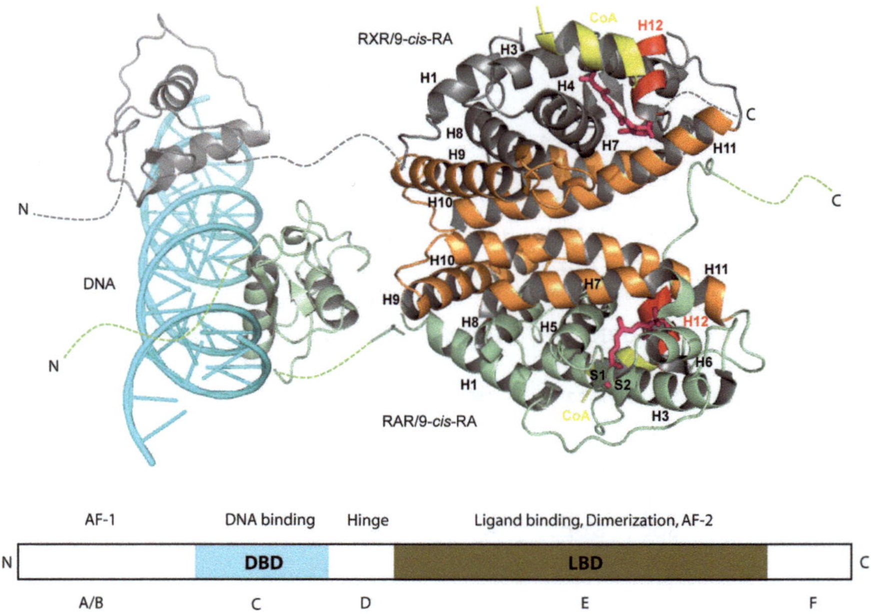

Fig. 3.3 Structural and functional organization of nuclear receptors. Like other nuclear receptors, retinoic acid receptors (RAR*s*) and retinoid X receptors (RXR*s*) consist of six structural and functional domains (*A–F*) based on regions of conserved sequence and function. The DNA-binding domain (DBD) is composed of two zinc-finger motifs and mediates sequence-specific DNA recognition. The ligand-binding domain (LBD) mediates ligand binding, dimerization and a ligand-dependent transactivation function (AF-2). The N-terminal *A/B* region contains a ligand-independent activation function (AF-1). The region *D* corresponds to a linker allowing the proper orientation of DBDs and LBDs within DNA-bound dimers. The *F* region is not present in all receptors and its function is poorly understood. The model of the full-length RXR-RAR heterodimer bound to DNA was virtually reconstituted from the crystal structures of the DNA-bound and 9-*cis*-retinoic acid-bound RAR-RXR DBD (Protein Data Bank code 1dsz) and LBD (Protein Data Bank code 1xdk) heterodimers, respectively. LBDs and DBDs are represented as ribbons. The dimerization interface comprising helices H7, H9, H10 and H11 as well as loops L8-9 and L9-10 of each LBD is colored in *orange*. The rest of RXR LBD and RAR LBD are colored in *gray* and *green*, respectively. Helix H12 of each monomer is highlighted in *red*. The coactivator peptides interacting with the heterodimer are drawn in *yellow* and the agonist ligand (9-*cis*-RA) is drawn as *pink sticks*. *Dotted lines* denote regions with unresolved structures

Structural Basis for RAR-Coactivator Interaction

Based on available crystal structures, recruitment of coactivators by RAR occurs when a ligand-induced receptor surface is formed that accommodates coactivator binding via interactions between RAR and the protein's LxxLL NR signature box motif [41, 57, 62, 79]. Structural analyses of RAR bound to an agonist and a coactivator peptide fragment containing the LxxLL motif indicate that the C-terminal, H12 helix seals the LBP (Fig. 3.4a). This particular position of H12

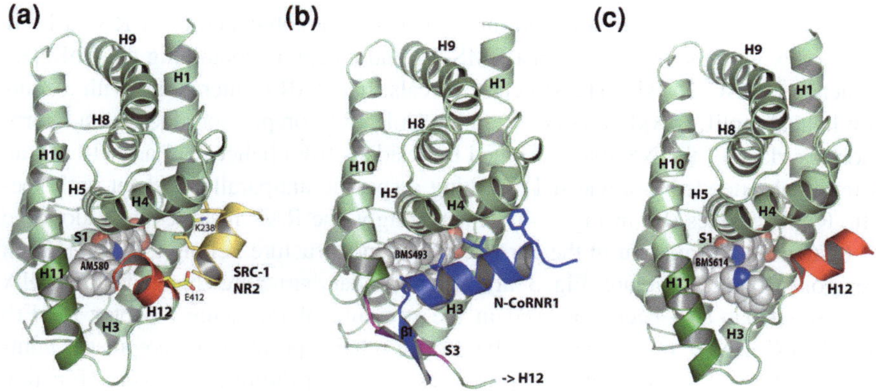

Fig. 3.4 Crystal structures of RAR LBD in various functional states. **a** Overall structure of the RAR LBD/SRC-1 NR2/AM580 (agonist) complex (Protein Data Bank code 3kmr) [41]. Helix H12 (*red*) is positioned into the so-called active position, allowing the formation of a surface specifically recognized by short LXXLL motifs (*yellow sticks*) contained in coactivators (CoA, *yellow*). Lys238 (H3) and Glu412 (H12) which generate a charge clamp interacting with the LXXLL helical motif of CoA are highlighted as *yellow* sticks. **b** Overall structure of the RAR LBD/N-CoRNR1/BMS493 (inverse agonist) complex (Protein Data Bank code 3kmz) [41]. The inverse agonist ligand stabilizes the interaction between RAR and the corepressor (CoR, *blue*) that involves an antiparallel β-sheet interface comprising S3 from RAR (*purple*) and β1 from the CoR (*blue*). The remaining of N-CoRNR1 folds as a four-turn α-helix which docks into the coregulator groove of RAR through the conserved LXXXIXX(I/V)IXXX(Y/F) motif (*blue sticks*). **c** Overall structure of the RAR LBD/BMS614 (antagonist) complex (Protein Data Bank code 1dkf) [8]. Neutral antagonist binding forces a displacement of H12 helix (*red*) that prevents coactivator recruitment and maintains the receptor in an inactive state

defines a hydrophobic surface comprising the C-terminal part of helix H3, helix H4, and the L3–4 loop linking H3 and H4. This region accommodates the short LxxLL-containing helical motif of coactivator proteins. The coactivator helix is held in place by interactions of the leucine residues with the hydrophobic groove and by hydrogen bonds with a lysine at the C-terminus of H3 and a glutamate in H12 which together form a "charge clamp" (Fig. 3.4a). It is noteworthy that this so-called holo-LBD conformation is displayed by all agonist-bound NRs and thus, corresponds to the canonical active form of these receptors.

Structural Basis for RAR-Corepressor Interaction and the Coregulator Exchange

In the absence of ligand or in the presence of so-called inverse agonists, RAR exhibits strong repressive activity that is brought about by the recruitment and binding of corepressors and establishment of a corepressor complex in the promoter region of target genes [26, 48] (Fig. 3.1b, c). Subsequent condensation of chromatin is believed to cause the gene repression. The molecular basis of the repression function of RAR has been revealed through crystallographic studies of RARα LBD in complex with the inverse agonist BMS493 and a peptide containing the CoRNR1 sequence of N-CoR [41]. The structure reveals that RARα, interacting with a four-turn helical motif, LxxxIxxxIxxxF/Y, in the CoRNR1 corepressor, docks to a hydrophobic surface of the NR that is formed by residues from helices H3 and H4 and an N-terminal extended β-strand (β1) forming a specific antiparallel β-sheet with specific RAR residues (S3 in Fig. 3.4b). Interestingly, the RAR residues that adopt the S3 β-strand conformation in the corepressor-bound structure belong to helix H11 in the agonist-bound structure (Fig. 3.4a). This secondary structure change from α-helix to β-strand has also been observed in the structure of the heme receptor RevErb bound to CoRNR1 of N-CoR [61], but not in other reported NR-corepressor complex structures obtained with receptors devoid of constitutive repression function [45, 78, 82]. These latter receptors have been shown to interact rather weakly with the shorter CoRNR2 motif of corepressors and only in the presence of an antagonist. The functional implication of the specific β-sheet interface between RAR and corepressors has been further validated experimentally by structure-guided mutagenesis. Mutations affecting specifically the integrity of either the β1 or S3 β-strands abrogated corepressor interaction and transcriptional repression by RAR [41].

A Working Model for RAR-Coregulator Complex Formation and Action

Comparing the structures of RAR in complex with coactivator and corepressor fragments provides a structural basis for the ligand-induced coregulator exchange. The repressive activity of RAR is conferred by a structural element consisting of an extended β-strand that forms an antiparallel β-sheet with specific corepressor residues. Agonist binding induces a β-strand to α-helix transition that allows helix H11 formation, which in turn provokes corepressor release, the repositioning of helix H12, and coactivator recruitment. Thus, it appears that the secondary-structure transition from β-strand S3 to α-helix H11 is the master regulator of corepressor dissociation from RAR, whereas H12 is primarily involved in the NR interaction with coactivators.

Coregulator Recruitment to RAR-RXR Heterodimers

RARs bind to DNA as heterodimers with RXRs [8, 62, 71]. Several studies, describing RAR and RXR LBDs in the presence of several combinations of RAR and RXR agonists and antagonists [8, 62, 71], have identified the structural organization of RAR-RXR heterodimers and the differential involvement of each subunit in the modulation of their activity. These structures (Fig. 3.3) show that the dimeric arrangements of these heterodimers are closely related, with residues from helices H7, H9, H10, as well as loops L8–9 and L9–10 of each protomer forming an interface comprising a network of complementary hydrophobic and charged residues that are further stabilized by neutralized basic and acidic surfaces.

In vitro, RAR and RXR ligands are able to individually bind to their corresponding receptors and as such activate RAR-RXR heterodimers. However, it appears that in the usual cellular environment, and in the absence of RAR agonist, RXR-specific ligands (rexinoids) are unable to do so [24, 51, 62]. Thus, in heterodimers, RAR agonists autonomously induce the recruitment of coactivators and activate transcription (Fig. 3.5a, b), while RXR agonists only induce activation when an agonist occupies the RAR LBD (Fig. 3.5c). The biological significance of this RXR "subordination" or "silencing" is presumably to avoid confusion between multiple RXR-partner signaling pathways. In the case of RAR-RXR, it is due to the inability of rexinoids to induce corepressor dissociation from heterodimers [24] and thus to induce the recruitment of coactivators because of the inaccessibility of the mutually exclusive binding site of the two coregulator types (Figs. 3.4a, b and 3.5f). Indeed, corepressors interact essentially with the RAR subunit through formation of the specific β-sheet interface described above (Fig. 3.5a). Thus, the conformational change from β-strand S3 to α-helix H11 upon RAR agonist binding switches the RAR-RXR heterodimers from an "off" to an "on" state which can be further activated by addition of rexinoids (Fig. 3.5b, c) [41].

The only way for RXR to modulate transactivation in response to its own ligand in RAR-RXR heterodimers is through synergy with RAR ligands. A structural and functional study of the RAR-RXR heterodimer in the presence of 9-*cis* retinoic acid and a coactivator fragment containing three LxxLL motifs [62] suggested that the synergy between receptor agonists would result from the enlargement of the contact area between the heterodimer and the coactivator through formation of one interacting surface on each heterodimer subunit (Fig. 3.5c). Thus, two LxxLL motifs would mediate the optimal assembly of one coactivator onto heterodimeric receptors. However, a study with a fragment of the coactivator Med1 containing several LxxLL motifs argued against this 'deck model' [65]. The data obtained using small angle X-ray scattering (SAXS) suggested that Med1 binds exclusively to the RAR partner (Fig. 3.5d). This observation was further supported by the report of an allosteric mechanism controlling coactivators recruitment by NRs whereby ligand and/or coregulator binding to one monomer could affect ligand and coregulator binding to the second monomer of a dimer [57]. In this model, binding of a coactivator LxxLL motif to one subunit would promote formation of a nonsymmetrical dimer in which the cofactor binding site of the unbound receptor would be negatively affected through the dimer interface. This phenomenon would generate a negative cooperativity between the two binding sites of the dimer resulting in the formation of a single contact point between the dimer and the coactivator. Interestingly, RAR antagonists can also synergize with rexinoid agonists and can activate transcription of endogenous target genes [41]. In fact, neutral antagonists (e.g. BMS614) prevent binding of both coactivators and corepressors to RAR by relocalizing helix H12 in the coregulator groove (see text below and Fig. 3.4c) and thus may allow some activation by rexinoids [24] (Fig. 3.5g).

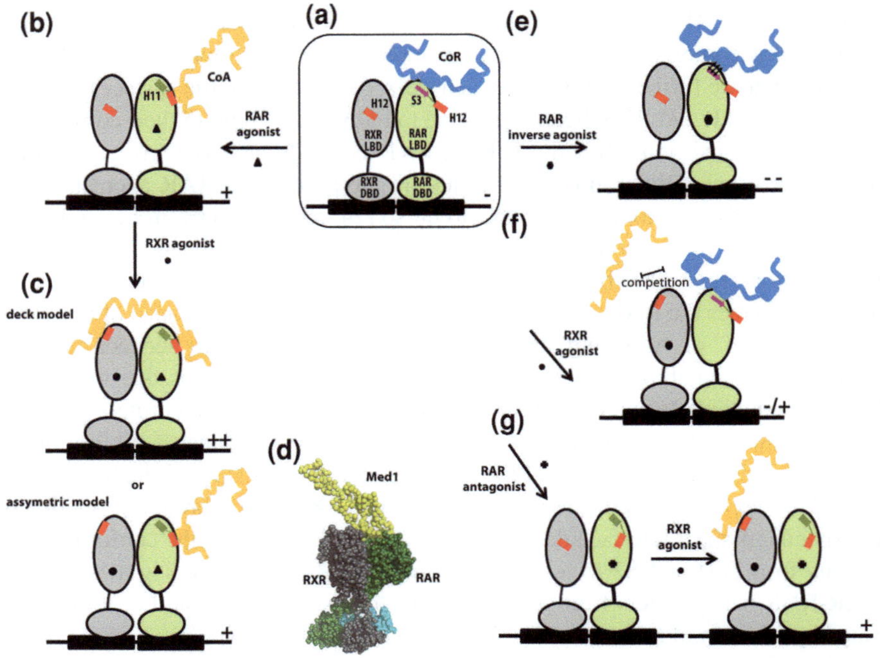

Fig. 3.5 Modulation of RAR-RXR coregulator interactions by ligands. **a** In the absence of ligand, RAR/RXR exists as a complex with a corepressor (CoR, *blue*) bound to RAR subunit through one CoRNR box (*blue square*). **b** A RAR agonist induces an allosteric change of RAR such that the CoR binding is disrupted and a novel surface is created to which a coactivator (CoA, *yellow*) can bind through its NR box (*yellow square*). **c** Further addition of a RXR agonist generates a CoA binding surface at the RXR subunit. Then in the so-called deck model, it could allow the CoA to bind cooperatively to the heterodimer using two of its NR box inducing a higher activation of transcription. On the contrary, in the asymmetric model, the CoA could not bind to the RXR subunit. **d** This asymmetric model was confirmed by the solution structure of the RXR (*gray*)/RAR (*green*)/DR5 (*cyan*)/Med1 (*yellow*) complex solved using SAXS experiment [65]. **e** A RAR inverse agonist stabilizes the heterodimer-CoR interaction through RAR subunit, thus leading to enhanced repression. **f** A RXR agonist induces a potential CoA binding surface at RXR subunit but the rexinoid does not, under normal conditions, induce CoR dissociation. Owing to steric interference, CoA cannot bind to the heterodimer, accounting for the phenomenon of RXR subordination. In some particular circumstances (see text), a rexinoid alone can transactivate. **g** A RAR neutral antagonist destabilizes the CoR interface without generating the surface for CoA binding, leading to derepression. Subsequent addition of RXR agonist allows the recruitment of CoA by RXR subunit, resulting in transcriptional activity

Relevance of the Field to Our Understanding of Retinoid Biology and Chemistry

Over the years, a plethora of synthetic ligands with a wide range of activities have been developed for potential therapeutic use. These compounds include agonists, antagonists, inverse agonists and partial agonists [1, 6, 14–17, 74, 75]. Structural studies on RXR-and RAR-coregulator interactions have provided deep insight into ligand action, revealing how different classes of ligands allosterically remodel the receptor surface to increase or reduce interaction with a given type of coregulator. The structures have indicated that ligand interactions with helices H11 and H12, or residues in their proximity, are primary determinants of helix H12 position, and that H12 can adopt not only the active and inactive positions, but also several intermediary positions and dynamics. This implies that relatively subtle ligand modifications have the capacity to significantly alter the conformation of the LBD, thereby generating distinct coregulator binding surfaces. Based on this knowledge, it is now possible to rationally design and synthesize function-specific ligands with tissue-selective activities originating from the divergent expression levels of coactivators and corepressors.

Full and Partial Agonists

As described, agonists induce the repositioning of H12 which contributes in a critical manner to the surface recognized by the LxxLL boxes of coactivators. In the active conformation, H12 docks against helices H3 and H11 to form one side of the coactivator-binding site (Fig. 3.4a). In contrast to full agonists, "weak agonists" are able to bind to retinoid receptors with high affinity, but even so, they incompletely stabilize the active form. The "partial" or "mixed" agonist/antagonist activity of these ligands is a consequence of the lowering of the interaction strength between H12 in the active position and the H3/H11 surface that renders the activation helix more dynamic. The consequence in this case, is that the association with coactivators is weaker than the agonist-bound situation [55, 58]. The presence of coactivators helps stabilize the active conformation so that the transcriptional outcome of partial agonist binding depends on the intracellular concentration of coregulators. Hence such ligands can act as cell-selective modulators with agonist or antagonist properties depending on the cellular context.

Neutral Antagonists and Inverse Agonists

Both neutral antagonists and inverse agonists bind to the same cavity as all-*trans*-retinoic acid, but these ligands prevent RARs from adopting an active conformation. The crystal structure of the BMS614-bound RARα LBD complex [8] revealed that the neutral antagonist BMS614 harbors a large 'antagonistic' extension that cannot be contained within the buried LBP so that helix H12 is displaced to allow the bulky extension of the ligand to protrude between H3 and H11 (Fig. 3.4c). Due to its amphipathic character, H12 adopts an "antagonist conformation" by docking its hydrophobic inner face into the coregulator recognition cleft, thereby preventing interaction with both coactivators and corepressors. In contrast, inverse agonists, such as BMS493, inhibit coactivator binding to RAR, but favor interaction with corepressors and enhance transcriptional silencing [23, 24, 38, 70]. The structural basis for inverse agonism was recently provided by resolution of the structure of RAR in complex with BMS493 and the CoRNR1 sequence of N-CoR [41] (Fig. 3.4b). This structural study revealed that strengthening of the RAR-corepressor interaction relies on the stabilization of the β-sheet S3/β1 interface. In contrast with agonists that disrupt the RAR-corepressor interface by inducing S3 destabilization and formation of helix H11, inverse agonists act by stabilizing the β-strand S3 conformation. Moreover, contrary to what is seen in the BMS614-containing structure, helix H12 in the BMS493-bound complex has no defined position, thus rendering the coregulator recognition groove of RAR available for docking of the helical motif of CoRNR1.

Future Directions: What Is Still Left to Do

Progress in NR structure research has revolutionized our view on how NRs in general and retinoid receptors in particular act as essential regulators of fundamental cellular processes. The structural principles of the interaction of receptors with their DNA response elements, ligands and coregulators have been decoded through crystallographic and other biophysical approaches revealing the link between NR conformation, ligand-induced allosteric changes and the resulting abilities of RARs and RXRs to communicate with the intracellular components. Notably, these studies have revealed the chemical and structural features of agonists, neutral antagonists, inverse agonists and partial agonists that all induce particular receptor-coregulator interactions and specific biological outcomes. Since NRs are primary drug targets, this information can be directly translated into the design of novel selective modulators with impact on drug development. However, despite this enormous gain in knowledge, substantial challenges remain.

Whereas detailed structural information is available on the isolated DBD and LBD domains, little is known about the interdomains that connect the DBD and LBD, or about the overall topology of the full-length receptors. The recently reported low resolution solution structure of RARα-RXRα heterodimer bound to its cognate response element [65] displays an elongated conformation with separated DBD and LBD domains lacking interdomain contacts. This model somehow contradicts the crystal structure of PPARγ-RXRα heterodimer in which a more compact organization is observed, with the relative arrangement of receptor domains positioning the PPARγ LBD so that it contacts both receptor DBDs [9]. Therefore, there is still a need for high-resolution structures of full-length RAR-RXR heterodimers in different functional states in order to unravel the relevant conformational states and the role of DNA in the heterodimer organization. In the same vein, high-resolution structural information on bound coactivators and corepressors is so far limited to peptides containing one NR or CoRNR box. Here again, there is a need for structural determination of RAR-RXR heterodimers in presence of full-length interaction domains of coactivators and corepressors.

Indeed, the relative contribution of RAR and RXR in the recruitment of coactivators also requires some clarification. Although both functional and structural data agree well on the fact that RAR is the prime contact point with corepressors within the RAR-RXR heterodimers, the involvement of RXR in the recruitment of coactivators is still in debate. This is a particularly important question especially in the case of heterodimers whose activity can, in principle, be fine-tuned by combinatorial sets of RAR and RXR ligands, thus providing interesting pharmacological opportunities. Many interaction and transcription regulation studies have provided evidence for a synergistic effect of RAR and RXR ligands in the recruitment of coactivators [12, 24, 62, 69, 81]. Because heterodimers contain two coactivator binding sites (one on RAR and one on RXR), it was hypothesized that the synergy between receptor agonists would result from the enlargement of the contact area between the heterodimer and the coactivator through formation of one LxxLL interacting surface on each heterodimer subunit. However, the only structural study reported to date suggests that the agonist-bound RXR subunit plays a minor role in the interaction [65]. Additional structural and biophysical experiments are thus needed to reconcile these seemingly conflicting observations.

In contrast to the DBD and the LBD, there are still no high-resolution structures available for the N-terminal AF-1 domain (NTD) of RARs. Several biochemical and structural studies coupled to structure prediction algorithms suggested that the NTDs of RARs, as well as any member of the nuclear receptor family, are of naturally-disordered structure [39, 80]. It has emerged that unstructured proteins or domains may be functional, undergoing transitions to more ordered states or folding into stable secondary or tertiary structures upon binding to DNA response elements or to coregulatory proteins [19]. Moreover, disordered domains provide the flexibility that is needed for modification by enzymes such as kinases and ubiquitin-ligases [19]. Such modifications may induce changes in the structural properties of the domain with profound impacts on its interactions with coregulators and/or on the dynamics of adjacent structural domains. Indeed, although the biological activity of RARs is primarily regulated by binding of a ligand, these receptors also serve as integrating platforms for a variety of post-translational modifications. As an example, phosphorylations of both the AF-1 and AF-2 of RARs have been shown to modulate transcriptional activity [21, 66], but it is still unknown whether or not these protein modifications induce local or global conformational changes in the receptors. Further investigation will be required to define the structural mechanism underlying this signal propagation. Finally, very little is known about the possible effects of ligand binding on the N-terminal activation function AF-1, while its existence and synergy with AF-2 has been confirmed [52].

References

1. Altucci L, Leibowitz MD, Ogilvie KM, de Lera AR, Gronemeyer H (2007) RAR and RXR modulation in cancer and metabolic disease. Nat Rev Drug Discov 6:793–810
2. Baniahmad A, Leng X, Burris TP, Tsai SY, Tsai MJ, O'Malley BW (1995) The tau 4 activation domain of the thyroid hormone receptor is required for release of a putative corepressor(s) necessary for transcriptional silencing. Mol Cell Biol 15:76–86

3. Baniahmad A, Tsai SY, O'Malley BW, Tsai MJ (1992) Kindred S thyroid hormone receptor is an active and constitutive silencer and a repressor for thyroid hormone and retinoic acid responses. Proc Natl Acad Sci USA 89:10633–10637
4. Baniahmad C, Nawaz Z, Baniahmad A, Gleeson MA, Tsai MJ, O'Malley BW (1995) Enhancement of human estrogen receptor activity by SPT6: a potential coactivator. Mol Endocrinol 9:34–43
5. Bocquel MT, Kumar V, Stricker C, Chambon P, Gronemeyer H (1989) The contribution of the N-and C-terminal regions of steroid receptors to activation of transcription is both receptor and cell-specific. Nucleic Acids Res 17:2581–2595
6. Bourguet W, de Lera AR, Gronemeyer H (2010) Inverse agonists and antagonists of retinoid receptors. Methods Enzymol 485:161–195
7. Bourguet W, Germain P, Gronemeyer H (2000) Nuclear receptor ligand-binding domains: three-dimensional structures, molecular interactions and pharmacological implications. Trends Pharmacol Sci 21:381–388
8. Bourguet W, Vivat V, Wurtz J-M, Chambon P, Gronemeyer H, Moras D (2000) Crystal structure of a heterodimeric complex of RAR and RXR ligand-binding domains. Mol Cell 5:289–298
9. Chandra V, Huang P, Hamuro Y, Raghuram S, Wang Y, Burris TP, Rastinejad F (2008) Structure of the intact PPAR-gamma-RXR-nuclear receptor complex on DNA. Nature 456:350–356
10. Chen H, Lin RJ, Xie W, Wilpitz D, Evans RM (1999) Regulation of hormone-induced histone hyperacetylation and gene activation via acetylation of an acetylase. Cell 98:675–686
11. Chen JD, Evans RM (1995) A transcriptional co-repressor that interacts with nuclear hormone receptors. Nature 377:454–457
12. Chen JY, Clifford J, Zusi C, Starrett J, Tortolani D, Ostrowski J, Reczek PR, Chambon P, Gronemeyer H (1996) Two distinct actions of retinoid-receptor ligands. Nature 382:819–822
13. Darimont BD, Wagner RL, Apriletti JW, Stallcup MR, Kushner PJ, Baxter JD, Fletterick RJ, Yamamoto KR (1998) Structure and specificity of nuclear receptor-coactivator interactions. Genes Dev 12:3343–3356
14. Dawson MI (2004) Synthetic retinoids and their nuclear receptors. Curr Med Chem Anticancer Agents 4:199–230
15. Dawson MI, Xia Z (2012) The retinoid X receptors and their ligands. Biochim Biophys Acta 1821:21–56
16. Dawson MI, Zhang XK (2002) Discovery and design of retinoic acid receptor and retinoid X receptor class-and subtype-selective synthetic analogs of all-trans-retinoic acid and 9-cis-retinoic acid. Curr Med Chem 9:623–637
17. de Lera AR, Bourguet W, Altucci L, Gronemeyer H (2007) Design of selective nuclear receptor modulators: RAR and RXR as a case study. Nat Rev Drug Discov 6:811–820
18. Ding XF, Anderson CM, Ma H, Hong H, Uht RM, Kushner PJ, Stallcup MR (1998) Nuclear receptor-binding sites of coactivators glucocorticoid receptor interacting protein 1 (GRIP1) and steroid receptor coactivator 1 (SRC-1): multiple motifs with different binding specificities. Mol Endocrinol 12:302–313
19. Dyson HJ, Wright PE (2005) Intrinsically unstructured proteins and their functions. Nat Rev Mol Cell Biol 6:197–208
20. Feng W, Ribeiro RC, Wagner RL, Nguyen H, Apriletti JW, Fletterick RJ, Baxter JD, Kushner PJ, West BL (1998) Hormone-dependent coactivator binding to a hydrophobic cleft on nuclear receptors. Science 280:1747–1749
21. Gaillard E, Bruck N, Brelivet Y, Bour G, Lalevee S, Bauer A, Poch O, Moras D, Rochette-Egly C (2006) Phosphorylation by PKA potentiates retinoic acid receptor alpha activity by means of increasing interaction with and phosphorylation by cyclin H/cdk7. Proc Natl Acad Sci USA 103:9548–9553
22. Gehin M, Vivat V, Wurtz JM, Losson R, Chambon P, Moras D, Gronemeyer H (1999) Structural basis for engineering of retinoic acid receptor isotype-selective agonists and antagonists. Chem Biol 6:519–529

23. Germain P, Gaudon C, Pogenberg V, Sanglier S, Van Dorsselaer A, Royer CA, Lazar MA, Bourguet W, Gronemeyer H (2009) Differential action on coregulator interaction defines inverse retinoid agonists and neutral antagonists. Chem Biol 16:479–489
24. Germain P, Iyer J, Zechel C, Gronemeyer H (2002) Co-regulator recruitment and the mechanism of retinoic acid receptor synergy. Nature 415:187–192
25. Germain P, Kammerer S, Perez E, Peluso-Iltis C, Tortolani D, Zusi FC, Starrett J, Lapointe P, Daris JP, Marinier A, de Lera AR, Rochel N, Gronemeyer H (2004) Rational design of RAR-selective ligands revealed by RARbeta crystal stucture. EMBO Rep 5:877–882
26. Glass CK, Rosenfeld MG (2000) The coregulator exchange in transcriptional functions of nuclear receptors. Genes Dev 14:121–141
27. Gronemeyer H, Gustafsson JA, Laudet V (2004) Principles for modulation of the nuclear receptor superfamily. Nat Rev Drug Discov 3:950–964
28. Heery DM, Kalkhoven E, Hoare S, Parker MG (1997) A signature motif in transcriptional co-activators mediates binding to nuclear receptors. Nature 387:733–736
29. Heinzel T, Lavinsky RM, Mullen TM, Soderstrom M, Laherty CD, Torchia J, Yang WM, Brard G, Ngo SD, Davie JR, Seto E, Eisenman RN, Rose DW, Glass CK, Rosenfeld MG (1997) A complex containing N-CoR, mSin3 and histone deacetylase mediates transcriptional repression. Nature 387:43–48
30. Horlein AJ, Naar AM, Heinzel T, Torchia J, Gloss B, Kurokawa R, Ryan A, Kamei Y, Soderstrom M, Glass CK et al (1995) Ligand-independent repression by the thyroid hormone receptor mediated by a nuclear receptor co-repressor. Nature 377:397–404
31. Hu X, Lazar MA (1999) The CoRNR motif controls the recruitment of corepressors by nuclear hormone receptors. Nature 402:93–96
32. Hu X, Li Y, Lazar MA (2001) Determinants of CoRNR-dependent repression complex assembly on nuclear hormone receptors. Mol Cell Biol 21:1747–1758
33. Imhof A, Yang XJ, Ogryzko VV, Nakatani Y, Wolffe AP, Ge H (1997) Acetylation of general transcription factors by histone acetyltransferases. Curr Biol 7:689–692
34. Jung SY, Malovannaya A, Wei J, O'Malley BW, Qin J (2005) Proteomic analysis of steady-state nuclear hormone receptor coactivator complexes. Mol Endocrinol 19:2451–2465
35. Kato S, Yokoyama A, Fujiki R (2011) Nuclear receptor coregulators merge transcriptional coregulation with epigenetic regulation. Trends Biochem Sci 36:272–281
36. Klaholz BP, Mitschler A, Belema M, Zusi C, Moras D (2000) Enantiomer discrimination illustrated by high-resolution crystal structures of the human nuclear receptor hRARgamma. Proc Natl Acad Sci USA 97:6322–6327
37. Klaholz BP, Renaud JP, Mitschler A, Zusi C, Chambon P, Gronemeyer H, Moras D (1998) Conformational adaptation of agonists to the human nuclear receptor RAR gamma. Nat Struct Biol 5:199–202
38. Klein ES, Pino ME, Johnson AT, Davies PJ, Nagpal S, Thacher SM, Krasinski G, Chandraratna RA (1996) Identification and functional separation of retinoic acid receptor neutral antagonists and inverse agonists. J Biol Chem 271:22692–22696
39. Lavery DN, McEwan IJ (2005) Structure and function of steroid receptor AF1 transactivation domains: induction of active conformations. Biochem J 391:449–464
40. Douarin B, Baur E, Zechel C, Heer D, Heine M, Vivat V, Gronemeyer H, Losson R, Chambon P (1996) Ligand-dependent interaction of nuclear receptors with potential transcriptional intermediary factors (mediators). Philos Trans R Soc Lond B Biol Sci 351:569–578
41. le Maire A, Teyssier C, Erb C, Grimaldi M, Alvarez S, de Lera AR, Balaguer P, Gronemeyer H, Royer CA, Germain P, Bourguet W (2010) A unique secondary-structure switch controls constitutive gene repression by retinoic acid receptor. Nat Struct Mol Biol 17:801–807
42. Li D, Yamada T, Wang F, Vulin AI, Samuels HH (2004) Novel roles of retinoid X receptor (RXR) and RXR ligand in dynamically modulating the activity of the thyroid hormone receptor/RXR heterodimer. J Biol Chem 279:7427–7437
43. Lonard DM, O'Malley BW (2007) Nuclear receptor coregulators: judges, juries, and executioners of cellular regulation. Mol Cell 27:691–700

44. Lonard DM, O'Malley BW (2006) The expanding cosmos of nuclear receptor coactivators. Cell 125:411–414
45. Madauss KP, Grygielko ET, Deng SJ, Sulpizio AC, Stanley TB, Wu C, Short SA, Thompson SK, Stewart EL, Laping NJ, Williams SP, Bray JD (2007) A structural and in vitro characterization of asoprisnil: a selective progesterone receptor modulator. Mol Endocrinol 21:1066–1081
46. Malovannaya A, Lanz RB, Jung SY, Bulynko Y, Le NT, Chan DW, Ding C, Shi Y, Yucer N, Krenciute G, Kim BJ, Li C, Chen R, Li W, Wang Y, O'Malley BW, Qin J (2011) Analysis of the human endogenous coregulator complexome. Cell 145:787–799
47. McInerney EM, Rose DW, Flynn SE, Westin S, Mullen TM, Krones A, Inostroza J, Torchia J, Nolte RT, Assa-Munt N, Milburn MV, Glass CK, Rosenfeld MG (1998) Determinants of coactivator LXXLL motif specificity in nuclear receptor transcriptional activation. Genes Dev 12:3357–3368
48. McKenna NJ, Lanz RB, O'Malley BW (1999) Nuclear receptor coregulators: cellular and molecular biology. Endocr Rev 20:321–344
49. McKenna NJ, Xu J, Nawaz Z, Tsai SY, Tsai MJ, O'Malley BW (1999) Nuclear receptor coactivators: multiple enzymes, multiple complexes, multiple functions. J Steroid Biochem Mol Biol 69:3–12
50. Meyer ME, Gronemeyer H, Turcotte B, Bocquel MT, Tasset D, Chambon P (1989) Steroid hormone receptors compete for factors that mediate their enhancer function. Cell 57:433–442
51. Minucci S, Leid M, Toyama R, Saint-Jeannet JP, Peterson VJ, Horn V, Ishmael JE, Bhattacharyya N, Dey A, Dawid IB, Ozato K (1997) Retinoid X receptor (RXR) within the RXR-retinoic acid receptor heterodimer binds its ligand and enhances retinoid-dependent gene expression. Mol Cell Biol 17:644–655
52. Nagpal S, Friant S, Nakshatri H, Chambon P (1993) RARs and RXRs: evidence for two autonomous transactivation functions (AF-1 and AF-2) and heterodimerization in vivo. EMBO J 12:2349–2360
53. Nagy L, Kao HY, Chakravarti D, Lin RJ, Hassig CA, Ayer DE, Schreiber SL, Evans RM (1997) Nuclear receptor repression mediated by a complex containing SMRT, mSin3A, and histone deacetylase. Cell 89:373–380
54. Nagy L, Kao HY, Love JD, Li C, Banayo E, Gooch JT, Krishna V, Chatterjee K, Evans RM, Schwabe JW (1999) Mechanism of corepressor binding and release from nuclear hormone receptors. Genes Dev 13:3209–3216
55. Nahoum V, Perez E, Germain P, Rodriguez-Barrios F, Manzo F, Kammerer S, Lemaire G, Hirsch O, Royer CA, Gronemeyer H, de Lera AR, Bourguet W (2007) Modulators of the structural dynamics of the retinoid X receptor to reveal receptor function. Proc Natl Acad Sci USA 104:17323–17328
56. Onate SA, Tsai SY, Tsai MJ, O'Malley BW (1995) Sequence and characterization of a coactivator for the steroid hormone receptor superfamily. Science 270:1354–1357
57. Osz J, Brelivet Y, Peluso-Iltis C, Cura V, Eiler S, Ruff M, Bourguet W, Rochel N, Moras D (2012) Structural basis for a molecular allosteric control mechanism of cofactor binding to nuclear receptors. Proc Natl Acad Sci USA 109:E588–E594
58. Santin EP, Germain P, Quillard F, Khanwalkar H, Rodriguez-Barrios F, Gronemeyer H, de Lera AR, Bourguet W (2009) Modulating retinoid X receptor with a series of (E)-3-[4-hydroxy-3-(3-alkoxy-5,5,8,8-tetramethyl-5,6,7,8-tetrahydronaphthalen-2-y l)phenyl]acrylic acids and their 4-alkoxy isomers. J Med Chem 52:3150–3158
59. Perissi V, Jepsen K, Glass CK, Rosenfeld MG (2010) Deconstructing repression: evolving models of co-repressor action. Nat Rev Genet 11:109–123
60. Perissi V, Staszewski LM, McInerney EM, Kurokawa R, Krones A, Rose DW, Lambert MH, Milburn MV, Glass CK, Rosenfeld MG (1999) Molecular determinants of nuclear receptor-corepressor interaction. Genes Dev 13:3198–3208
61. Phelan CA, Gampe RT Jr, Lambert MH, Parks DJ, Montana V, Bynum J, Broderick TM, Hu X, Williams SP, Nolte RT, Lazar MA (2010) Structure of Rev-erbalpha bound to N-CoR reveals a unique mechanism of nuclear receptor-co-repressor interaction. Nat Struct Mol Biol 17:808–814

62. Pogenberg V, Guichou JF, Vivat-Hannah V, Kammerer S, Perez E, Germain P, de Lera AR, Gronemeyer H, Royer CA, Bourguet W (2005) Characterization of the interaction between retinoic acid receptor/retinoid X receptor (RAR/RXR) heterodimers and transcriptional coactivators through structural and fluorescence anisotropy studies. J Biol Chem 280:1625–1633
63. Ptashne M, Gann AA (1990) Activators and targets. Nature 346:329–331
64. Renaud JP, Rochel N, Ruff M, Vivat V, Chambon P, Gronemeyer H, Moras D (1995) Crystal structure of the RAR-gamma ligand-binding domain bound to all-trans retinoic acid. Nature 378:681–689
65. Rochel N, Ciesielski F, Godet J, Moman E, Roessle M, Peluso-Iltis C, Moulin M, Haertlein M, Callow P, Mely Y, Svergun DI, Moras D (2011) Common architecture of nuclear receptor heterodimers on DNA direct repeat elements with different spacings. Nat Struct Mol Biol 18:564–570
66. Rochette-Egly C (2003) Nuclear receptors: integration of multiple signalling pathways through phosphorylation. Cell Signal 15:355–366
67. Roeder RG (1998) Role of general and gene-specific cofactors in the regulation of eukaryotic transcription. Cold Spring Harb Symp Quant Biol 63:201–218
68. Rosenfeld MG, Lunyak VV, Glass CK (2006) Sensors and signals: a coactivator/corepressor/epigenetic code for integrating signal-dependent programs of transcriptional response. Genes Dev 20:1405–1428
69. Roy B, Taneja R, Chambon P (1995) Synergistic activation of retinoic acid (RA)-responsive genes and induction of embryonal carcinoma cell differentiation by an RA receptor alpha (RAR alpha)-, RAR beta-, or RAR gamma-selective ligand in combination with a retinoid X receptor-specific ligand. Mol Cell Biol 15:6481–6487
70. Sanglier S, Bourguet W, Germain P, Chavant V, Moras D, Gronemeyer H, Potier N, Van Dorsselaer A (2004) Monitoring ligand-mediated nuclear receptor-coregulator interactions by noncovalent mass spectrometry. Eur J Biochem 271:4958–4967
71. Sato Y, Ramalanjaona N, Huet T, Potier N, Osz J, Antony P, Peluso-Iltis C, Poussin-Courmontagne P, Ennifar E, Mely Y, Dejaegere A, Moras D, Rochel N (2010) The, "Phantom Effect" of the Rexinoid LG100754: structural and functional insights. PLoS ONE 5:e15119
72. Tasset D, Tora L, Fromental C, Scheer E, Chambon P (1990) Distinct classes of transcriptional activating domains function by different mechanisms. Cell 62:1177–1187
73. Torchia J, Rose DW, Inostroza J, Kamei Y, Westin S, Glass CK, Rosenfeld MG (1997) The transcriptional co-activator p/CIP binds CBP and mediates nuclear-receptor function. Nature 387:677–684
74. Vaz B, de Lera AR (2012) Advances in drug design with RXR modulators. Expert Opin Drug Discov 7:1003–1016
75. Vivat-Hannah V, Zusi FC (2005) Retinoids as therapeutic agents: today and tomorrow. Mini Rev Med Chem 5:755–760
76. Voegel JJ, Heine MJ, Tini M, Vivat V, Chambon P, Gronemeyer H (1998) The coactivator TIF2 contains three nuclear receptor-binding motifs and mediates transactivation through CBP binding-dependent and -independent pathways. EMBO J 17:507–519
77. Voegel JJ, Heine MJ, Zechel C, Chambon P, Gronemeyer H (1996) TIF2, a 160 kDa transcriptional mediator for the ligand-dependent activation function AF-2 of nuclear receptors. EMBO J 15:3667–3675
78. Wang L, Zuercher WJ, Consler TG, Lambert MH, Miller AB, Orband-Miller LA, McKee DD, Willson TM, Nolte RT (2006) X-ray crystal structures of the estrogen-related receptor-gamma ligand binding domain in three functional states reveal the molecular basis of small molecule regulation. J Biol Chem 281:37773–37781
79. Wang S, Wang Z, Lin S, Zheng W, Wang R, Jin S, Chen J, Jin L, Li Y (2012) Revealing a natural marine product as a novel agonist for retinoic acid receptors with a unique binding mode and inhibitory effects on cancer cells. Biochem J 446:79–87
80. Warnmark A, Treuter E, Wright AP, Gustafsson JA (2003) Activation functions 1 and 2 of nuclear receptors: molecular strategies for transcriptional activation. Mol Endocrinol 17:1901–1909

81. Westin S, Kurokawa R, Nolte RT, Wisely GB, McInerney EM, Rose DW, Milburn MV, Rosenfeld MG, Glass CK (1998) Interactions controlling the assembly of nuclear-receptor heterodimers and co-activators. Nature 395:199–202
82. Xu HE, Stanley TB, Montana VG, Lambert MH, Shearer BG, Cobb JE, McKee DD, Galardi CM, Plunket KD, Nolte RT, Parks DJ, Moore JT, Kliewer SA, Willson TM, Stimmel JB (2002) Structural basis for antagonist-mediated recruitment of nuclear co-repressors by PPARα. Nature 415:813–817

Chapter 4
Evolution of Retinoic Acid Receptors and Retinoic Acid Signaling

Juliana Gutierrez-Mazariegos, Michael Schubert and Vincent Laudet

Abstract Retinoic acid (RA) is a vitamin A-derived morphogen controlling important developmental processes in vertebrates, and more generally in chordates, including axial patterning and tissue formation and differentiation. In the embryo, endogenous RA levels are controlled by RA synthesizing and degrading enzymes and the RA signal is transduced by two retinoid receptors: the retinoic acid receptor (RAR) and the retinoid X receptor (RXR). Both RAR and RXR are members of the nuclear receptor superfamily of ligand-activated transcription factors and mainly act as heterodimers to activate the transcription of target genes in the presence of their ligand, all-trans RA. This signaling pathway was long thought to be a chordate innovation, however, recent findings of gene homologs involved in RA signaling in the genomes of a wide variety of non-chordate animals, including ambulacrarians (sea urchins and acorn worms) and lophotrochozoans (annelids and mollusks), challenged this traditional view and suggested that the RA signaling pathway might have a more ancient evolutionary origin than previously thought. In this chapter, we discuss the evolutionary history of the RA signaling pathway, and more particularly of the RARs, which might have experienced independent gene losses and duplications in different animal lineages. In sum, the available data reveal novel insights into the origin of the RA signaling pathway as well as into the evolutionary history of the RARs.

J. Gutierrez-Mazariegos (✉) · V. Laudet
Molecular Zoology Team, Institut de Génomique Fonctionnelle de Lyon, Université de Lyon, Université Lyon 1, CNRS, INRA, Ecole Normale Supérieure de Lyon, 46 allée d'Italie, 69364 Lyon Cedex 07, France
e-mail: juliana.gutierrez_mazariegos@ens-lyon.fr

V. Laudet
e-mail: vincent.laudet@ens-lyon.fr

M. Schubert
Laboratoire de Biologie du Développement de Villefranche-sur-Mer (UMR 7009—CNRS/UMPC), Observatoire Océanologique, 181 Chemin du Lazaret, BP 28, 06230 Villefranche-sur-Mer, France
e-mail: michael.schubert@obs-vlfr.fr

M. A. Asson-Batres and C. Rochette-Egly (eds.), *The Biochemistry of Retinoic Acid Receptors I: Structure, Activation, and Function at the Molecular Level*, Subcellular Biochemistry 70,
DOI: 10.1007/978-94-017-9050-5_4,

Abbreviations

ADH	Alcohol dehydrogenase
AF	Activation function
BCO	β-carotene 15,15′-monooxygenase
BMP-1	Bone morphogenetic protein-1
CRABP	Cellular retinoic acid binding protein
CRBP	Cellular retinol binding protein
CYP	Cytochrome P450
CYP26	Cytochrome P450 subfamily 26
DBD	DNA binding domain
DR	Direct repeat
FRET	Fluorescence resonance energy transfer
HPLC	High performance liquid chromatography
LBD	Ligand binding domain
RA	Retinoic acid
RALDH	Retinal dehydrogenase
RAR	Retinoic acid receptor
RARE	Retinoic acid response element
RDH10	Retinol dehydrogenase 10
RXR	Retinoid X receptor
SDR	Short-chain dehydrogenase/reductase
USP	Ultraspiracle

Introduction

Retinoic acid (RA) is a fat-soluble morphogen derived from vitamin A that, in vertebrates, controls the organization of the anteroposterior axis and the formation and differentiation of various tissues during development. It is well known that one of the major actors of the RA signaling pathway is the retinoic acid receptor (RAR), a member of the nuclear receptor superfamily of ligand-activated transcription factors. RARs mainly act as heterodimers with the retinoid x receptors (RXRs) to activate the transcription of target genes in the presence of their ligand, all-*trans* RA (reviewed in [36]) (Fig. 4.1). In most vertebrates, including human and mouse, there are three *rar* genes (*rarα*, *rarβ* and *rarγ*) and three *rxr* genes (*rxrα*, *rxrβ* and *rxrγ*) (Fig. 4.2), each encoding several isoforms (reviewed in [51]). In fact, at the origin of vertebrates two whole genome duplications took place [4, 21, 57, 82] giving rise to the multiple RAR and RXR paralogs in vertebrates. Thus, a number of different heterodimers can be formed between these receptors and it is believed that there are more than thirty different RAR-RXR heterodimer associations, taking into account the many RAR and RXR isoforms produced, that transduce signals in the presence of all-*trans* RA [51, 59]. It is

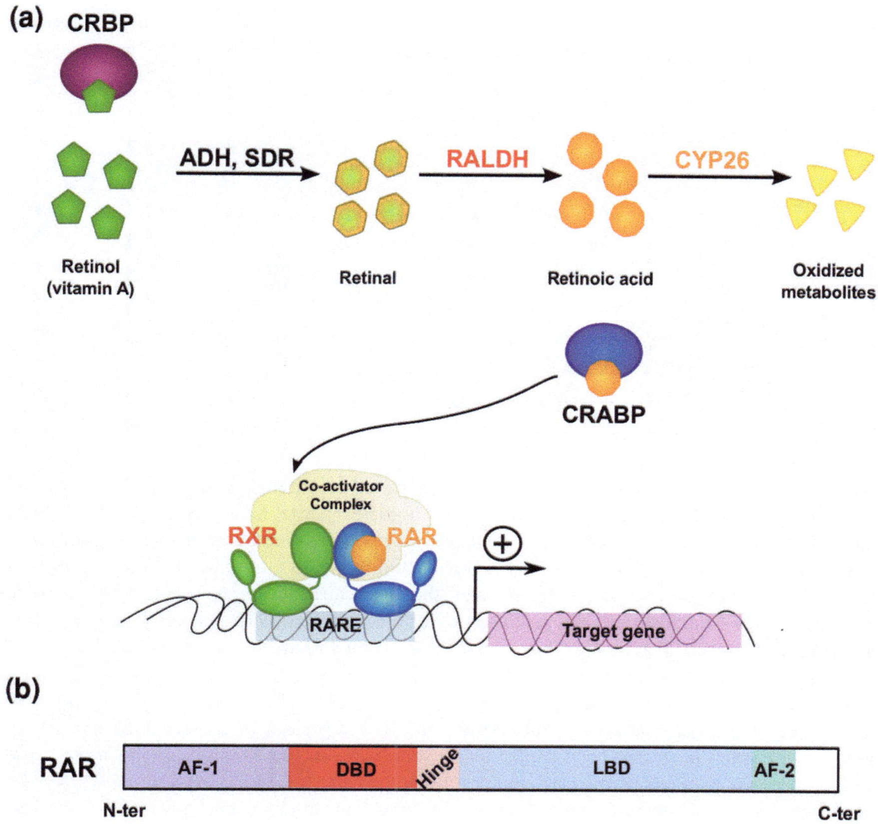

Fig. 4.1 RA signaling pathway. **a** Schematic representation of RA metabolism and signaling. Retinol is converted to RA in two steps, first retinol (vitamin A) is oxidized into retinal by a reversible reaction and then retinal is oxidized into RA. RA is transported into the nucleus where it binds to the retinoic acid receptor (RAR), which forms a heterodimer with the retinoid X receptor (RXR). Upon binding of RA, the RAR-RXR heterodimer will recruit co-activators and activate the transcription of target genes. The colors represent the evolutionary conservation: RXR and RALDH are in *red* because they are present in all metazoans and CYP26 and RAR are in *orange* because they are absent in ecdysozoans and cnidarians. ADH alcohol dehydrogenase; CRBP cellular retinol binding protein; CRABP cellular retinoic acid binding protein; CYP26 cytochrome P450 subfamily 26; RA retinoic acid; RALDH retinaldehyde dehydrogenase; RAR retinoic acid receptor; RARE retinoic acid response element; RXR retinoid X receptor; SDR short chain dehydrogenase/reductase. **b** Overview of the structure of the RAR protein: N-terminal ligand-independent transcriptional activation domain (AF-1), a centrally located DNA-binding domain (DBD) consisting of a highly conserved core region, which contains two zinc finger modules, a hinge that allows flexibility between the N- and C-terminal portion of the molecule and a C-terminal ligand binding domain (LBD), which interacts with the ligand, allowing receptor dimerization and additionally serving as a ligand-activated transcriptional activation function (AF-2) domain

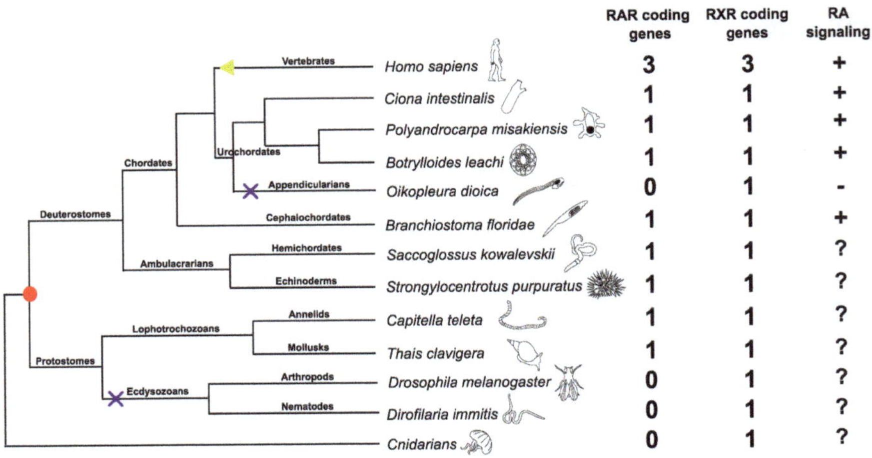

Fig. 4.2 Evolution of RAR. Schematic phylogenetic distribution of RARs and RXRs in bilaterian animals: RARs are generally present in chordates (vertebrates, tunicates and cephalochordates) and have recently been identified in the genomes of ambulacrarians (hemichordates and echinoderms) and lophotrochozoans (annelids and mollusks). In general, RARs seem to be absent from ecdysozoan genomes. RAR might thus have already been present in the last common ancestor of deuterostomes and protostomes, called Urbilateria [12]. Given the absence from ecdysozoan genomes, RAR might have secondarily been lost in this lineage. Moreover, RAR has probably also been secondarily lost in a tunicate lineage, the appendicularians. In vertebrates, the *rar* genes have been duplicated, with, for example, *Homo sapiens* possessing three *rar* paralogs (*rarα*, *rarβ* and *rarγ*). In contrast, RXRs are present in all metazoan lineages. In addition, *rxr* genes have also been duplicated in vertebrates as, for example, *Homo sapiens* possessing three *rxr* paralogs (*rxrα*, *rxrβ* and *rxrγ*). The presence of a functional retinoic acid (RA) signaling pathway is indicated by a + or − sign, however, in many lineages this has not been determined (?). Numbers indicate the number of RAR and RXR coding genes in a given lineage. The *red dot* indicates the probable origin of rar and the *yellow triangle* indicates the whole genome duplication that took place in the vertebrate lineage and the purple *x* indicates the loss of RAR. Protein accession numbers: (*source* http://www.ncbi.nlm.nih.gov/, excepting the *Capitella teleta* sequence, which is accessible at: http://genome.jgi-psf.org/Capca1/Capca1.home.html): *Homo sapiens* RARα (P10276), RARβ (P10826), RARγ (P13631), RXRα (P19793.1), RXRβ (P28702.2) and RXRγ (P48443.1), *Ciona intestinalis* RAR (NP_001072037) and RXR (NP_001071809), *Polyandrocarpa misakiensis* RAR (BAA25569) and RXR (BAA82618), *Botrylloides leachi* RAR (DQ523226) and RXR, *Branchiostoma floridae* RAR (AAM46149) and RXR (AAM46151), *Saccoglossus kowalevskii* RAR (XP_002742241) and RXR (ADB22634), *Strongylocentrotus purpuratus* RAR (XP_779976) and RXR (XP_784246), *Capitella teleta* RAR (168520) and RXR (186691), *Thais* (*Reisha*) *clavigera* RAR (BAN82614.1) and RXR (E9RHD8), *Drosophila melanogaster* USP (AAF45707), *Dirofilaria immitis* USP (AAM08269) and *Tripedaliacystophora* RXR (AAC80008)

also important to mention that RXR is believed to be a receptor for a specific RA isomer, 9-*cis* RA, although the in vivo relevance of this observation is still debated ([11, 47, 68] and references therein). Moreover, other ligands such as docohexaenoic acid or phytanic acid, among others, have been proposed as ligands for RXR [22, 60]. The functional role of RXR in the heterodimer is outside the scope of this chapter, but it is nonetheless important to point out that RXR is not a passive partner, and the true nature of its in vivo role remains to be described (see [32] and references therein).

History: Production, Metabolism and Signaling by Retinoids in Vertebrates

The Structure of the RAR-RXR Heterodimer

Like other members of the nuclear receptor superfamily, RARs are modular proteins, possessing several domains, which carry out specific functions required for their activities as ligand-regulated transcription factors (Fig. 4.1). The RAR-RXR heterodimer binds to DNA on specific sequences called retinoic acid response elements (RAREs), which, most frequently, consist of direct repeats (DRs) [2, 15] (Fig. 4.1).

In the apo form, which corresponds to the receptor without the ligand, RAR, together with RXR, interacts with transcriptional co-repressors, such as NCoR or SMRT, and represses the transcription of target genes (reviewed in [36]). Following ligand binding, the LBD undergoes a conformational change during which the most C-terminal helix (H12) forms a lid that closes the hydrophobic pocket in which the ligand is buried (reviewed in [36]). It has been shown that this conformational change significantly alters the composition of the proteins that interact with the LBD. The conformational change causes dissociation of the co-repressors and the binding of co-activators, such as the members of the p160 proteins (SRC1, 2 and 3) that induce histone acetylation and subsequently, transcriptional activation (reviewed in [66]. This is, however, an oversimplified presentation of RAR-RXR functions, as many different types of co-repressors, co-activators and chromatin remodeling complexes work together to orchestrate the remodeling of chromatin linked to transcriptional regulation [3, 34]. Recently, older models of unliganded RAR action have been challenged by new genome-wide analyses of RAR target genes (see Chaps. 9 and 10). It has become clear that ligand binding generally increases the ability of RAR-RXR to interact with the regulatory regions of at least some target genes [10, 67, 69].

Regulation of Endogenous RA Levels

Many studies have thoroughly characterized the function of RAR in vertebrates, focusing mainly on developmental and physiological aspects [35, 86]. Other studies have characterized the biochemical pathways that produce RA in vivo. These data have revealed that the RA precursor, retinol (which is vitamin A), is catabolized to retinal in a reversible manner by alcohol dehydrogenase (ADH) [8, 54, 103] or the short-chain dehydrogenase/reductase (SDR) enzymes, such as RDH10 [102]. This reaction is followed by an irreversible oxidation step of retinal to RA by retinal dehydrogenases (RALDHs), which is the rate-limiting step of this pathway [25]. There are three RALDH enzymes in most vertebrates, and RALDH2 is the most relevant RALDH during embryonic development, with its expression allowing us to infer, at least grossly, the regions of the embryo characterized by high endogenous RA levels (see [78]). Importantly, this zone of high endogenous RA levels gives rise to two RA gradients, one oriented anteriorly and one posteriorly in the embryo, as has recently been revealed in developing zebrafish using a visualization technique based on fluorescence resonance energy transfer (FRET) [91].

Within a target cell, retinol is associated with cellular retinol binding protein (CRBP), and RA is associated with the cellular retinoic acid binding protein (CRABP) [72]. Cellular and tissue concentrations of RA are generally regulated through homeostatic processes involving both RA production and degradation, with enzymes of the cytochrome P450 subfamily 26 (CYP26) principally mediating the catalysis of RA to inactive products like 4-oxo RA and 4-hydroxy RA [76]. Three genes encoding CYP26 enzymes (CYP26A1, CYP26B1 and CYP26C1) exist in vertebrates [99]. Moreover, in vitro experiments have identified further RA-degrading CYP enzymes that may also be involved in the RA catalysis process, such as CYP1A2, 2A4, 2A6, 1B1, 2B1, 2B6, 2C3, 2C7, 2C8, 2C9, 2D6, 2E1, 2E2, 2G1, 3A4/5, 3A6, 3A7, and 4A11. However, the in vivo functions of these enzymes remain unclear [37, 61, 81].

Development of the Field: Evolutionary Origins of RA Signaling

In contrast to the situation in vertebrates, where the roles and mode of action of RA are relatively well understood, very little is known about the origins of the RA signaling pathway. Only a few studies have focused on the functions of RAR in non-vertebrate animals. Among non-vertebrates, only representative RAR receptors found in other chordates, i.e. in tunicates and cephalochordates, have been functionally characterized [27, 28, 30, 31]. However, the recent identification of some components of the RA machinery in the genomes of non-chordate animals has provided some useful hints that shed new light on the action of RA and the presence of a RA signaling pathway in metazoan animals. For example, homologs of *rar*, *cyp26* and *raldh* have been identified in the genomes of ambulacrarians (sea urchins and acorn worms) and lophotrochozoans (annelids and mollusks) (Fig. 4.2) [1, 12, 13, 65, 92]. Our current understanding of the origin of RAR suggests the evolutionary history of RAR was influenced by specific losses and duplications in particular animal lineages. Altogether, this analysis allows us to conclude that, evolutionarily speaking, the RA signaling pathway is more ancient than originally thought.

Current State of the Field: RA Signaling in Non-vertebrate Animals

Presence or Absence of RA-Signaling in Metazoan Genomes: Chordates

The RA signaling pathway was long thought to be vertebrate-specific. However, it is now well established that most of the components of the RA machinery are present and functional in all chordate phyla. More particularly, single *rar* homologs have been identified and functionally characterized in the following chordates (Fig. 4.2): *Ciona intestinalis* [31], *Botrylloides leachi* [80] and *Polyandrocarpa misakiensis* [40] (all of which are ascidian tunicates), as well as in the cephalochordate *Branchiostoma floridae* also referred to as amphioxus [28].

As previously mentioned, vertebrates possess at least three RARs, RARα, RARβ and RARγ (Fig. 4.2), but as indicated, only one RAR is present in chordates (cephalochordates and ascidian tunicates) (Fig. 4.2). The evolution of the chordate RA signaling pathway has been studied extensively [27, 30, 31, 88]. In amphioxus it has been shown that RAR and RXR are able to heterodimerize, bind RA and activate the transcription of genes upon RA binding [28]. Moreover, it was shown that the amphioxus RAR has a ligand-binding pocket structure similar to RARβ [27]. RA functions have been described in the embryonic development of amphioxus. For example, it has been shown that RA regulates the anteroposterior patterning of the CNS and neuronal specification in a *hox*-dependent manner [88]. RA also mediates the patterning of the ectoderm and the formation of the tail fin in this specie [14]. In addition, orthologs of *cyp26*, *rdh* and *raldh* have been identified in the genome of amphioxus, however, their function remains to be assessed. In contrast, there is no evidence of the existence of genes coding for proteins responsible for retinol storage, transport and cellular uptake, suggesting that these might have appeared specifically in the vertebrate lineage [14].

Components of the RA machinery have been identified in tunicates. It was shown that RARs are present in *Ciona intestinalis* [31], *Botrylloides leachi* [80] and *Polyandrocarpa misakiensis* [40] that bind and are activated by RA. However, the function of the enzymes involved in RA synthesis and metabolism has not been assessed yet in these animals [14]. In addition, RA signaling has been secondarily modified in different lineages [14]. For example, there is a partial loss of the RA-dependent regulation of the *hox* code in ascidians [14], but the appendicularian tunicate, *Oikopleura dioica*, stands out as an exception because it lacks the genes coding for the main actors of the RA signaling pathway, such as, *rar*, *aldh1* and *cyp26* [13]. In the course of evolution, these genes were very likely secondarily lost in this animal since these gene are present in all the other urochordate species studied, such as *Ciona intestinalis* [31], *Botrylloides leachi* [80] and *Polyandrocarpa misakiensis* [40]. Of interest, it seems that the anterioposterior patterning of *O. dioica* during embryonic development may be RA-independent, although detailed characterization of the developmental pathways used by this species is still lacking. These observations suggest that anteroposterior patterning can be achieved using many different signaling pathways and is not necessarily dependent on RA.

In conclusion, the RA signaling pathway is present and functional in chordates, however, it should be noted that, compared to the situation in vertebrates, very little is known about the regulation of this signaling cascade and its function in invertebrate-chordates.

Presence or Absence of RA-Signaling in Metazoan Genomes: Non-chordates

Bioinformatic analyses have revealed that genes involved in the RA signaling machinery are present in the genomes of a variety of non-chordates (ambulacrarians and lophotrochozoans) [1, 12, 13, 65, 92]. Based on phylogenetic analysis of the nuclear hormone receptor superfamily, it was hypothesized that a proto-RAR might have been present in Urbilateria, the last common ancestor of all bilaterians [4].

Homologous genes of *rar* have so far been identified in the genomes of the following non-chordate invertebrates (Fig. 4.2): the ambulacrarians *Saccoglossus kowalevskii* (a hermichordate) and *Strongylocentrotus purpuratus* (an echinoderm) [13, 65] and the lophotrochozoans *Lottia gigantea* (a mollusk) and *Capitella teleta* (an annelid, formerly *Capitella capitata*) [12]. The phylogenetic analyses performed and the surprisingly high degree of amino acid conservation of the RARs found in these species indicate that these receptors may be *bona-fide* RARs that might be able to bind RA [12]. These findings allow us to conclude that the gene encoding RAR has a much more ancient evolutionary origin than previously believed. In particular, the fact that a clear rar ortholog is found in lophotrochozoans (annelids and mollusks) strongly suggests that a *rar* gene was already present in the ancestor of all bilaterians. This has important implications for the evolution and diversification of metazoan animals, since it indicates that functional roles for RA in developmental patterning might be evolutionarily much more ancient than originally thought (reviewed in [12]). However, recently, the characterization of a RAR from the mollusk *Thais clavigera* showed that this receptor is unable to bind RA [97], raising the question of the function of RAR and RA in non-chordate animals. Thus, the unambiguous experimental demonstration of a functional RAR and of a functional RA signaling cascade outside the chordate phylum still remains elusive.

In contrast, previous in silico analyses did not identify *rar* homologs in appendicularian tunicates (see above), ecdysozoans (arthropods and nematodes), and cnidarians (Fig. 4.2). These results allow two tentative conclusions: (1) if the absence of *rar* from cnidarian genomes is confirmed, for example by the analysis of other genomes, the origin of RAR and RA signaling can probably firmly be placed at the base of bilaterian animals and not at the base of metazoan animals. Given that anteroposterior patterning is an important feature of bilaterians, it is tempting to speculate that RAR might have played a role in the emergence of bilaterian body axes and, particularly, in the evolution of anteroposterior patterning systems, but this remains speculative given the very limited amount of data currently available on the role of RA outside chordates (see below); (2) the absence of *rar* in appendicularian tunicates and ecdysozoans suggests that these genes have been independently lost in these animal lineages and indicate that a complex anteroposterior patterning system can be established in the absence of RAR.

Homologs of the enzymes involved in RA metabolism, such as RALDH and CYP26 have also been identified in non-chordate animals. For example, *raldh* homologs have been identified in the genome of ambulacrarians (*Strongylocentrotus purpuratus* and *Saccoglossus kowalevskii*), lophotrochozoans (*Lottia gigantea* and *Capitella teleta*), ecdyzosoans (*Caenorhabditis elegans*, *Daphnia pulex* and *Drosophila melanogaster*) and cnidarian (*Nematostella vectensis*) [1, 12, 13, 65, 92]. In contrast, *cyp26* homologs have only been identified in lophotrochozoans and ambulacrarians [1, 12, 13, 65, 92]. Altogether, these data reinforce the notion that the RA signaling pathway might have been present in Urbilateria, the last common ancestor of protostomes and deuterostomes. However, the unambiguous experimental demonstration of a functional RAR and of a functional RA signaling cascade outside the chordate phylum still remains elusive.

In contrast, *rxr* homologs have been identified in the genomes of various non-chordate phyla including lophotrochozoans [9], ecdysozoans [104] and basal metazoan species, such as cnidarians [29, 55] and sponges [100]. Therefore, RXR seems to have a more ancient evolutionary origin than RAR. Apparently, the RXR receptors have also experienced a complex evolutionary history. For example, in vertebrates, several paralogs of RXR were acquired following whole genome duplication events [77] and in insects, ligand recognition by USP (i.e. RXR) has been altered independently in different lineages [16, 44, 46]. For instance, it was shown that the USP from *Tribolium castaneum* does not have a ligand-binding pocket and, thus, does not bind and is not activated by RXR ligands despite the high degree of identity between these receptors [46]. In addition, in mecopteridan USPs, the ligand-binding pocket is very large and is occupied by a phospholipid [6, 17]. These data reveal the evolutionary plasticity of the ligand binding pocket of nuclear hormone receptors and highlight the requirement for experimental analyses of divergent receptors.

RA Derivatives in Non-chordate Invertebrates

In vertebrates, RAR is able to bind at least two different isomers of RA, all-*trans* RA and 9-*cis* RA, whereas RXR only binds to 9-*cis* RA in vitro. Moreover, many studies have shown that treatment of vertebrate species with all-*trans* RA or 9-*cis* RA causes developmental defects and malformations [79]. In an effort to better understand the biological roles of each RA isomer, efforts have been made to measure the presence of these compounds in different organs of vertebrates. The presence of all-*trans* RA during development was detected, for example, in mouse, rat and human serum, liver, kidney, brain and testis [39, 48, 87]. The concentration of all-*trans* RA in these tissues was between 7 and 10 nM, however, it was lower in the serum [48].

In contrast, far less is known about the metabolism of 9-*cis* RA and the presence of 9-*cis* RA in vivo. Until recently, the biological significance of this isomer remained controversial. In fact, 9-*cis* RA was first detected by high performance liquid chromatography (HPLC) in mouse liver and kidney [39] and rat epididymal tissue [75] at concentrations two or three times lower than those of all-*trans* RA. Subsequent analytical assays, however, failed to detect this isomer in liver, kidney and other murine tissues (reviewed in [47]). This failure to detect 9-*cis* RA in vivo has contributed to the controversy about the biological relevance of this RA isomer. In contrast, by using liquid chromatography/tandem mass spectrometry a recent study has revealed the presence of 9-*cis* RA in mouse pancreas [49]. This work further established that 9-*cis* RA is able to attenuate glucose-stimulated insulin secretion in the pancreas [49]. These data not only show that 9-*cis* RA is a naturally-occurring RA isomer in the pancreas, but also that 9-*cis* RA is biologically active. In fact, it is possible, and even likely, that 9-*cis* RA can simply not be detected in other tissues, because its levels are below the limit of detection of the employed analysis methods.

In addition to mice, rats and humans, endogenous RA levels have also been measured in a number of other animals. But before reviewing these data, it has to be mentioned that derivatives of retinol, generally termed retinoids, are also involved in the visual cycle and that in particular 11-*cis* retinaldehyde plays a crucial role in vision. This aspect is not considered in the following discussion, but it is nonetheless important to be aware of the retinoid derivatives that participate in vision as it helps to explain the wide distribution in metazoan animals of various enzymes able to act on retinoid compounds. For example, 9-*cis* RA was detected in the regenerating limb of urodele amphibians [98], all-*trans* RA was detected in budding tunicates [52] and both all-*trans* RA and 9-*cis* RA were detected in the chordate amphioxus [20].

In non-chordates, endogenous all-*trans* RA and 9-*cis* RA were detected by HPLC in the central nervous system and the hemolymph of the mollusk *Lymnea stagnalis* [23]. All-*trans* RA concentrations were higher than those of 9-*cis* RA with all-*trans* RA concentrations estimated at 693 nM in the central nervous system and 155 nM in the hemolymph, whereas, for 9-*cis* RA, the estimated concentrations were 380 nM in the central nervous system and 120 nM in the hemolymph [23]. These values are thus relatively high, when compared to the values typically reported in mammals, but the significance of this difference is still unclear. Similarly, in the locust embryo, both all-*trans* RA and 9-*cis* RA were detected [74]. The all-*trans* RA concentration in the whole embryo was estimated to be 3 nM and the 9-*cis* RA concentration in the locust embryo was around 1.5 nM [74]. All-*trans* RA and 9-*cis* RA were also detected in limb blastemas of the fiddler crab during regeneration. Here, all-*trans* RA concentrations were 19 pg/μg protein/blastema and 9-*cis* RA concentrations were 83 pg/μg protein/blastema [41].

There are other RA isomers, such as 9,13-di-*cis* RA and 13-*cis* RA, that, at least in vertebrates, are characterized by reduced or absent biological activity [43, 95]. Interestingly, these RA isomers are distributed relatively widely in different mammalian tissues, for example, serum, brain, liver, kidney, adipose tissue, muscle, spleen and testis [50]. Moreover, they have also been detected in invertebrates, including sponges, which are basal metazoans. Sponges contain both retinyl esters and 13-*cis* RA and are able to accumulate β-carotene [5, 71]. In addition, retinol and retinyl esters have been detected in mollusks, which also possess some kind of capacity to endogenously store retinoids [33].

In nematodes, the only evidence of a possible action of RA has been reported in the parasitic nematodes *Brugia malayi* and *Onchocerca volvulus* [94, 101]. In both worm species only retinol could be detected [94, 101] and it was shown that *Brugia malayi* worms were able to take up and accumulate all-*trans* RA in their tissues [101]. In addition, it was shown that the development of the parasitic nematode *Litomosoides carinii* in vitamin A-deficient cotton rats was delayed [93]. Together, these data suggest that retinoids may play a role in the development of parasitic nematodes, however, it is important to stress that there is no RAR in nematodes and that a homolog of USP (i.e. RXR) has only been identified in a single species of parasitic nematodes [90].

Role of RA Derivatives in Non-chordate Invertebrates

As discussed above, in vertebrates, both RA isomers, all-*trans* RA and 9-*cis* RA, are not only present, but also biologically active. In fact, it is known that RA controls many developmental processes of vertebrates [79]. Indeed, exposure of developing embryos to exogenous RA causes severe defects in vertebrates and more generally in chordates. However, outside chordates, evidence for biological functions of RA is scarce and only a few studies have addressed the impact of RA on embryonic development. For example, in many mollusk species, such as, *Lymnaea stagnalis*, *Physa fontinalis* and *Bithynia tentaculata*, constant treatment starting at gastrulation with all-*trans* RA (100 nM) affects eye formation [19]. In addition, the use of higher concentrations of all-*trans* RA (1 μM) on *Lymnaea stagnalis* embryos causes various malformations, for example, eye and shell defects and, at times, growth arrest [19]. Moreover, it was shown that RA treatments were only effective during gastrulation as treatment of embryos after gastrulation did not cause any evident developmental defects [19]. This suggests that RA might control specific developmental events in this species, which is reminiscent of the role played by RA during vertebrate development.

Treatment of cultured neurons from the mollusk *Lymnaea stagnalis* with all-*trans* RA (100 nM) induces neurite outgrowth and growth cone turning [23, 24]. A comparable effect was observed in vertebrates with a similar RA concentrations [18]. 9-*cis* RA treatments also have an effect on mollusks. For example, in the mollusk *Thais clavigera*, the injection of 9-*cis* RA (1 μM) causes imposex, i.e. the development of male organs in females [73]. Taken together, although the available evidence is still very fragmentary and based mainly on morphological observations, these data on effects of RA treatments in mollusks are nonetheless indicative of possible functional roles of RA derivatives in mollusks. However, recently a RAR from the mollusk *Thais clavigera* has been characterized and it was shown that this receptor is unable to bind RA [97], thus suggesting that effects of RA on mollusks might be RAR independent.

Data are also available that describe the effects of treatments with RA derivatives in echinoderms. For example, treatment of micromere-derived cells from the sea urchin *Hemicentrotus pulcherrimus* with all-*trans* RA (100 nM) induces pseudopodial cable growth [56]. Intriguingly, no effect was observed after treatment of embryos of the sea urchin *Paracentrotus lividus* with all-*trans* RA (20 μM). The embryos seem to develop normally, although their development is delayed [89].

Finally, there is growing evidence for a role of RA derivatives in tissue formation and regeneration of a wide variety of animals, including platyhelminthes [83], insects [38], crustaceans [42], cnidarians [70] and sponges [45]. This role might be related to the known effect of retinoids in controlling fin regeneration in zebrafish [7] and limb regeneration in urodeles [98]. In addition, retinoids appear to be implicated in regulating tissue regeneration in the fruit fly *Drosophila melanogaster* [38]. Mutations in the genes coding for enzymes responsible of retinoid metabolism, such as BCO (β-carotene 15,15′-monooxygenase) and a retinaldehyde dehydrogenase, generated a delay in tissue regeneration in this species. A similar effect was observed when flies were deprived of caroteinoids, a natural source of retinoids, and the normal timing was restored by supplementation of retinoids [38]. These results are very surprising because in the fruit fly retinoids were thought to be involved exclusively in vision. Moreover, there is no RAR in the fruit fly genome and USP (i.e. RXR) is unable to bind retinoids (see [46] and references therein).

Moreover, in the platyhelminthe *Girardia tigrina*, treatment with exogenous all-*trans* RA (0.5 mM) disrupts anterior but not posterior regeneration [83]. In the fiddler crab *Uca pugilator*, treatment of regenerating limb buds with high concentrations of all-*trans* RA (50 μM) causes growth delay and malformations of the limb bud [42]. In the cnidarian *Hydractinia echinata*, treatments with retinol, retinaldehyde and all-*trans* RA (1 nM to 1 μM each) increase the regeneration rate, number of tentacles and budding rate [70]. Sponges are also affected by treatment with all-*trans* RA (50 μM), which causes morphogenetic malformations of buds and gemmules [45]. In addition, it was shown that treatment with all-*trans* RA (1 mM) upregulates the expression of the gene encoding the BMP-1 (bone morphogenetic protein-1) homolog in the sponge *Suberites domuncula* [71] and downregulates the expression of the *c-myb* gene in the sponge *Geodia cydonium* [5]. Altogether, these observations suggest that RA might have various pleiotropic effects in metazoan animals that may range from organ formation to regeneration.

Relevance

It is important to point out that from these experiments we cannot formally conclude that endogenous retinoids play physiological roles in these different animals. In fact, some of the effects observed upon RA exposure might be caused exclusively by the toxicity of the molecule at high dose or by its transformation into a different active compound. In addition, even if, in a given animal, the molecule acts on a specific target at a relatively weak concentration, one cannot necessarily conclude that the relevant metabolic pathway exists endogenously in that animal [62]. The case of bisphenol A is exemplary for this situation: bisphenol A is an artificially produced molecule that has been shown to exert very strong hormonal effects in mammals by regulating the activity of estrogen receptors [84]. Although these effects inadvertently exist in vivo, bisphenol A is certainly not an endogenously-produced hormone.

Coming back to RA, its derivatives and their potential biological functions, the solution of the problem seems rather straightforward: the receptors need to be characterized in full, including both molecular capacities and physiological functions, and the data hence obtained need to be correlated with the in vivo retinoid content of a given species. Of course, this work approach will not yield insights into retinoid functions that are not mediated by RAR and/or RXR. Even so, to date we completely lack a functional characterization of non-chordate RARs and we have thus no indication whatsoever of the possible ligand of these RARs. It could be all-*trans* RA, another RA isomer or some other compound of unknown nature. Therefore, we cannot conclude that the effects observed upon RA treatment are indeed RAR-dependent or even RA signaling-dependent.

Future Directions

With the sequencing of the genomes of many different animal species, it has become evident that the RA signaling pathway, that was initially thought to be chordate specific, has a much wider phylogenetic distribution than originally thought. In fact, some of the components of this pathway are found in the genomes of a wide variety of non-chordate taxa. For example, many components of this signaling pathway, such as the homologs of *rar*, *raldh* and *cyp26* have been identified in the genomes of ambulacrarians and lophotrochozoans [1, 12, 13, 65, 92]. Although an experimental validation of the biological functions of these receptors and metabolic enzymes is still lacking, these data nonetheless suggests that some kind of RA signaling pathway was probably already present in the last common ancestor of all bilaterian animals: Urbilateria [12].

Recent data have highlighted that RAR, in addition to being a classical nuclear receptor regulated by the binding of its ligand is also a protein whose activity can be controlled by phosphorylation [86]. Several studies have shown that these phosphorylation events are important for the full regulatory potential of RARs. It is still unclear how this level of regulation evolved but a recent phylogenetic analysis has shown that some, but not all, of those sites, are conserved between RARs from various taxa, including vertebrates and invertebrates [85]. This is, in particular, the case of the serine phosphorylation site located in the N-terminal A/B region of the receptor that is present in the RARs of vertebrates, amphioxus and ascidian tunicates (such as *Ciona intestinalis*) and that may also be present in protostome

RARs. It will therefore be of great interest to study how the regulatory logic controlling these invertebrate receptors has been set up. Similarly, recent work has revealed the evolutionary conservation in vertebrates of RAR targets through the conservation of direct repeats DR5, DR2 and related elements in target gene promoter regions [58]. Again, almost nothing is known about RAR target genes outside of vertebrates, but it can be anticipated that studies in non-vertebrates will yield exciting insights into the evolutionary diversification of the gene regulatory network controlled by RA and its receptors.

The RA signaling pathway has evolved differently in chordates, lophotrochozoans, ambulacrarians and ecdysozoans. The receptor controlling this pathway in vertebrates, RAR, is present in all these groups except in ecdysozoans where it seems to have been lost. However, it should be noted that, given the limited number of species studied and the intriguing biological effects of RA derivatives in insects and crustaceans, one should not conclude that this pathway was lost altogether in ecdysozoans. Moreover, although lophotrochozoans and ambulacrarians posses an evident *rar* homolog, there is still no experimental evidence showing that these genes encode *bona-fide* RARs that are capable of binding RA and of activating transcription in a ligand-dependent fashion.

Thus, it is important to be cautious and to remember that the presence of the ortholog of a given vertebrate gene in a distant organism does not necessarily imply the presence of the identical function known for the product of this gene in vertebrates [62]. For example, an ortholog of the estrogen receptor is present in annelids and mollusks, but several studies have shown that the mollusk orthologs do not bind estrogens [96], whereas the annelid othologs are able to bind estrogens at high concentrations [53]. In addition, it seems that estrogens might not even be present endogenously in these species [26, 63, 64]. Therefore, future research should focus on a detailed characterization of the functions of the components of the RA signaling pathway in a wide variety of different metazoan species.

Acknowledgment This work was supported by funds from ANR (ANR-09-BLAN-0129-01, ANR-09-BLAN-0262-02, ANR-11-JSV2-002-01) and CNRS to Michael Schubert and Vincent Laudet.

References

1. Albalat R, Cañestro C (2009) Identification of Aldh1a, Cyp26 and rar orthologs in protostomes pushes back the retinoic acid genetic machinery in evolutionary time to the bilaterian ancestor. Chem-Biol Interact 178:188–196
2. Balmer JE, Blomhoff R (2005) A robust characterization of retinoic acid response elements based on a comparison of sites in three species. J Steroid Biochem Mol Biol 96:347–354
3. Baek SH, Rosenfeld MG (2004) Nuclear receptor coregulators: their modification codes and regulatory mechanism by translocation. Biochem Biophys Res Co 319:707–714
4. Bertrand S, Brunet FG, Escriva H, Parmentier G, Laudet V, Robinson-Rechavi M (2004) Evolutionary genomics of nuclear receptors: from twenty-five ancestral genes to derived endocrine systems. Mol Biol Evol 21:1923–1937

5. Biesalski HK, Doepner G, Tzimas G, Gamulin V, Schroder HC, Batel R, Nau H, Muller WE (1992) Modulation of myb gene expression in sponges by retinoic acid. Oncogene 7:1765–1774
6. Billas IM, Moulinier L, Rochel N, Moras D (2001) Crystal structure of the ligand-binding domain of the ultraspiracle protein USP, the ortholog of retinoid X receptors in insects. J Biol Chem 276:7465–7474
7. Blum N, Begemann G (2012) Retinoic acid signaling controls the formation, proliferation and survival of the blastema during adult zebrafish fin regeneration. Development 139:107–116
8. Boleda MD, Saubi N, Farrés J, Parés X (1993) Physiological substrates for rat alcohol dehydrogenase classes: aldehydes of lipid peroxidation, omega-hydroxyfatty acids, and retinoids. Arch Biochem Biophys 307:85–90
9. Bouton D, Escriva H, de Mendonça RL, Glineur C, Bertin B, Noël C, Robinson-Rechavi M, de Groot A, Cornette J, Laudet V, Pierce RJ (2005) A conserved retinoid X receptor (RXR) from the mollusk *Biomphalaria glabrata* transactivates transcription in the presence of retinoids. J Mol Endocrinol 34:567–582
10. Bruck N, Vitoux D, Ferry C, Duong V, Bauer A, de The H, Rochette-Egly C (2009) A coordinated phosphorylation cascade initiated by p38MAPK/MSK1 directs RARα to target promoters. EMBO J 28:34–47
11. Calleja C, Messaddeq N, Chapellier B, Yang H, Krezel W, Li M, Metzger D, Mascrez B, Ohta K, Kagechika H, Endo Y, Mark M, Ghyselinck NB, Chambon P (2006) Genetic and pharmacological evidence that a retinoic acid cannot be the RXR-activating ligand in mouse epidermis keratinocytes. Genes Dev 20:1525–1538
12. Campo-Paysaa F, Marlétaz F, Laudet V, Schubert M (2008) Retinoic acid signaling in development: tissue-specific functions and evolutionary origins. Genesis 46:640–656
13. Cañestro C, Postlethwait JH, Gonzàlez-Duarte R, Albalat R (2006) Is retinoic acid genetic machinery a chordate innovation? Evol Dev 8:394–406
14. Carvalho JE, Schubert M (2013) Retinoic acid: metabolism, developmental functions and evolution. In: Dakshinamurti K, Dakshinamurti S (eds) Vitamin-binding proteins: their functional consequences. CRC Press/Taylor & Francis Group, Boca Raton, Florida, USA (in press)
15. Chambon P (1996) A decade of molecular biology of retinoic acid receptors. FASEB J 10:940–954
16. Chaumot A, da Lage JL, Maestro O, Martin D, Iwema T, Brunet F, Belles X, Laudet V, Bonneton F (2012) Molecular adaptation and resilience of the insect's nuclear receptor USP. BMC Evol Biol 12(1):199
17. Clayton GM, Peak-Chew SY, Evans RM, Schwabe JW (2001) The structure of the ultraspiracle ligand-binding domain reveals a nuclear receptor locked in an inactive conformation. Proc Natl Acad Sci USA 98:1549–1554
18. Corcoran J, Maden M (1999) Nerve growth factor acts via retinoic acid synthesis to stimulate neurite outgrowth. Nat Neurosci 2:307–308
19. Creton R, Zwaan G, Dohmen R (1993) Specific developmental defects in molluscs after treatment with retinoic acid during gastrulation. Dev Growth Differ 35:357–364
20. Dalfó D, Albalat R, Molotkov A, Duester G, Gonzàlez-Duarte R (2002) Retinoic acid synthesis in the prevertebrate amphioxus involves retinol oxidation. Dev Genes Evol 212:388–393
21. Dehal P, Boore JL (2005) Two rounds of whole genome duplication in the ancestral vertebrate. PLoS Biol 3:e314
22. de Urquiza AM, Liu S, Sjöberg M, Zetterström RH, Griffiths W, Sjövall J, Perlmann T (2000) Docosahexaenoic acid, a ligand for the retinoid X receptor in mouse brain. Science 290:2140–2144
23. Dmetrichuk JM, Carlone RL, Jones TRB, Vesprini ND, Spencer GE (2008) Detection of endogenous retinoids in the molluscan CNS and characterization of the trophic and tropic actions of 9-*cis* retinoic acid on isolated neurons. J Neurosci 28:13014–13024

24. Dmetrichuk JM, Carlone RL, Spencer GE (2006) Retinoic acid induces neurite outgrowth and growth cone turning in invertebrate neurons. Dev Biol 294:39–49
25. Duester G (2008) Retinoic acid synthesis and signaling during early organogenesis. Cell 134:921–931
26. Eick GN, Thornton JW (2011) Evolution of steroid receptors from an estrogen-sensitive ancestral receptor. Mol Cell Endocrinol 334:31–38
27. Escriva H, Bertrand S, Germain P, Robinson-Rechavi M, Umbhauer M, Cartry J, Duffraisse M, Holland L, Gronemeyer H, Laudet V (2006) Neofunctionalization in vertebrates: the example of retinoic acid receptors. PLoS Gene 2(7):e102
28. Escriva H, Holland ND, Gronemeyer H, Laudet V, Holland LZ (2002) The retinoic acid signaling pathway regulates anterior/posterior patterning in the nerve cord and pharynx of amphioxus, a chordate lacking neural crest. Development 129:2905–2916
29. Escriva H, Safi R, Hänni C, Langlois MC, Saumitou-Laprade P, Stehelin D, Capron A, Pierce R, Laudet V (1997) Ligand binding was acquired during evolution of nuclear receptors. PNAS 94:6803–6808
30. Fujiwara S (2006) Retinoids and nonvertebrate chordate development. J Neurobiol 66:645–652
31. Fujiwara S, Kawamura K (2003) Acquisition of retinoic acid signaling pathway and innovation of the chordate body plan. Zool Sci 20:809–818
32. Germain P, Iyer J, Zechel C, Gronemeyer H (2002) Co-regulator recruitment and the mechanism of retinoic acid receptor synergy. Nature 415:187–192
33. Gesto M, Castro LF, Reis-Henriques MA, Santos MM (2012) Retinol metabolism in the mollusk *Osilinus lineatus* indicates an ancient origin for retinyl ester storage capacity. PLoS One 7:e35138
34. Glass CK, Rosenfeld MG (2000) The coregulator exchange in transcriptional functions of nuclear receptors. Genes Dev 14:121–141
35. Glover JC, Renaud JS, Rijli FM (2006) Retinoic acid and hindbrain patterning. J Neurobiol 66:705–725
36. Gronemeyer H, Gustafsson JA, Laudet V (2004) Principles for modulation of the nuclear receptor superfamily. Nat Rev Drug Discov 3:950–964
37. Guengerich FP, Cheng Q (2011) Orphans in the human cytochrome P450 superfamily: approaches to discovering functions and relevance in pharmacology. Pharmacol Rev 63:684–699
38. Halme A, Cheng M, Hariharan IK (2010) Retinoids regulate a developmental checkpoint for tissue regeneration in Drosophila. Curr Biol 20:458–463
39. Heyman RA, Mangelsdorf DJ, Dyck JA, Stein RB, Eichele G, Evans RM, Thaller C (1992) 9-cis retinoic acid is a high affinity ligand for the retinoid X receptor. Cell 68:397–406
40. Hisata K, Fujiwara S, Tsuchida Y, Ohashi M, Kawamura K (1998) Expression and function of a retinoic acid receptor in budding ascidians. Dev Genes Evol 208:537–546
41. Hopkins P (2001) Limb regeneration in the fiddler crab, *Uca pugilator*: hormonal and growth factor control. Amer Zool 398:389–398
42. Hopkins PM, Durica DS (1995) Effects of all-trans retinoic acid on regenerating limbs of the fiddler crab. *Uca pugilator*. J Exp Zool 272:455–463
43. Horst RL, Reinhardt TA, Goff JP, Nonnecke BJ, Gambhir VK, Fiorella PD, Napoli JL (1995) Identification of 9-*cis*,13-*cis*-retinoic acid as a major circulating retinoid in plasma. Biochemistry 34:1203–1209
44. Hult EF, Tobe SS, Chang BS (2011) Molecular evolution of ultraspiracle protein (USP/RXR) in insects. PLoS One 6:e23416
45. Imsiecke G, Borojevic R, Custudio M, Müller WEG (1994) Retinoic acid acts as a morphogen in freshwater sponges. Invertebr Reprod Dev 26:89–98
46. Iwema T, Billas IM, Beck Y, Bonneton F, Nierengarten H, Chaumot A, Richards G, Laudet V, Moras D (2007) Structural and functional characterization of a novel type of ligand-independent RXR-USP receptor. EMBO J 26:3770–3782

47. Kane MA (2012) Analysis, occurrence, and function of 9-cis-retinoic acid. Biochim Biophys Acta 1821:10–20
48. Kane MA, Chen N, Sparks S, Napoli JL (2005) Quantification of endogenous retinoic acid in limited biological samples by LC/MS/MS. Biochem J 388:363–369
49. Kane MA, Folias AE, Pingitore A, Perri M, Obrochta KM, Krois CR, Cione E, Ryu JY, Napoli JL (2010) Identification of 9-*cis*-retinoic acid as a pancreas-specific autacoid that attenuates glucose-stimulated insulin secretion. PNAS 107:21884–21889
50. Kane MA, Folias A, Napoli J (2008) HPLC/UV quantitation of retinal, retinol, and retinyl esters in serum and tissues. Anal Biochem 378:71–79
51. Kastner P, Leid M, Chambon P (1994) The role of nuclear retinoic acid receptors in the regulation of gene expression. Vitamin A in health and disease. Marcel Dekker, New York, pp 189–238
52. Kawamura K, Hara K, Fujiwara S (1993) Developmental role of endogenous retinoids in the determination of morphallactic field in budding tunicates. Development 117:835–845
53. Keay J, Thornton JW (2009) Hormone-activated estrogen receptors in annelid invertebrates: implications for evolution and endocrine disruption. Endocrinology 150:1731–1738
54. Kim CI, Leo MA, Lieber CS (1992) Retinol forms retinoic acid via retinal. Arch Biochem Biophys 294:388–393
55. Kostrouch Z, Kostrouchova M, Love W, Jannini E, Piatigorsky J, Rall JE (1998) Retinoic acid X receptor in the diploblast, Tripedalia cystophora. PNAS 95:13442–13447
56. Kuno S, Kawamoto M, Okuyama M, Yasumasu I (1999) Outgrowth of pseudopodial cables induced by all-trans retinoic acid in micro- mere-derived cells isolated from sea urchin embryos. Dev Growth Differ 41:193–199
57. Kuraku S (2011) Hox gene clusters of early vertebrates: do they serve as reliable markers for genome evolution? Genomics Proteomics Bioinform 9:97–103
58. Lalevée S, Anno YN, Chatagnon A, Samarut E, Poch O, Laudet V, Benoit G, Lecompte O, Rochette-Egly C (2011) Genome-wide in silico identification of new conserved and functional retinoic acid receptor response elements (direct repeats separated by 5 bp). J Biol Chem 286:33322–33334
59. Laudet V, Gronemeyer H (2002) The nuclear receptors factsbook. Academic Press, London
60. Lemotte PK, Keidel S, Apfel CM (1996) Phytanic acid is a retinoid X receptor ligand. Eur J Biochem 236:328–333
61. Marill J, Capron CC, Idres N, Chabot GG (2002) Human cytochrome P450s involved in the metabolism of 9-*cis*- and 13-*cis*-retinoic acids. Biochem Pharmacol 63:933–943
62. Markov GV, Paris M, Bertrand S, Laudet V (2008) The evolution of the ligand/receptor couple: a long road from comparative endocrinology to comparative genomics. Mol Cell Endocrinol 293:5–16
63. Markov GV, Laudet V (2011) Origin and evolution of the ligand-binding ability of nuclear receptors. Mol Cell Endocrinol 334:21–30
64. Markov GV, Tavares R, Dauphin-Villemant C, Demeneix BA, Baker ME, Laudet V (2009) Independent elaboration of steroid hormone signaling pathways in metazoans. PNAS 106:11913–11918
65. Marlétaz F, Holland L, Laudet V, Schubert M (2006) Retinoic acid signaling and the evolution of chordates. Int J Biol Sci 2:38–47
66. McKenna NJ, O'Malley BW (2002) Combinatorial control of gene expression by nuclear receptors and coregulators. Cell 108:465–474
67. Mendoza-Parra MA, Walia M, Sankar M, Gronemeyer H (2011) Dissecting the retinoid-induced differentiation of F9 embryonal stem cells by integrative genomics. Mol Syst Biol 7:538
68. Mic FA, Molotkov A, Benbrook DM, Duester G (2003) Retinoid activation of retinoic acid receptor but not retinoid X receptor is sufficient to rescue lethal defect in retinoic acid synthesis. PNAS 100:7135–7140
69. Moutier E, Ye T, Choukrallah MA, Urban S, Osz J, Chatagnon A, Delacroix L, Langer D, Rochel N, Moras D, Benoit G, Davidson I (2012) Retinoic acid receptors recognize the

mouse genome through binding elements with diverse spacing and topology. J Biol Chem 287:26328–26341
70. Müller W (1984) Retinoids and pattern formation in a hydroid. J Embryol Exp Morph 271:253–271
71. Müller WE, Binder M, von Lintig J, Guo YW, Wang X, Kaandorp JA, Wiens M, Schröder HC (2011) Interaction of the retinoic acid signaling pathway with spicule formation in the marine sponge *Suberites domuncula* through activation of bone morphogenetic protein-1. Biochim Biophys Acta 1810:1178–1194
72. Napoli JL (1999) Retinoic acid: its biosynthesis and metabolism. Prog Nucleic Acid Res Mol Biol 63:139–188
73. Nishikawa J, Mamiya S, Kanayama T, Nishikawa T, Shiraishi F, Horiguchi T (2004) Involvement of the retinoid X receptor in the development of imposex caused by organotins in gastropods. Environ Sci Technol 38:6271–6276
74. Nowickyj SM, Chithalen JV, Cameron D, Tyshenko MG, Petkovich M, Wyatt GR, Jones G, Walker VK (2008) Locust retinoid X receptors: 9-Cis-retinoic acid in embryos from a primitive insect. PNAS 105:9540–9545
75. Pappas RS, Newcomer ME, Ong DE (1993) Endogenous retinoids in rat epididymal tissue and rat and human spermatozoa. Biol Reprod 48:235–247
76. Petkovich PM (2001) Retinoic acid metabolism. J Am Acad Dermatol 45:136–142
77. Philip S, Castro LF, da Fonseca RR, Reis-Henriques MA, Vasconcelos V, Santos MM, Antunes A (2012) Adaptive evolution of the retinoid X receptor in vertebrates. Genomics 99:81–89
78. Pittlik S, Begemann G (2012) New sources of retinoic acid synthesis revealed by live imaging of an Aldh1a2-GFP reporter fusion protein throughout zebrafish development. Dev Dyn 241:1205–1216
79. Rhinn M, Dollé P (2012) Retinoic acid signalling during development. Development 139:843–858
80. Rinkevich Y, Paz G, Rinkevich B, Reshef R (2007) Systemic bud induction and retinoic acid signaling underlie whole body regeneration in the urochordate *Botrylloides leachi*. PLoS Biol 5:e71
81. Roberts ES, Vaz AD, Coon MJ (1992) Role of isozymes of rabbit microsomal cytochrome P-450 in the metabolism of retinoic acid, retinol, and retinal. Mol Pharmacol 41:427–433
82. Robinson-Rechavi M, Boussau B, Laudet V (2004) Phylogenetic dating and characterization of gene duplications in vertebrates: the cartilaginous fish reference. Mol Biol Evol 21:580–586
83. Romero R, Bueno D (2001) Disto-proximal regional determination and intercalary regeneration in planarians, revealed by retinoic acid induced disruption of regeneration. Int J Dev Biol 45:669–673
84. Rubin BS (2011) Bisphenol A: an endocrine disruptor with widespread exposure and multiple effects. J Steroid Biochem 127:27–34
85. Samarut E, Amal I, Markov GV, Stote R, Dejaegere A, Laudet V, Rochette-Egly C (2011) Evolution of nuclear retinoic acid receptor alpha (RARα) phosphorylation sites. Serine gain provides fine-tuned regulation. Mol Biol Evol 7:2125–2137
86. Samarut E, Rochette-Egly C (2011) Nuclear retinoic acid receptors: conductors of the retinoic acid symphony during development. Mol Cell Endocrinol 348:348–360
87. Schmidt CK, Brouwer A, Nau H (2003) Chromatographic analysis of endogenous retinoids in tissues and serum. Anal Biochem 315:36–48
88. Schubert M, Holland ND, Laudet V, Holland LZ (2006) A retinoic acid-hox hierarchy controls both anterior/posterior patterning and neuronal specification in the developing central nervous system of the cephalochordate amphioxus. Dev Biol 296:190–202
89. Sciarrino S, Matranga V (1995) Effects of retinoic acid and dimethylsulfoxide on the morphogenesis of the sea urchin embryo. Cell Biol Int 19:675–680

90. Shea C, Hough D, Xiao J, Tzertzinis G, Maina CV (2004) An rxr/usp homolog from the parasitic nematode, *Dirofilaria immitis*. Gene 324:171–182
91. Shimozono S, Iimura T, Kitaguchi T, Higashijima SI, Miyawaki A (2013) Visualization of an endogenous retinoic acid gradient across embryonic development. Nature 496:363–366
92. Simões-Costa MS, Azambuja AP, Xavier-Neto J (2008) The search for non-chordate retinoic acid signaling: lessons from chordates. J Exp Zoo B Mol Dev Evol 310:54–72
93. Storey DM (1982) Vitamin A deficiency and the development of *Litomosoides carinii* (Nematoda, Filarioidea) in cotton rats. Z Parasitenk 67:309–315
94. Sturchler D, Wyss F, Hanck A (1981) Retinol, onchocerciasis and *Onchocerca volvulus*. T Roy Soc Trop Med H 75:617
95. Tang GW, Russell RM (1990) 13-*cis*-retinoic acid is an endogenous compound in human serum. J Lipid Res 31:175–182
96. Thornton JW, Need E, Crews D (2003) Resurrecting the ancestral steroid receptor: ancient origin of estrogen signaling. Science 301:1714–1717
97. Urushitani H, Katsu Y, Ohta Y, Shiraishi H, Iguchi T, Horiguchi T (2013) Cloning and characterization of the retinoic acid receptor-like protein in the rock shell, *Thais clavigera*. Aquat Toxicol 142–143:403–413
98. Viviano CM, Horton C, Maden M, Brockes JP (1995) Synthesis and release of 9-cis retinoic acid by the urodele wound epidermis. Development 121:3753–3762
99. White RJ, Schilling TF (2008) How degrading: Cyp26 s in hindbrain development. Dev Dyn 237:2775–2790
100. Wiens M (2003) Retinoid X receptor and retinoic acid response in the marine sponge *Suberites domuncula*. J Exp Biol 206:3261–3271
101. Wolff KM, Scott AL (1995) *Brugia malayi*: retinoic acid uptake and localization. Exp Parasitol 80:282–290
102. Wu BX, Moiseyev G, Chen Y, Rohrer B, Crouch RK, Ma JX (2004) Identification of RDH10, an all-*trans* retinol dehydrogenase, in retinal Müller cells. IOVS 45:3857–3862
103. Yang ZN, Davis GJ, Hurley TD, Stone CL, Li TK, Bosron WF (1994) Catalytic efficiency of human alcohol dehydrogenases for retinol oxidation and retinal reduction. Alcohol Clin Exp Res 18:587–591
104. Yao TP, Segraves WA, Oro AE, McKeown M, Evans RM (1992) Drosophila ultraspiracle modulates ecdysone receptor function via heterodimer formation. Cell 71:63–72

Chapter 5
RXRs: Collegial Partners

Federica Gilardi and Béatrice Desvergne

Abstract Retinoid X Receptors (RXR) were initially identified as nuclear receptors binding with stereo-selectivity the vitamin A derivative *9-cis* retinoic acid, although the relevance of this molecule as endogenous activator of RXRs is still elusive. Importantly, within the nuclear receptor superfamily, RXRs occupy a peculiar place, as they are obligatory partners for a number of other nuclear receptors, thus integrating the corresponding signaling pathways. In this chapter, we describe the structural features allowing RXR to form homo- and heterodimers, and the functional consequences of this unique ability. Furthermore, we discuss the importance of studying RXR activity at a genome-wide level in order to comprehensively address the biological implications of their action that is fundamental to understand to what extent RXRs could be exploited as new therapeutic targets.

Abbreviations

RXR	Retinoid X receptor
NR	Nuclear receptor
DBD	DNA-binding domain
LBD	Ligand binding domain
NCOR1	Nuclear receptor corepressor 1

F. Gilardi · B. Desvergne (✉)
Center for Integrative Genomics, University of Lausanne, Genopode Building, 1015 Lausanne, Switzerland
e-mail: Beatrice.Desvergne@unil.ch

F. Gilardi
e-mail: Federica.Gilardi@unil.ch

M. A. Asson-Batres and C. Rochette-Egly (eds.), *The Biochemistry of Retinoic Acid Receptors I: Structure, Activation, and Function at the Molecular Level*, Subcellular Biochemistry 70,
DOI: 10.1007/978-94-017-9050-5_5, © Springer Science+Business Media Dordrecht 2014

SMRT	Silencing mediator of Retinoid acid and Thyroid hormone receptor
RA	Retinoic acid
ER	Estrogen receptor
RAR	Retinoid acid receptor
TR	Thyroid hormone receptor
VDR	Vitamin D receptors
EMSA	Electrophoretic mobility shift assay
COUP-TFs	Chicken ovalbumin upstream promoter transcription factors
PPAR	Peroxisome proliferator-activated receptor
FXR	Farnesoid X receptor
LXR	Liver X receptor
NGFI-B	Nerve growth factor-induced protein I-B
Nurr1	Nuclear receptor related 1
AR	Androgen receptor
MR	Mineralcorticoid receptor
GR	Glucocorticoid receptor
PXR	Pregnane X receptor
CAR	Constitutive androstane receptor
HNF4	Hepatocyte nuclear factor 4
SHP	Small heterodimer partner
DR	Direct repeat
EM	Cryo electron microscopy
NMR	Nuclear magnetic resonance
SAXS	Small angle X-ray
SANS	Small single neutron scattering
FRET	Fluorescence resonance energy transfer
HDX	Hydrogen/Deuterium exchange mass spectrometry
ChIP	Chromatin immunoprecipitation
SRC-1	Steroid receptor coactivator 1
TIF2	Transcriptional intermediary factor 2
AF2	Activating function 2
MAPK	Mitogen-activated kinase
PKC	Protein kinase C
FABP4	Fatty acid binding protein 4
BSEP1	Bile salt export pump 1
ER	Everted repeat
IR	Inverted repeat
HMCS2	3-Hydroxy-3-Methylglutaryl-CoA synthase 2
DHA	Docohexanoic acid
PUFA	Polyunsaturated fatty acids
TBT	Trybutyltin chloride
TSS	Transcription start site
TZD	Thiazolidinedione
SNuRM	Selective nuclear receptor modulator

Introduction

In living organisms, cells need to adapt their intracellular activities to environmental conditions in order to control development, homeostasis and metabolism. One family of proteins playing a key role in the cellular response to internal and external stimuli is the nuclear hormone receptor superfamily, which includes Retinoid X Receptors (RXRs). RXRs occupy a central position within the nuclear receptor (NR) superfamily, serving as obligatory partners for numerous other NRs. This property sets RXRs at the crossroads of multiple signaling pathways where NRs play a coordinating role. The functionality of NRs is facilitated by two conserved domains in the proteins: the central DNA-Binding Domain (DBD) and the C-terminal Ligand-Binding Domain (LBD). The DBD is required for NR binding to specific DNA sequences and is folded into two small structural motifs called "zinc fingers". In the absence of a ligand, NRs physically interact with transcriptional corepressors, such as Nuclear receptor Co-Repressors 1 and Silencing Mediator of Retinoid acid and Thyroid hormone receptor (NCOR1 and SMRT) [26, 52], that trigger the repression of gene expression when they are bound to DNA. Upon ligand binding, the receptors undergo a conformational change that can influence their intracellular localization and/or their affinity for coregulator proteins, thus increasing the recruitment of transcriptional coactivators and the release of corepressors. These events promote interaction of NRs with the transcription initiation complex which leads to the transcription of target genes. In this way, NRs are able to modulate gene expression in response to a variety of stimuli.

History: RXRs—Central Players Within the NR Superfamily

First Evidence for a New Receptor for Retinoids

The cloning of the first NR, the glucocorticoid hormone receptor, dates back to almost three decades ago [46, 101, 136]. Thereafter, many other hormone-dependent transcription factors sharing similar structure and properties have progressively enriched this superfamily of proteins. Humans express 48 NR genes, some of them generating more than one isoform (Table 5.1). RXRs were first described in 1990 by Mangelsdorf et al. [93] as NRs able to respond to vitamin A metabolites with a peculiar ligand specificity for *9-cis* retinoic acid (*9cis*-RA).

Early work in this field showed that NRs, such as the estrogen receptor (ER), bind to target DNA in the form of homodimers. New evidence generated in the early 1990s suggested there might also be heterodimer interactions between different NRs [44]. In particular, it became evident that nuclear extracts from different cell types contained proteins that were necessary for NRs, including Retinoic Acid Receptors (RARs) [45], Thyroid hormone Receptors (TRs) [18, 103], and Vitamin

Table 5.1 The human 48 nuclear receptors, categorized according to their ligand binding properties

	Receptor	Full name	Sub-type	Official symbol	Monomer/Homodimers/ Heterodimer	LIGAND
Nuclear hormone receptors	ER	Estrogen receptors	α	NR3A1	Homodimer	Estradiol
			β	NR3A2		
	GR	Glococorticoid receptor		NR3C1	Homodimer	Glucocorticoids
	MR	Mineralocorticoid receptor		NR3C2	Homodimer	Aldosterone
	PR	Progesterone receptor		NR3C3	Homodimer	Progesterone
	AR	Androgen receptor		NR3C4	Homodimer	Testosterone
	TR	Thyroid hormone receptors	α	NR1A1	Heterodimer/Monomer	Thyroid hormone
			β	NR1A2		
	RAR	Retinoic acid receptors	α	NR1B1	Heterodimer	Retinoic acid
			β	NR1B2		
			γ	NR1B3		
	VDR	Vitamin D receptor		NR1I1	Heterodimer	Vitamin D
	RXR	Retinoid X receptors	α	NR2B1	Homodimer/Heterodimer	9-cis retinoic acid
			β	NR2B2		
			γ	NR2B3		
Metabolic sensors	PPAR	Peroxisome proliferator-activated receptors	α	NR1C1	Heterodimer	Polyunsaturated fatty acids, prostaglandins
			β	NR1C2		
			γ	NR1C3		
	LXR	Liver X receptors	α	NR1H3	Heterodimer	Oxysterols
			β	NR1H2		
	FXR	Farnesoid X receptor		NR1H4	Heterodimer	Bile acids
	PXR	Pregnane X receptor		NR1I2	Heterodimer	Xenobiotics
	CAR	Constitutive androstane receptor		NR1I3	Heterodimer	Androstane

(continued)

Table 5.1 continued

	Receptor	Full name	Sub-type	Official symbol	Monomer/Homodimers/ Heterodimer	LIGAND
"Adopted" orphan nuclear receptors	SF1	Steroidogenic factor 1		NR5A1	Monomer	Oxysterols
	LRH-1	Liver receptor homolog 1		NR5A2	Monomer	Phosphatidilcholine derivatives
	RevErb	Reverse Erb A	α	NR1D1	Monomer/Homodimer	Heme
			β	NR1D2		
	ROR	RAR-related orphan receptors	α	NR1F1	Monomer	Cholesterol, all *trans* retinoic acid
			β	NR1F2		
			γ	NR1F2		
	HNF4	Hepatocyte nuclear factor 4	α	NR2A1	Homodimer	Fatty acids
			γ	NR2A2		
Orphan nuclear receptors	COUP-TF	Chicken ovalbumine upstream promoter transcription factors	I	NR2F1	Homodimer/Heterodimer	Unknown
			II	NR2F2		
	ERR	Estrogen related receptors	α	NR3B1	Monomer/Homodimer	Unknown
			β	NR3B2		
			γ	NR3B3		
	TR2	Testicular receptors	2	NR2C1	Homodimer/Heterodimer	Unknown
			4	NR2C2		
	TLX	Tailless-related receptors		NR2E1	Monomer/Homodimer	Unknown
	PNR	Photoreceptor-specific nuclear receptor		NR2E2	Monomer/Homodimer	Unknown
	NGFIB	Nerve growth factor IB		NR4A1	Monomer/Homodimer/ Heterodimer	Unknown
	Nurr1	Nuclear receptor related 1		NR4A2		
	Nor1	Neuron-derived orphan receptor 1		NR4A3		
	GCNF	Germ cell nuclear factor		NR6A1	Homodimer	Unknown
	DAX1	Dosage-sensitive sex reversal, adrenal hypoplasia critical region, on chromosome X, gene 1		NR0B1	Heterodimer	Unknown
	SHP	Small heterodimer partner		NR0B2		Unknown

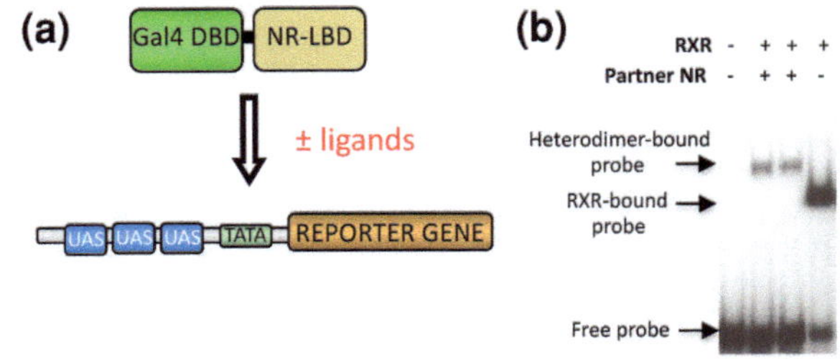

Fig. 5.1 Traditional techniques employed to evaluate RXR functionality. **a** In transactivation assays, the LBD of RXR or other NR is fused with the DBD of Gal4, a yeast-derived transcription factor. This construct is transfected in cells together with a vector expressing a reporter gene under the control of a series of Gal4 responsive elements (UAS). In the presence of the NR ligand, the transcriptional activity of Gal4 is enhanced and the expression reporter gene is increased. **b** In electrophoretic mobility shift assays (*EMSA*), a radiolabeled DNA probe is incubated with the NR of interest and loaded on a gel. If the NR binds to the DNA probe, the migration of the DNA is delayed and we observe the shift (RXR-bound probe). The formation of a heterodimer complex is also revealed (heterodimer-bound probe)

D Receptors (VDRs) [86], to bind DNA with high affinity. After the identification and first characterization of RXRs, several groups independently demonstrated that RXRs were indeed the common missing factors that were forming heterodimeric complexes with RARs, TRs and VDRs [17, 70, 82, 95, 133, 144, 147].

The ability of RXRs to form heterodimers was determined using Electrophoretic Mobility Shift Assay (EMSA). This technique evaluates the ability of protein complexes to bind to specific DNA sequences in an in vitro context (Fig. 5.1). Continued reports showed that RXRs are also able to form homodimers [54, 147] as well as to heterodimerize with Chicken Ovalbumin Upstream Promoter Transcription Factors (COUP-TFs) [6, 69], Peroxisome Proliferator-Activated Receptors (PPARs) [5, 40, 64, 71], Farnesoid X Receptors (FXRs) [37], Liver X Receptors (LXRs) [138], Nerve Growth Factor-induced protein I-B (NGFI-B) and transcriptional inducible NUclear Receptor Related (Nurr1) [111]. Thus, RXRs appear to play a unique role in integrating the action of several NRs by the forming heterodimers with multiple partners. Other work has shown that RXRs can also form homodimers [54, 147].

Classification of Nuclear Receptors

To unify the nomenclature of all NRs, a classification according to their phylogeny was adopted in 1999 (Nuclear Receptors Nomenclature [29]). However other criteria, more related to their functionality, are often used to categorize NRs [4]. For instance, when NRs are ordered based on ligand binding properties, they fall into three groups (Table 5.1). The first class is composed of the "classic" hormone receptors, including ERs, Androgen Receptor (AR), Mineralocorticoid Receptor

(MR), Glucocorticoid Receptor (GR), TRs, VDR and RARs. The receptors belonging to this group can bind DNA in form of homodimers, like the ER and the GR, or in form of heterodimers with RXR, like RARs, VDR and TRs. The peculiarity of these NRs is their ability to bind a narrow range of small molecules with very high affinity, thus mediating a corresponding endocrine function.

By contrast, receptors belonging to the so-called metabolic sensors bind, with relatively low affinity, a large number of molecules, which are often intermediate or final metabolites of different metabolic pathways. Within this group are PPARs, LXRs and FXRs, which are activated by a broad spectrum of fatty acids, cholesterol and bile acid derivatives, respectively. Due to the nature of their activators and due to their target genes, these receptors regulate energy metabolism at multiple levels, but also cell proliferation, cell differentiation and cell survival (reviewed in [20, 30, 35]). Other members of this group are the pregnane X receptor (PXR) and the constitutive androstane receptor (CAR), which both induce cellular pathways important for detoxification [58, 140]. Most of the metabolic sensors bind DNA in the form of heterodimers with RXRs.

A third group is composed of "orphan" receptors that so far have no identified ligand. A number of studies have proposed putative ligands for some of these receptors [19, 68, 78, 92], that, thereafter, are now referred as "adopted" NRs. In contrast, for some others, such as Nurr1, the lack of a ligand binding pocket large enough to accommodate a ligand suggests that such receptors may have no ligand at all [135]. The orphan NRs are the most varied in terms of DNA-binding properties: indeed, they can act in the form of homodimers, such as the Hepatocyte Nuclear Factor 4 (HNF4), heterodimers with RXR, like COUP-TFs, or monomers, such as NGFI-B. Furthermore, the atypical receptor, Small Heterodimer Partner (SHP), lacks a DBD and cannot bind DNA.

The central role of RXRs as heterodimerization partners for NRs belonging to all three classes sets them apart from the above-described functional classification scheme. Complete occupation of the ligand-binding pocket by high-affinity ligands is a typical feature of classic NRs, a property exhibited by RXR, which has a small ligand binding cavity (between 400 and 500 Å) [51] that tightly accommodates *9-cis* RA. Nonetheless, it is not possible to strictly categorize the RXRs as classic NRs because a physiological role for *9-cis* RA has yet to be confirmed in vivo, and other natural ligands have much lower affinities for the receptor (see "Relevance: Endogenous and Synthetic RXR Ligands").

RXR Isotypes

In both humans and mice, there are three RXR isotypes, RXRα (NR2B1), RXRβ (NR2B2) and RXRγ (NR2B3), that are encoded by separate genes sharing strong sequence homology [82, 91]. Each gene can further give rise to different isoforms by means of alternative splicing and/or alternative promoter utilization. In mice, three RXRα isoforms (RXRα1, 2 and 3), differing in the N-terminal domain, have been functionally characterized [16], and a fourth possible isoform has been predicted. Four RXRβ and two RXRγ isoforms have been identified [80, 89].

RXR isotypes and isoforms have a specific tissue distribution with partially redundant functions. In mice, RXRα1 is most abundant in the liver, but is also highly expressed in kidney, spleen, placenta and epidermis. RXRα2 and 3 are found in adult testis [16]. RXRβ is ubiquitous, with higher levels present in the central nervous system. RXRγ1 is expressed in skeletal muscle, olfactory bulb and pituitary gland, while RXRγ2 appears in both cardiac and skeletal muscle [10]. In humans, the tissue expression profile is more ubiquitous for RXRα and RXRβ, while RXRγ is predominantly in pineal gland (http://biogps.org [139]).

Development of the Field: RXR Behavior as Partners of Different Functional Complexes

How RXR-Containing Heterodimers Bind DNA

Many factors contribute to determining the recognition of the response element by the NR dimer in a specific target promoter, including the properties of the DNA consensus motif, its polarity, the NR ligand, and the DNA itself. The ligand may alter the DNA binding properties of a given NR heterodimer [146], and reciprocally, the DNA may itself play a role in the allosteric regulation of NR activity [54, 114, 120, 146].

NRs bind to specific DNA sequences, called consensus motifs. The DNA sequence recognized by NR dimers consists of two copies of a derivative of the consensus hexamer AGGTCA. Orientation and spacing of these two motifs are important determinants of their specificity toward each receptor dimer. Most of the consensus sequences for homo- or heterodimers containing RXRs are direct repeats of the core motif, with one to five nucleotides spacing (DR1 to DR5). In particular, RXR-homodimers and PPAR/RXR heterodimers preferentially bind to DR1, RAR/RXR complexes can bind to DR2 or DR5, whereas VDR/RXR and TR/RXR favor binding to DR3 and DR4 respectively. However, the rule underlying the recognition of a response element by NR heterodimers is difficult to decipher, due to some heterogeneity. For instance, RAR/RXR was also observed to associate with DR1 motifs, in some specific promoter contexts [119]. TR/RXR [34] and FXR/RXR [75] may also bind to inverted repeats (IR).

Another crucial feature of the heterodimer interaction with DNA is its polarity (see Chap. 3 and [77]). DR motifs are asymmetrical and can be read correctly only from one direction [90, 104]. This property provides a further level of specificity as gene regulatory responses may differ depending on whether RXRs occupy the 5′ upstream or 3′ downstream half-site [71, 73, 74, 94]. For instance, when complexed with RARs, RXRs always occupy the 3′ half-site in a DR1 but the 5′ half-site in a DR5. This difference in position influences the ability of the NR complex to be activated by the ligand. In contrast, PPARs always binds to the 5′ extended half-site of the DR1, while RXRs occupy the 3′ half-site [53]. This was confirmed by the crystal structure of the heterodimer PPARγ/RXRα complexed with its DNA response element [25].

Many homo/heterodimers compete for the same DR elements (for example, PPAR/RXR, HNF4/HNF4, and RXR/RXR bind to DR1; TR/RXR and LXR/RXR bind to DR4). This may have a strong functional impact depending on the protein abundance of RXRs and/or their partners and the heterodimer associating with DNA. As a consequence, the resulting expression of target genes may be significantly different. Alternately and not exclusively, the importance of promoter context may be a key determinant.

Structural Basis of Hetero- and Homo-Dimer Formation

As described, RXR is a peculiar member of the NR family since it can form homodimers or heterodimers with many different NRs. The first crystal structure of a RXR-dimer to be obtained was the RXR-LBD homodimer [12]. The analysis of its structure showed that both DBD and LBD are involved in dimerization. However, the LBD is more crucial due to a larger and stronger dimerization interface that is composed of amino acids arranged in a hydrophobic cluster and surrounded by polar residues. For example, the short region of RXRα between amino acids 387 and 429, called the I-box, has been shown to be required for heterodimerization [112]. A similar I-box, with a number of highly conserved amino acids, is also present in all RXR heterodimerization partners. Other crystal structures of RXR/partner LBD have been resolved, including RAR/RXR [13, 113], CAR/RXR [126, 142], LXR/RXR [59, 127] and PPAR/RXR [39, 47, 141]. The overall structure of these different heterodimers is highly similar, but within the molecules, there is variability between the different heterodimer interfaces, suggesting that some partners have a higher affinity for RXR than others.

More recently, thanks to tremendous progress in development of the techniques used to analyze protein structure and protein-protein interactions [i.e. Cryo Electron Microscopy (EM), nuclear magnetic resonance (NMR), small angle X-ray and neutron scattering (SAXS and SANS) and FRET (fluorescence resonance energy transfer)], the structures of intact heterodimers bound to DNA in their native environment have been reported (see Chaps. 2 and 3 and reviewed in [15]).

In 2008, the crystal structure of the full-length PPARγ /RXRα complex bound to ligands, coactivator peptides, and DNA was resolved, fully highlighting the domain–domain interactions involving these intact NRs. The heterodimer formed by PPARγ and RXRα is asymmetric, allowing the PPARγ-LBD to lie at the center of the complex and to make direct contacts with multiple domains in both proteins. Three distinct heterodimerization interfaces are formed between RXRα and PPARγ and, interestingly, some of them are DNA-dependent [25].

A different conformation was observed by Rochel et al. [117], who noticed that, in solution, the structure of several heterodimer complexes bound to DNA (VDR/RXR bound to a DR3, RAR/RXR to a DR5, RAR/RXR to a DR1 and PPAR/RXR to a

DR1) is characterized by an asymmetric open conformation with separate DBDs and LBDs, connected by extended hinges. Subsequent analyses of the full VDR/RXR visualized the hinge domains of both receptors that stabilize the complex in a precise conformation [109]. Besides these physical interactions, an extensive communication throughout the heterodimer, influenced by the recognized sequence of DNA, has been documented by Hydrogen/Deuterium Exchange (HDX) experiments [146].

Based on these results, the currently accepted model for heterodimerization is that, first, RXR and the partner NR form a complex through their dimerization interfaces, and second, some remodeling of the formed heterodimer occurs once the complex recognizes the response element.

Functional Behavior of RXR-Containing Complexes

In the last twenty years, a great effort has been made to understand the behavior of the different RXR-containing complexes, namely homotetramers, homodimers and heterodimers, and the specific role of RXR proteins within the heterodimer. Unexpectedly considering the behavior of most members of the nuclear hormone receptor family, RXRs are able to self-associate in solution and form transcriptionally inactive homotetramers [38, 65, 66]. Upon ligand binding, these tetramers dissociate, allowing the formation of homo- and heterodimers, exhibiting transcriptional activity [28]. The function of RXR homotetramers is not completely understood. A first hypothesis regarding the role of these complexes was that they could act as structural elements able to fold the DNA, in order to approach distant RXR response elements [143]. Subsequently, it was proposed that homotetramers could function as a way for RXR molecules to be stored and then be mobilized for formation of heterodimers, depending on the ligand availability for the partner [36, 131]. No further evidence has been reported and we are left with only hypotheses.

Besides aggregating as homotetramers, RXR can homodimerize. *9-cis* RA induces RXR homodimer formation in vitro, suggesting the existence of a retinoid response pathway distinct from that activated by the heterodimer RAR/RXR [147]. In cells and in animals treated with 9-*cis* RA, the actual presence of RXR homodimers on specific DNA response elements was subsequently demonstrated by in vivo chromatin immunoprecipitation (ChIP) studies that are extremely powerful in indicating whether a protein is recruited to specific DNA binding sites in a given cell [54] (the principles of the technique are presented in Fig. 5.2). Additionally, analysis of fluorescence fluctuation brightness in living cells expressing RXRs labeled with molecules that emit fluorescence showed that homodimerization is triggered by *9-cis* RA binding to RXR [27] (Fig. 5.3). This event induces the recruitment of coactivators, such as Steroid Receptor Coactivator 1 (SRC-1) and Transcriptional Intermediary Factor 2 (TIF2), thus stabilizing the association of the complex to DR1 elements on the DNA. However and as mentioned, all these experimental studies were performed in the presence of *9-cis* RA. Thus, the physiological role of RXR homodimers, which seems to highly depend

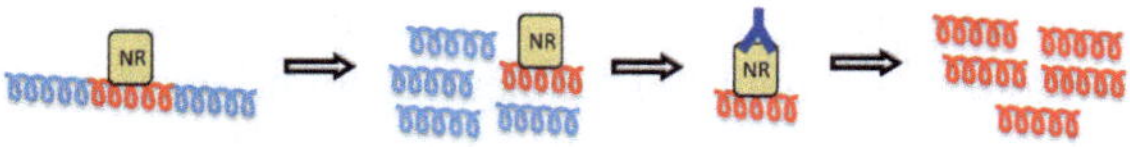

Fig. 5.2 Chromatin Immunoprecipitation (*ChIP*). In ChIP assays, NR proteins are cross-linked with formaldehyde to regions of DNA that have recognition sites (*red*) for the proteins (NR). In the following step, DNA is fragmented in short pieces (300–1,000 bases). Fragments of DNA that are associated with the protein of interest are then captured using an antibody (*blue* inverted Y) that specifically recognizes the NR. In the last step, cross-linking is reversed in the captured DNA-NR fragments: the DNA is detached from the NR and the captured DNA fragments are identified by amplifing in a PCR or by DNA sequencing

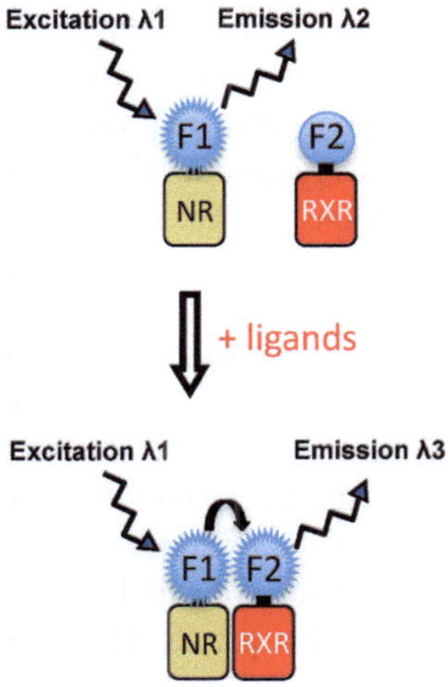

Fig. 5.3 Fluorescence resonance energy transfer (*FRET*). FRET assays evaluate the interaction between two proteins (i.e. heterodimer partners). The two NR are linked to two different fluorophores (F1 and F2). If the two partners do not interact, the excitation of the F1 fluorophore at the appropriate wavelength (λ1) results in energy emission (λ2). If the two partners interact, the interaction brings the fluorophores in tight proximity, and, in this case, excitation of the F1 fluorophore results in an energy transfer to the F2 fluorophore, which gets excited and emits energy at wavelength λ3

on the presence of RXR ligand, will remain an open question until the existence and nature of endogenous ligands for RXR is revealed [54].

First attempts to understand the functional consequences of heterodimerization were based mostly on transient transfection experiments in which RXR proteins and their partners were overexpressed and activated by their respective ligands (Fig. 5.1a). These studies highlighted two categories of functionally distinct RXR partners, so-called permissive and non-permissive partners. The activity of a non-permissive NR partner is enhanced by its own agonist, but is unaffected by the RXR-agonist. Non-permissive partners include hormone receptors with a high affinity for their ligands, such as TRs, RARs and VDR. In contrast, permissive RXR heterodimers can be activated by either an RXR and/or the NR partner agonist.

Permissive partners are typically metabolic sensors like PPARs, LXRs, FXRs, PXR and CAR, that sense mainly small lipid molecules with low affinity. One attractive hypothesis to explain the appearance during evolution of these two classes of receptors is that non-permissive RXR partners diverged from permissive receptors as they acquired the ability to recognize high-affinity hormonal ligands. In parallel, the amino acid sequence of permissive partners evolved to allow them to communicate trough allosteric mechanisms with RXR. Thus, the evolutionary process allowed NRs to differentially couple metabolic and endocrine regulation [123].

The molecular mechanisms underlying the allosteric coupling of partners in permissive heterodimers are still poorly understood. A crucial element underlying the functional synergy of activation could reside in the recruitment of coactivators. Several reports indicate that, upon ligand binding, a single coactivator molecule binds to receptor heterodimers [84]. Each heterodimer partner can be autonomously anchored to one of the LXXLL signature motifs that are present in the coactivator [50, 137]. In the presence of both ligands, synergy would originate from the cooperative binding of one coactivator molecule with the two heterodimer partners, through the simultaneous establishment of two NR-coactivator interfaces [41]. For permissive heterodimers, such as LXR/RXR, the existence of allosteric interactions, which can influence the activity of the protein complex in response to the binding of the RXR ligand, was demonstrated by recent structural data [124]. The RXR region mediating these interactions can be different depending on the partner. As an example, the RXR-AF2 region is necessary for RXR-mediated activation of PPARβ, but not of PPARγ [21]. Such allosteric interactions seem to be excluded within the complex RAR/RXR, in agreement with the non-permissiveness of this heterodimer [113]. In contrast, and quite unexpectedly, the non-permissive partner TR is needed to specifically activate the prolactin promoter in response to 9-*cis* RA in cells [24]. Taken together, these observations suggest that the mechanisms underlying the molecular crosstalk between the partners are strongly specific for each heterodimer and, for a given heterodimer, these mechanisms may differ depending on the evaluated target gene.

Therefore, to deeply understand RXR biology, it will be fundamentally important to couple the innovative techniques used to analyze protein structure and interactions with new genome-wide approaches as this will allow comprehensive evaluation of RXR binding and its effects on the expression of all target genes. Noteworthy, the physiological relevance of the distinction between non-permissive and permissive partners may have to be revisited once the nature of the natural ligand for RXR in vivo is elucidated.

Regulation of RXR Activity Through Post-Translational Modifications

The dynamic assembly/disassembly of retinoid receptors to promoters of target genes is regulated by proteasome-mediated degradation. Both partners within DNA-bound RXRα/RAR heterodimers can be degraded in response to retinoids acting either on RXR or on RAR (9-*cis* RA or derivatives and *all-trans*-retinoic acid) [11, 42, 43, 72, 130]. The impact of the ubiquitin-proteosome machinery on heterodimer activity appears to be isoform specific. It has been shown in different cell lines and adipose tissue that RXRα is degraded by this process, but RXRβ is not [79].

Proteasome-mediated degradation of RXRα is linked to its phosphorylation at specific serine residues by mitogen-activated kinases (MAPK) and/or protein kinase C (PKC) that can interfere with its function [1, 43, 97]. These post-translational modifications seem to be important for the potential involvement of RXRs in cell proliferation and in cancer development. RXR in tumors can be degraded by cathepsin L and calpain producing a fragment of 44KDa which goes to the cytosol and activates a cancer cell survival pathway [129, 148]. Thus, RXRα may also have extra-nuclear action and mechanisms governing its cellular localization could control its transcriptional activity, further complicating RXR biology.

Current State of the Field: Addressing RXR Functions on a Global Level

Targeted Gene Disruption: RXR-Null Mice

The generation of mice carrying RXR gene deletions or mutations has been a valuable tool to investigate the impact that RXR has on the biological functions of other NRs and the specific role, if any, of each RXR isotype. As expected, phenotypic characterization of the mutant mice demonstrated that RXR is involved in a plethora of developmental and physiological pathways. The analysis of RXRβ mutants revealed a reduction of spermatid formation in the mice [63]. Specific ablation of RXRγ impaired cholinergic responses in nigrostriatal pathways [118] and caused important metabolic effects, including an increased metabolic rate with an accompanying resistance to gain fat mass in response to high-fat feeding [49].

Whereas the RXRβ and RXRγ null mice had only mild developmental defects, ubiquitous inactivation of RXRα resulted in embryonic lethality at mid-gestation due to hypoplastic development of the ventricular myocardium [62, 125]. To bypass this lethality, several conditional, tissue-specific, RXRα knockout mice have been generated. Conditional hepatic ablation of RXRα results in alteration of fatty acid oxidation and hepatocyte lifespan suggesting that the absence of RXRα in the liver cannot be compensated for by RXRβ or RXRγ. This supports the hypothesis that each isotype may have specific functions [56, 134]. Tissue-specific deletion of RXRα in adipose tissue resulted in resistance to diet-induced obesity in these mice, due to impaired adipocyte differentiation, likely reflecting an altered PPARγ function [55].

More recently, characterization of mice lacking RXRα in myeloid cells has demonstrated an important role of RXRα in the innate immune response to inflammatory stimuli. Interestingly, this function seems to be mediated by RXR homodimers, evidence that supports the existence of an RXR signaling pathway in vivo [108].

In most cases, the phenotypes observed in RXR mutant mice can be related to alterations in pathways regulated by other NRs (reviewed in [31]). It remains difficult to unambiguously define the functions of individual heterodimers in regulating signaling pathways that influence the expression of specific genes via direct and/or indirect regulatory loops. Thus, much is left to be learned about the functional roles of RXRs in vivo. Besides a thorough investigation using tissue-specific knockout models, the systematic analysis of transcriptional programs regulated by agonists for RXRs or their partner is essential to bring new insight into the complex field of RXR biology.

Global Analyses of RXR-Dependent Transcriptional Programs

With the rapid development of quantitative PCR and microarray technologies, the evaluation of how the expression of large gene sets is affected by different experimental conditions has become routine. These approaches are suitable in order to identify potential target genes of transcription factors, including NRs. The application of such techniques has elucidated tissue- and gene-specific differences in the permissiveness of RXR-heterodimers, by using synthetic ligands.

One study performed in rats showed that, in adipose tissue, the effect of the RXR-specific agonist LG268 was not the same as that of the PPARγ-specific ligand rosiglitazone, suggesting that in this tissue, the PPARγ/RXR heterodimer is not permissive [2]. Differently, in mouse liver, the rexinoid bexarotene induced the expression of LXR-target genes that participate in lipogenesis, but did not affect those involved in cholesterol homeostasis. The latter findings support the concept that the permissivity of LXR/RXR heterodimers also depends on the target gene [76].

A systematic analysis of the different transcriptional programs induced by permissive and non-permissive heterodimers was performed in monocyte-derived dendritic cells, as a case study [128]. To comprehensively address all aspects of RXR signaling, the global gene expression profile was evaluated in cells treated with RXR ligands (LG628 and 9-*cis* RA), ligands for the non-permissive partners RAR and VDR (AM580 and 1,25-vitamin D, respectively), or ligands for the permissive partners PPARγ, PPARβ/δ and LXRs (rosiglitazone, GW1516 and GW3965, respectively). The results showed that RXR ligands regulated only a few of the target genes of the non-permissive partners, while most of the LXR and PPAR responsive genes were also induced by RXR-agonists, albeit to a lesser degree than was observed with LXR and PPAR agonists. While confirming the different functional behaviors of permissive and non-permissive heterodimers, this report also emphasized the fact that RXR-mediated activation on permissive heterodimers may have specific outcomes. For example, the combination of RXR and PPARγ agonists had a differential effect depending on the PPARγ targets: either synergism (most of the PPARγ/RXR responsive genes) or negative interaction, resulting in a reduced induction compared to that triggered by the PPARγ ligand alone (for example, the *Fatty acid binding protein* 4 (FABP4) gene), were observed [128]. Similar observations were made for other permissive heterodimers, such as FXR/RXR [61], strongly indicating that the heterodimer activation state is strongly dependent not only on the tissue, but also on the gene.

The molecular mechanisms underlying this differential tissue- and gene-dependent regulation are far from being understood. Presently, the most likely hypothesis is that cofactor availability in different cell types could be the determinant in such activities. The observation that the coactivator recruited on PPARγ/RXR heterodimers is different if induced by RXR or PPAR agonists [84] supports this hypothesis. This suggests that the amount of each cofactor in given cells and the nature of the available ligand influence RXR effects on target gene activation and expression.

Taken together, these observations underscore the existence of very intricate RXR signaling mechanisms. Further, this highlights the importance of performing analyses in different tissues or cell-types to fully decipher RXR behavior, especially in view of potential therapeutic uses of RXR-specific activators.

RXR Cistrome

In recent years, key technological advancements have allowed chromatin immune-precipitation to be coupled, initially with tiled oligonucleotide microarray (ChIP-chip) and later on with ultra high-throughput sequencing (ChIP-seq), providing a valuable tool to identify the NR *cistrome*, namely all DNA binding sites (referred to as *cis* as these sites are located on the DNA strand) recognized by a particular DNA-binding factor or complex (referred to as *trans*-acting factors), analyzed at the whole genome level. The first NR whose DNA binding sites were characterized genome-wide was the ER [22]. Thereafter, dozens of studies reported the cistrome of NRs in different cell lines or tissues. Currently, the landscape of RXR binding sites has been elucidated in combination with PPARγ during adipogenesis [106], with VDR in osteoblasts [99], with the oncofusion protein PML-RARα in an acute promyelitic leukemia cell line [96], with LXRβ in human keratinocytes [121] and with PPARα and LXR in the liver [9]. Most of these studies were primarily aimed at gaining a better understanding of the biology of the partner NR, rather than RXR, but nonetheless, they did highlight several features that are relevant to RXR, both as a NR and as a common heterodimerization partner.

One surprising finding that came out of analyzing the RXR-heterodimer cistrome was that the number of DNA binding sites was one, or even two, orders of magnitude greater than the number of effective target genes (Fig. 5.4). This difference can be partially explained by an underestimation of the ligand-regulated gene number due to confounding effects, such as RNA abundance and stability, which can influence

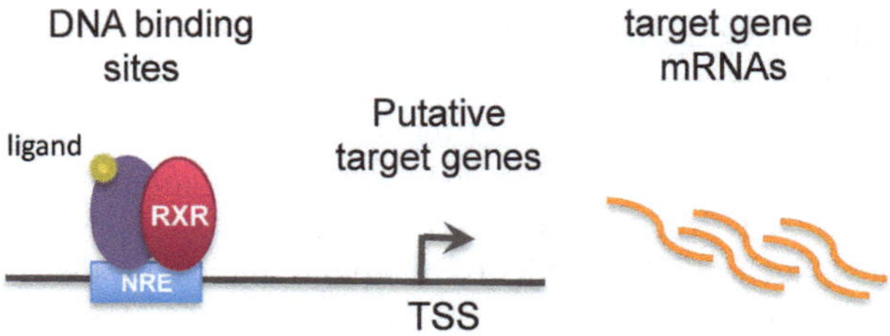

Heterodimer	Ligand	RXR/partner common binding sites	Putative target genes	Effective target genes	Reference
VDR/RXR	1,25(OH)$_2$D$_3$	1845	1445	141	Meyer et al. 2010
LXR/RXR	T0901317	12806	5524	161	Boergesen et al. 2012
PPARγ/RXR	Differentiation cocktail	5236	ND	861	Nielsen et al. 2008

Fig. 5.4 The number of NR binding sites to DNA is one or two orders of magnitude higher than the number of their target genes. Upon ligand treatment, ChIP on chip or ChIP-seq experiments were performed to identify genome-wide DNA binding sites of RXR with VDR in murine osteoblasts, with LXR in mouse liver, and with PPARγ in murine adipocytes. Compared to a set of thousands of *putative target genes* that were proposed for each heterodimer based on several criteria, including proximity of the RXR/partner binding site to the transcription start site (*TSS*), the global gene expression analysis indicated that only hundreds of genes (*effective target genes*) were modulated by the treatment

the sensitivity of microarrays. However, the disparity is pronounced and raises the question of the functional significance of RXR, and more generally, NR, binding to these numerous binding sites that are apparently not related to transcriptional regulation. Unexpectedly, the analysis of NR binding sites localization showed that most of them (>60 %) are located in distal intergenic regions or within introns, whereas only a few sites are found in the proximal promoter region. This observation correlates well with the fact that only 16 % of the chromatin regions that are open and more accessible to transcription factors are located in promoter regions, as demonstrated by DNAse I hypersensitivity assays coupled to DNA sequencing [14]. Accordingly, the identified binding sites for RXR and LXR in the liver, in the absence of specific ligands, are located within open chromatin regions, whereas treatment with ligands enables LXR and RXR to bind also to less accessible sites [9].

As expected, RXR binding sites are more numerous than those for a particular partner. Intriguingly, RXR binding to sites recognized by one heterodimer is not strictly dependent on the presence of the partner. For instance, many binding sites that are co-occupied by RXR and PPARγ in mature adipocytes, are already bound by RXR in preadipocytes, possibly in the form of a homodimer or as a heterodimer with PPARβ [106] or another unidentified partner. Similarly, VDR binding sites pre-marked by the presence of RXR have been identified in osteoblasts [99].

The constant recruitment of RXR on a number of DNA binding sites that are shared with other NRs was observed also in mouse liver [9]. The characterization of RXR binding sites in LXR double knockout mice livers reveals that, despite LXR absence, RXR is still binding to the majority of the shared LXR/RXR binding sites identified in the wild type mice. Interestingly, when a motif search was performed on sequences under the LXR/RXR peaks, it turned out that they were not only enriched in DR4 motifs, which are the canonical binding sites for LXR/RXR and TR/RXR heterodimers, but they were also enriched in other DRs, Inverted Repeats (IR1) and Everted Repeats (ER2), which suggests that different homo/heterodimers may bind to these regions. Indeed, a sub-group of these sites can be bound also by PPARα, HNF4α, FXR, and Reverbα, suggesting the existence of NR "hot spot" sequences, to which multiple NRs bind depending on the precise signaling acting within the cell at a certain moment. Notably, while LXR binds to sequences containing DR1 elements in vivo, it fails to bind the same sequences in vitro, indicating the chromatin context plays a major role in determining the recognition of a certain response element. The functional outcome of this extensive crosstalk appears to also be determined by the context of the single genomic binding sites. The observation that a very low percentage of RXR binding sites are overlapping in different tissues (i.e. only 12 % of RXR binding sites are common between adipocytes and liver) further supports the importance of the chromatin context in determining NR recruitment on specific DNA regions. However, the molecular mechanisms underlying the promiscuity and the regulation of the activity of individual receptors at particular sites are not yet fully understood.

The global maps of RXR binding sites obtained to date have been valuable tools for identifying novel transcription factor binding sites. With respect to RXR biology, the comparison of different global genome-wide binding profiles has clarified genomic regions occupied by different partners. Future work will reveal new levels of regulation by individual heterodimers that are strictly dependent on the chromatin context.

Relevance: Endogenous and Synthetic RXR Ligands

Given the widespread relevance of the superfamily of NRs to multiple aspects of human physiology and etiopathogenesis and the role of RXR as partner for many of these receptors, a detailed characterization of RXR ligands and their effects has major implications for the understanding of retinoid biology, as well as for the development of new potential drug treatments.

The physiological ligand for RXR is still unknown. A study performed in mouse epidermal keratinocytes demonstrated the concomitant presence of a transcriptional repressor complex composed of RXRα heterodimerized with unliganded RARγ and an active heterodimer PPARβ(δ)/RXRα. The induced expression of the target gene *3-Hydroxy-3-Methylglutaryl-CoA Synthase 2* (*hmgcs2*) indicates that the RXR agonist must have been present in these cells [21]. These results imply that the ligand for RXR must be different from the ligand for RAR. This excludes *9-cis* RA from the possible candidates since this molecule also activates RAR.

More recently, other vitamin A metabolites acting as ligand for RXRs have been detected in vivo, such as dihydroretinoids [102], *all-trans*-retinaldehyde [149] and b-apo-14′-carotenal [150]. A group of flexible unsaturated fatty acids, either endogenously produced or derived from the diet, can act as RXR ligand. Two monocyclic terpenoid compounds, methoprenic acid and phytanic acid have been shown to be highly selective RXR ligands, albeit at much higher concentrations than *9-cis*-RA [48, 67, 83]. In addition, the structural analysis of RXR-RAR LBD led to the identification of an oleic acid molecule into the RXR ligand binding pocket [13]. Docohexanoic acid (DHA) can also be accommodated in the LBD of RXRs and activate the receptors, similarly to other n-3 PUFA, such as docosatetraenoic acid (C22: 4 all-*cis*-Δ7,10,13,16) or arachidonic acid (C20: 4 all-*cis*-Δ5,8,11,14) [33, 132]. These findings suggest that RXRs, besides regulating metabolic homeostasis in association with their heterodimerization partners, can also directly act as intracellular sensors.

Besides natural ligands, many synthetic compounds target RXRs (reviewed in [23, 110]. Several environmental pollutants, such as the organotin tributyltin chloride (TBT), activate RXRs and other NRs, thus interfering with the physiological activity of these receptors [60].

An interest in RXR ligands as potential pharmacological treatments arose some time ago from the observation of their strong apoptotic effect [98, 105]. Accordingly, an RXR selective ligand, bexarotene (LGD1069, Targretin®) [8], has been successfully used in the therapy of refractory or persistent cutaneous T-cell lymphoma. Preclinical studies have shown that rexinoids are also promising drugs for acute promyelocytic and myeloid leukemias [3, 145], and clinical trials have demonstrated their efficacy in the treatment of non-small-cell lung cancer [7, 115], as well as in the prevention of both lung and mammary tumors [85, 87, 88].

Given their pro-apoptotic effects and their multi-partnerships with NRs that act as metabolic regulators, rexinoids are clearly molecules with potential to exert broad impact on metabolic regulation. LG1069 administration to patients results in hypertriglyceridemia, hypercholesterolemia and hypothyroidism [32, 116]. The characterization of rexinoid effects in vivo showed that they mimic PPARγ ligands such as thiazolidinediones (TZDs), improving insulin sensitivity. However, while TZD acts primarily at the level of adipose tissue, RXR ligands exert their activity mainly at hepatic and muscular levels [2, 122], suggesting that PPARα, PPARβ(δ) and LXRs are likely the partners accounting for rexinoid action.

The multiple metabolic effects of rexinoids highlight their potential as drugs for complex disorders, such as the metabolic syndrome. The striking gain in knowledge on how ligand binding influences RXR structure and the resulting ability of the complex to interact with cofactor proteins has made it possible to start designing rexinoids with heterodimer selective activities, such as *selective NR modulators* (SNuRM). Several reports describe rexinoids that are able to activate different combinations of heterodimers including HX630 (selective for PPARγ/RXR), PA024 (acting on both PPARγ/RXR and LXR/RXR) [107], and LG101506 (selectively activates PPARγ/RXR and not LXR/RXR or RAR/RXR) [81]. The distinct behavior is reflected in the different patterns of gene expression induced by HX630 and PA024 when used in combination with an RAR agonist [57]. Rats treated with LG101506, which activates only PPARγ/RXR, have improved insulin sensitivity but unchanged triglyceride levels, likely due to a lack of LXR/RXR activation [100].

Future Directions

RXRs can form heterodimers with a number of NRs involved in the regulation of multiple cellular functions. This peculiar ability represents a formidable example of how activity of one single receptor can broadly modulate a very complex regulatory network. In recent years, striking technological advancements have made it possible to improve our understanding of RXR biology.

Significant steps forward have been made in our understanding of how changes in RXR three-dimensional structure are induced by ligands and in how the DNA consensus sequence affects the ability of the receptor to interact with cofactor proteins. In parallel, the possibility to globally address the transcriptional effects of RXR activation, together with the genome-wide identification of DNA binding sites of RXRs and their partners, have highlighted an extremely intricate regulatory network in which multiple factors, such as cofactor availability, competition among possible partners, and ligand presence, play a role. All of this has led to an appreciation of the increasing levels of complexity that underly RXR biology.

There is a need for more in depth studies of the individual signaling pathways that are specifically activated by each *RXR/partner NR* heterodimer and/or single RXR isoforms. New approaches and new concepts, particularly based on the development of bioinformatics and modeling, are also needed to push the field forward and foster a way to grasp and control the inherent levels of complexity.

Such global analyses are essential in order to understand to what extent the unique properties of RXR signaling network can be exploited as a new therapeutic target for the treatment of diseases and metabolic disorders. Studies have paved the way for the design of pharmaceutical rexinoids with selective activities, and clinical trials have indicated their potential efficacy. But at the same time, the difficulty to keep in vivo responses to such promiscuous receptors under control speaks to a pressing need for additional research aimed at producing a deeper understanding of their direct and indirect actions.

References

1. Adachi S, Okuno M, Matsushima-Nishiwaki R, Takano Y, Kojima S, Friedman SL, Moriwaki H, Okano Y (2002) Phosphorylation of retinoid X receptor suppresses its ubiquitination in human hepatocellular carcinoma. Hepatology 35:332–340

2. Ahuja HS, Liu S, Crombie DL, Boehm M, Leibowitz MD, Heyman RA, Depre C, Nagy L, Tontonoz P, Davies PJA (2001) Differential effects of Rexinoids and Thiazolidinediones on metabolic gene expression in diabetic rodents. Mol Pharmacol 59:765–773
3. Altucci L, Rossin A, Hirsch O, Nebbioso A, Vitoux D, Wilhelm E, Guidez F, De Simone M, Schiavone EM, Grimwade D, Zelent A, De The H, Gronemeyer H (2005) Rexinoid-triggered differentiation and tumor-selective apoptosis of acute myeloid leukemia by protein kinase A-mediated desubordination of retinoid X receptor. Cancer Res 65:8754–8765
4. Aranda A, Pascual A (2001) Nuclear hormone receptors and gene expression. Physiol Rev 81:1269–1304
5. Bardot O, Aldridge TC, Latruffe N, Green S (1993) PPAR-RXR heterodimer activates a peroxisome proliferator response element upstream of the bifunctional enzyme gene. Biochem Biophys Res Commun 192:37–45
6. Berrodin TJ, Marks MS, Ozato K, Linney E, Lazar MA (1992) Heterodimerization among thyroid hormone receptor, retinoid X receptor, Chicken ovalbumin upstream promoter transcription factor, and an endogenous liver protein. Mol Endocrinol 6:1468–1478
7. Blumenschein GR, Khuri FR, Von Pawel J, Gatzemeier U, Miller WH, Jotte RM, Le Treut J, Sun S-L, Zhang JK, Dziewanowska ZE, Negro-Vilar A (2008) Phase III trial comparing carboplatin, paclitaxel, and bexarotene with carboplatin and paclitaxel in chemotherapy-naive patients with advanced or metastatic non-small-cell lung cancer: SPIRIT II. J Clin Oncol 26:1879–1885
8. Boehm MF, Zhang L, Zhi L, Mcclurg MR, Berger E, Wagoner M, Mais DE, Suto CM, Davies PJA, Heyman RA, Nadzan AM (1995) Design and synthesis of potent retinoid X receptor selective ligands that induce apoptosis in leukemia cells. J Med Chem 38:3146–3155
9. Boergesen M, Pedersen TA, Gross B, Van Heeringen SJ, Hagenbeek D, Bindesboll C, Caron S, Lalloyer F, Steffensen KR, Nebb HI, Gustafsson J-A, Stunnenberg HG, Staels B, Mandrup S (2012) Genome-wide profiling of liver X receptor, retinoid X receptor, and peroxisome proliferator-activated receptor alpha in mouse liver reveals extensive sharing of binding sites. Mol Cell Biol 32:852–867
10. Bookout AL, Jeong Y, Downes M, Yu R, Evans RM, Mangelsdorf DJ (2006) Anatomical profiling of nuclear receptor expression reveals a hierarchical transcriptional network. Cell 126:789–799
11. Boudjelal M, Wang Z, Voorhees JJ, Fisher GJ (2000) Ubiquitin/Proteasome pathway regulates levels of retinoic acid receptor Œ ≥ and retinoid X receptor Œ ± in human keratinocytes. Cancer Res 60:2247–2252
12. Bourguet W, Ruff M, Chambon P, Gronemeyer H, Moras D (1995) Crystal structure of the ligand binding domain of the human nuclear receptor RXR-alpha. Nature 375:377–382
13. Bourguet W, Vivat V, Wurtz J-M, Chambon P, Gronemeyer H, Moras D (2000) Crystal structure of a heterodimeric complex of RAR and RXR ligand-binding domains. Mol Cell 5:289–298
14. Boyle AP, Davis S, Shulha HP, Meltzer P, Margulies EH, Weng Z, Furey TS, Crawford GE (2008) High-resolution mapping and characterization of open chromatin across the genome. Cell 132:311–322
15. Brelivet Y, Rochel N, Moras D (2012) Structural analysis of nuclear receptors: from isolated domains to integral proteins. Mol Cell Endocrinol 348:466–473
16. Brocard J, Kastner P, Chambon P (1996) Two novel RXRalpha isoforms from mouse testis. Biochem Biophys Res Commun 229:211–218
17. Bugge TH, Pohl J, Lonnoy O, Stunnenberg HG (1992) RXR alpha, a promiscuous partner of retinoic acid and thyroid hormone receptors. EMBO J 11:1409–1418
18. Burnside J, Darling DS, Chin WW (1990) A nuclear factor that enhances binding of thyroid hormone receptors to thyroid hormone response elements. J Biol Chem 265:2500–2504
19. Burris TP, Busby SA, Griffin PR (2012) Targeting orphan nuclear receptors for treatment of metabolic diseases and autoimmunity. Chem Biol 19:51–59

20. Calkin AC, Tontonoz P (2012) Transcriptional integration of metabolism by the nuclear sterol-activated receptors LXR and FXR. Nat Rev Mol Cell Biol 13:213–224
21. Calleja C, Messaddeq N, Chapellier B, Yang H, Krezel W, Li M, Metzger D, Mascrez BND, Ohta K, Kagechika H, Endo Y, Mark M, Ghyselinck NB, Chambon P (2006) Genetic and pharmacological evidence that a retinoic acid cannot be the RXR-activating ligand in mouse epidermis keratinocytes. Genes Dev 20:1525–1538
22. Carroll JS, Liu XS, Brodsky AS, Li W, Meyer CA, Szary AJ, Eeckhoute J, Shao W, Hestermann EV, Geistlinger TR, Fox EA, Silver PA, Brown M (2005) Chromosome-wide mapping of estrogen receptor binding reveals long-range regulation requiring the forkhead protein FoxA1. Cell 122:33–43
23. Casals-Casas C, Desvergne B (2011) Endocrine disruptors: from endocrine to metabolic disruption. Annu Rev Physiol 73:135–162
24. Castillo AI, Sanchez-Martinez R, Moreno JL, Martinez-Iglesias OA, Palacios D, Aranda A (2004) A permissive retinoid X receptor/thyroid hormone receptor heterodimer allows stimulation of prolactin gene transcription by thyroid hormone and 9-cis-retinoic acid. Mol Cell Biol 24:502–513
25. Chandra V, Huang P, Hamuro Y, Raghuram S, Wang Y, Burris TP, Rastinejad F (2008) Structure of the intact PPAR-γ-RXR-α nuclear receptor complex on DNA. Nature 456:350–356
26. Chen JD, Evans RM (1995) A transcriptional co-repressor that interacts with nuclear hormone receptors. Nature 377:454–457
27. Chen Y, Wei LN, Muller JD (2005) Unraveling protein-protein interactions in living cells with fluorescence fluctuation brightness analysis. Biophys J 88:4366–4377
28. Chen Z-P, Iyer J, Bourguet W, Held P, Mioskowski C, Lebeau L, Noy N, Chambon P, Gronemeyer H (1998) Ligand- and DNA-induced dissociation of RXR tetramers. J Mol Biol 275:55–65
29. Committee NRN (1999) A unified nomenclature system for the nuclear receptor superfamily. Cell 97:161–163
30. Desvergne B, Michalik L, Wahli W (2006) Transcriptional regulation of metabolism. Physiol Rev 86:465–514
31. Desvergne B (2007) RXR: from partnership to leadership in metabolic regulations. In: Gerald L (ed) Vitamins & hormones. Academic Press, New York
32. Duvic M, Martin AG, Kim Y et al (2001) PHase 2 and 3 clinical trial of oral bexarotene (targretin capsules) for the treatment of refractory or persistent early-stage cutaneous t-cell lymphoma. Archives of Dermatology 137:581–593
33. Egea PF, Mitschler A, Moras D (2002) Molecular recognition of agonist ligands by RXRs. Mol Endocrinol 16:987–997
34. Farsetti A, Desvergne B, Hallenbeck PL, Robbins J, Nikodem VM (1992) Characterization of the myelin basic protein thyroid hormone response element and its function in the context of native and heterologous promoter. J Biol Chem 1992(267):15784–15788
35. Feige JN, Gelman L, Michalik L, Desvergne B, Wahli W (2006) From molecular action to physiological outputs: peroxisome proliferator-activated receptors are nuclear receptors at the croassroads of key cellular functions. Prog Lipid Res 45:120–159
36. Feige JRMN, Gelman L, Tudor C, Engelborghs Y, Wahli W, Desvergne B (2005) Fluorescence imaging reveals the nuclear behavior of peroxisome proliferator-activated receptor/retinoid X receptor heterodimers in the absence and presence of ligand. J Biol Chem 280:17880–17890
37. Forman BM, Goode E, Chen J, Oro AE, Bradley DJ, Perlmann T, Noonan DJ, Burka LT, Mcmorris T, Lamph WW, Evans RM, Weinberger C (1995) Identification of a nuclear receptor that is activated by farnesol metabolites. Cell 81:687–693
38. Gampe RT Jr, Montana VG, Lambert MH, Wisely GB, Milburn MV, Xu HE (2000) Structural basis for autorepression of retinoid X receptor by tetramer formation and the AF-2 helix. Genes Dev 14:2229–2241

39. Gampe RT, Montana VG, Lambert MH, Miller AB, Bledsoe RK, Milburn MV, Kliewer SA, Willson TM, Xu HE (2000) Asymmetry in the PPARgamma/RXRalpha crystal structure reveals the molecular basis of heterodimerization among nuclear receptors. Mol Cell 5:545–555
40. Gearing KL, Gottlicher M, Teboul M, Widmark E, Gustafsson JA (1993) Interaction of the peroxisome proliferator-activated receptor and retinoid receptor. Proc Natl Acad Sci USA 90:1440–1444
41. Germain P, Jaya I, Zechel C, Gronemeyer H (2002) Co-regulator recruitment and the mechanism of retinoic acid receptor synergy. Nature 415:187–192
42. Gianni M, Bauer A, Garattini E, Chambon P, Rochette-Egly C (2002) Phosphorylation by p38MAPK and recruitment of SUG-1 are required for RA-induced RAR[gamma] degradation and transactivation. EMBO J 21:3760–3769
43. Gianni M, Tarrade A, Nigro EA, Garattini E, Rochette-Egly C (2003) The AF-1 and AF-2 domains of RARγ2 and RXRα cooperate for triggering the transactivation and the degradation of RARγ2/RXRα heterodimers. J Biol Chem 278:34458–34466
44. Glass CK (1994) Differential recognition of target genes by nuclear receptor monomers, dimers and heterodimers. Endocr Rev 15:391–407
45. Glass CK, Devary OV, Rosenfeld MG (1990) Multiple cell-type specific proteins differentially regulate target sequence recognition by the alpha retinoic acid receptor. Cell 63:729–738
46. Govindan MV, Devic M, Green S, Gronemeyer H, Chambon P (1985) Cloning of the human glucocorticoid receptor cDNA. Nucleic Acids Res 13:8293–8304
47. Haffner CD, Lenhard JM, Miller AB, Mcdougald DL, Dwornik K, Ittoop OR, Gampe RT, Xu HE, Blanchard S, Montana VG, Consler TG, Bledsoe RK, Ayscue A, Croom D (2004) Structure-based design of potent retinoid X receptor alpha agonists. J Med Chem 47:2010–2029
48. Harmon MA, Boehm MF, Heyman RA, Mangelsdorf DJ (1995) Activation of mammalian retinoid X receptors by the insect growth regulator methoprene. Proc Natl Acad Sci 92:6157–6160
49. Haugen BR, Jensen DR, Sharma V, Pulawa LK, Hays WR, Krezel W, Chambon P, Eckel RH (2004) Retinoid X receptor gamma-deficient mice have increased skeletal muscle lipoprotein lipase activity and less weight gain when fed a high-fat diet. Endocrinology 145:3679–3685
50. Heery DM, Kalkhoven E, Hoare S, Parker M (1997) A signature motif in transcriptional coactivators mediates binding to nuclear receptors. Nature 387:733–736
51. Hollenberg SM, Weinberger C, Ong ES, Cerelli G, Oro A, Lebo R, Thompson EB, Rosenfeld MG, Evans RM (1985) Primary structure and expression of a functional human glucocorticoid receptor cDNA. Nature 318:635–641
52. Horlein A, Naar A, Heinzel T, Torchia J, Gloss B, Kurokawa R, Ryan A, Kamei Y, Soderstrom M, Glass C (1995) Ligand-independent repression by the thyroid hormone receptor mediated by a nuclear receptor co-repressor. Nature 377:397–404
53. Ijpenberg A, Jeannin E, Wahli W, Desvergne B (1997) Polarity and specific sequence requirements of peroxisome proliferator-activated receptor (PPAR)/retinoid X receptor heterodimer binding to DNA. J Biol Chem 272:20108–20117
54. Ijpenberg A, Tan NS, Gelman L, Kersten S, Seydoux J, Xu J, Metzger D, Canaple L, Chambon P, Wahli W, Desvergne B (2004) In vivo activation of PPAR target genes by RXR homodimers. EMBO J 23:2083–2091
55. Imai T, Jiang M, Chambon P, Metzger D (2001) Impaired adipogenesis and lipolysis in the mouse upon selective ablation of the retinoid X receptor α mediated by a tamoxifen-inducible chimeric Cre recombinase (Cre-ERT2) in adipocytes. PNAS 98:224–228
56. Imai T, Jiang M, Kastner P, Chambon P, Metzger D (2001) Selective ablation of retinoid X receptor alpha in hepatocytes impairs their lifespan and regenerative capacity. Proc Natl Acad Sci 98:4581–4586
57. Ishida S, Shigemoto-Mogami Y, Kagechika H, Shudo K, Ozawa S, Sawada J-I, Ohno Y, Inoue K (2003) Clinically potential subclasses of retinoid synergists revealed by gene expression profiling1. Mol Cancer Ther 2:49–58

58. Jacobs MN, Nolan GT, Hood SR (2005) Lignans, bacteriocides and organochlorine compound activate the human pregnane X receptor (PXR). Toxicol Appl Pharmacol 209:123–133
59. Jaye MC, Krawiec JA, Campobasso N, Smallwood A, Qiu C, Lu Q, Kerrigan JJ, De Los Frailes Alvaro M, Laffitte B, Liu W-S, Marino JP, Meyer CR, Nichols JA, Parks DJ, Perez P, Sarov-Blat L, Seepersaud SD, Steplewski KM, Thompson SK, Wang P, Watson MA, Webb CL, Haigh D, Caravella JA, Macphee CH, Willson TM, Collins JL (2005) Discovery of substituted maleimides as liver X receptor agonists and determination of a ligand-bound crystal structure. J Med Chem 48:5419–5422
60. Kanayama T, Kobayashi N, Mamiya S, Nakanishi T, Nishikawa J-I (2005) Organotin compounds promote adipocyte differentiation as agonists of the peroxisome proliferator-activated receptor Œ ≥/retinoid X receptor pathway. Mol Pharmacol 67:766–774
61. Kassam A, Miao B, Young PR, Mukherjee R (2003) Retinoid X receptor (RXR) agonist-induced antagonism of farnesoid X receptor (FXR) activity due to absence of coactivator recruitment and decreased DNA binding. J Biol Chem 278:10028–10032
62. Kastner P, Grondona JM, Mark M, Gansmuller A, Lemeur M, Decimo D, Vonesch J-L, Dolle P, Chambon P (1994) Genetic analysis of RXR alpha developmental function: convergence of RXR and RAR signaling pathways in heart and eye morphogenesis. Cell 78:987–1003
63. Kastner P, Mark M, Leid M, Gansmuller A, Chin W, Grondona JM, Decimo D, Krezel W, Dierich A, Chambon P (1996) Abnormal spermatogenesis in RXR beta mutant mice. Genes Dev 10:80–92
64. Keller H, Dreyer C, Medin J, Mahfoudi A, Ozato K, Wahli W (1993) Fatty acids and retinoids control lipid metabolism through activation of peroxisome proliferator-activated receptor-retinoid X receptor heterodimers. Proc Natl Acad Sci USA 90:2160–2164
65. Kersten S, Dong D, Lee W-Y, Reczek PR, Noy N (1998) Auto-silencing by the retinoid X receptor. J Mol Biol 284:21–32
66. Kersten S, Kelleher D, Chambon P, Gronemeyer H, Noy N (1995) Retinoid X receptor alpha forms tetramers in solution. Proc Natl Acad Sci 92:8645–8649
67. Kitareewan S, Burka LT, Tomer KB, Parker CE, Deterding LJ, Stevens RD, Forman BM, Mais DE, Heyman RA, Mcmorris T, Weinberger C (1996) Phytol metabolites are circulating dietary factors that activate the nuclear receptor RXR. Mol Biol Cell 7:1153–1166
68. Kliewer SA, Lehmann JM, Willson TM (1999) Orphan nuclear receptors: shifting endocrinology into reverse. Science 284:757–760
69. Kliewer SA, Umesono K, Heyman RA, Mangelsdorf DJ, Dyck JA, Evans RM (1992) Retinoid X receptor-COUP-TF interactions modulate retinoic acid signalling. Proc Natl Acad Sci USA 89:1448–1452
70. Kliewer SA, Umesono K, Mangelsdorf DJ, Evans RM (1992) Retinoid X receptors interacts with nuclear receptors in retinoic acid, thyroid hormone and vitamin D3 signalling. Nature 355:446–449
71. Kliewer SA, Umesono K, Nonan DJ, Heyman RA, Evans RM (1992) Convergence of 9-*cis* retinoic acid and peroxisome proliferator signalling pathways through heterodimer formation of their receptors. Nature 358:771–774
72. Kopf E, Plassat J-L, Vivat V, De The H, Chambon P, Rochette-Egly C (2000) Dimerization with retinoid X receptors and phosphorylation modulate the retinoic acid-induced degradation of retinoic acid receptors alpha and gamma through the ubiquitin-proteasome pathway. J Biol Chem 275:33280–33288
73. Kurokawa R, Dirienzo J, Boehm M, Sugarman J, Gloss B, Rosenfeld MG, Heyman RA, Glass CK (1994) Regulation of retinoid signalling by receptor polarity and allosteric control of ligand binding. Nature 371:528–531
74. Kurokawa R, Söderström M, Hörlein A, Halachmi S, Brown M, Rosenfeld MG, Glass CK (1995) Polarity-specific activities of retinoic acid receptors determined by a co-repressor. Nature 377:451–454
75. Laffitte BA, Kast HR, Nguyen CM, Zavacki AM, Moore DD, Edwards PA (2000) Identification of the DNA binding specificity and potential target genes for the farnesoid X-activated receptor. J Biol Chem 275:10638–10647

76. Lalloyer F, Pedersen TA, Gross B, Lestavel S, Yous S, Vallez E, Gustafsson J-A, Mandrup S, Fievet C, Staels B, Tailleux A (2009) Rexinoid bexarotene modulates triglyceride but not cholesterol metabolism via gene-specific permissivity of the RXR/LXR heterodimer in the liver. Arterioscler Thromb Vasc Biol 29:1488–1495
77. Lap Shu AC, Richard AW (2009) Cross-talk between PPARs and the partners of RXR: a molecular perspective. PPAR Res
78. Lee JM, Lee YK, Mamrosh JL, Busby SA, Griffin PR, Pathak MC, Ortlund EA, Moore DD (2011) A nuclear receptor-dependent phosphatidylcholine pathway with antidiabetic effects. Nature 474:506–510
79. Lefebvre B, Benomar Y, Guedin A, Langlois A, Hennuyer N, Dumont J, Bouchaert E, Dacquet C, Nicaud L, Casteilla L, Pattou F, Ktorza A, Staels B, Lefebvre P (2010) Proteasomal degradation of retinoid X receptor alpha reprograms transcriptional activity of PPAR gamma in obese mice and humans. J Clin Investig 120:1454–1468
80. Lefebvre P, Benomar Y, Staels B (2010) Retinoid X receptors: common heterodimerization partners with distinct functions. Trends Endocrinol Metab: TEM 21:676–683
81. Leibowitz MD, Ardecky RJ, Boehm MF, Broderick CL, Carfagna MA, Crombie DL, D'Äôarrigo J, Etgen GJ, Faul MM, Grese TA, Havel H, Hein NI, Heyman RA, Jolley D, Klausing K, Liu S, Mais DE, Mapes CM, Marschke KB, Michellys P-Y, Montrose-Rafizadeh C, Ogilvie KM, Pascual B, Rungta D, Tyhonas JS, Urcan MS, Wardlow M, Yumibe N, Reifel-Miller A (2006) Biological characterization of a heterodimer-selective retinoid X receptor modulator: potential benefits for the treatment of type 2 diabetes. Endocrinology 147:1044–1053
82. Leid M, Kastner P, Lyons R, Nakshatri H, Saunders M, Zacharewski T, Chen JY, Staub A, Garnier JM, Mader S, Chambon P (1992) Purification, cloning and RXR identity of the HeLa cell factor with which RAR or TR heterodimerizes to bind target sequences efficiently. Cell 68:377–395
83. Lemotte PK, Keidel S, Apfel CM (1996) Phytanic acid is a retinoid X receptor ligand. Eur J Biochem 236:328–333
84. Leo C, Yang X, Liu J, Li H, Chen JD (2001) Role of retinoid receptor coactivator pockets in cofactor recruitment and transcriptional regulation. J Biol Chem 276:23127–23134
85. Li Y, Zhang Y, Hill J, Shen Q, Kim H-T, Xu X, Hilsenbeck SG, Bissonnette RP, Lamph WW, Brown PH (2007) The rexinoid LG100268 prevents the development of preinvasive and invasive estrogen receptor negative tumors in MMTV-erbB2 mice. Clin Cancer Res 13:6224–6231
86. Liao J, Ozono K, Sone T, Mcdonnell DP, Pike JW (1990) Vitamin D receptor interaction with specific DNA requires a nuclear protein and 1,25-dihydroxyvitamin D3. Proc Natl Acad Sci USA 87:9751–9788
87. Liby K, Royce DB, Risingsong R, Williams CR, Wood MD, Chandraratna RA, Sporn MB (2007) A new rexinoid, NRX194204, prevents carcinogenesis in both the lung and mammary gland. Clin Cancer Res 13:6237–6243
88. Liby KT, Yore MM, Sporn MB (2007) Triterpenoids and rexinoids as multifunctional agents for the prevention and treatment of cancer. Nat Rev Cancer 7:357–369
89. Liu Q, Linney E (1993) The mouse retinoid-X receptor-gamma gene: genomic organization and evidence for functional isoforms. Mol Endocrinol 7:651–658
90. Mader S, Leroy P, Chen J-Y, Chambon P (1993) Multiple parameters control the selectivity of nuclear receptors for their response elements. J Biol Chem 268:591–600
91. Mangelsdorf DJ, Borgmeyer U, Heyman RA, Zhou J, Ong ES, Oro A, Kakizuka A, Evans RM (1992) Characterization of three RXR genes that mediate the action of 9-cis retinoic acid. Genes Dev 6:329–344
92. Mangelsdorf DJ, Evans RM (1995) The RXR heterodimers and orphan receptors. Cell 83:841–850
93. Mangelsdorf DJ, Ong ES, Dyck JA, Evans RM (1990) Nuclear receptor that identifies a novel retinoic acid response pathway. Nature 345:224–229

94. Mangelsdorf DJ, Umesono K, Kilewer SA, Borgmeyer U, Ong ES, Evans RM (1991) A direct repeat in the cellular retinol-binding protein type II gene confers differential regulation by RXR and RAR. Cell 66:555–561
95. Marks MS, Hallenbeck PL, Nagata T, Segars JH, Appella E, Nikodem VM, Ozato K (1992) H-2RIIBP (RXR beta) heterodimerization provides a mechanism for combinatorial diversity in the regulation of retinoic acid and thyroid hormone responsive genes. EMBO J 11:1419–1435
96. Martens JHA, Brinkman AB, Simmer F, Francoijs K-J, Nebbioso A, Ferrara F, Altucci L, Stunnenberg HG (2010) PML-RARa/RXR alters the epigenetic landscape in acute promyelocytic leukemia. Cancer Cell 17:173–185
97. Matsushima-Nishiwaki R, Okuno M, Adachi S, Sano T, Akita K, Moriwaki H, Friedman SL, Kojima S (2001) Phosphorylation of retinoid X receptor alpha at Serine 260 impairs its metabolism and function in human hepatocellular carcinoma. Cancer Res 61:7675–7682
98. Mehta K, Mcqueen T, Neamati N, Collins S, Andreeff M (1996) Activation of retinoid receptors RAR alpha and RXR alpha induces differentiation and apoptosis, respectively, in HL-60 cells. Cell Growth Differ 7:179–186
99. Meyer MB, Goetsch PD, Pike JW (2010) Genome-wide analysis of the VDR/RXR cistrome in osteoblast cells provides new mechanistic insight into the actions of the vitamin D hormone. J Steroid Biochem Mol Biol 121:136–141
100. Michellys PY, Ardecky RJ, Chen JH, Crombie DL, Etgen GJ, Faulkner AL, Grese TA, Heyman RA, Karanewsky DS, Klausing K, Leibowitz MD, Liu S, Mais DA, Mapes CM, Marschke KB, Reifel-Miller A, Ogilvie KM, Rungta D, Thompson AW, Tyhonas JS, Boehm MF (2003) Novel (2E,4E,6Z)-7-(2-Alkoxy-3,5-dialkylbenzene)-3-methylocta-2,4,6-trienoic acid retinoid X receptor modulators are active in models of type 2 diabetes. J Med Chem 46:2683–2696
101. Miesfeld R, Okret S, Wikstrom A-C, Wrange O, Gustafsson J-A, Yamamoto KR (1984) Characterization of a steroid hormone receptor gene and mRNA in wild-type and mutant cells. Nature 312:779–781
102. Moise AR, Alvarez S, Dominguez M, Alvarez R, Golczak M, Lobo GP, Von Lintig J, De Lera AR, Palczewski K (2009) Activation of retinoic acid receptors by dihydroretinoids. Mol Pharmacol 76:1228–1237
103. Murray MB, Towle HC (1989) Identification of nuclear factors that enhance binding of the htyroid hormone receptor to a thyroid hormone response element. Mol Endocrinol 3:1434–1442
104. Naar AM, Boutin J, Lipkin SM, Yu VC, Holloway JM, Glass CK, Rosenfeld MG (1991) The orientation and spacing of core DNA-binding motifs dictate selective transcriptional responses to three nuclear receptors. Cell 65:1267–1279
105. Nagy L, Thomazy VA, Shipley GL, Fesus L, Lamph W, Heyman RA, Chandraratna RA, Davies PJ (1995) Activation of retinoid X receptors induces apoptosis in HL-60 cell lines. Mol Cell Biol 15:3540–3551
106. Nielsen R, Pedersen T, Hagenbeek D, Moulos P, Siersbaek R, Megens E, Denissov S, Borgesen M, Francoijs K-J, Mandrup S, Stunnenberg HG (2008) Genome-wide profiling of PPARgamma: RXR and RNA polymerase II occupancy reveals temporal activation of distinct metabolic pathways and changes in RXR dimer composition during adipogenesis. Genes Dev 22:2953–2967
107. Nishimaki-Mogami T, Tamehiro N, Sato Y, Okuhira K-I, Sai K, Kagechika H, Shudo K, Abe-Dohmae S, Yokoyama S, Ohno Y, Inoue K, Sawada J-I (2008) The RXR agonists PA024 and HX630 have different abilities to activate LXR/RXR and to induce ABCA1 expression in macrophage cell lines. Biochem Pharmacol 76:1006–1013
108. Nunez V, Alameda D, Rico D, Mota R, Gonzalo P, Cedenilla M, Fischer T, Bosca L, Glass CK, Arroyo AG, Ricote M (2010) Retinoid X receptor alpha controls innate inflammatory responses through the up-regulation of chemokine expression. Proc Natl Acad Sci 107:10626–10631

109. Orlov I, Rochel N, Moras D, Klaholz BP (2012) Structure of the full human RXR/VDR nuclear receptor heterodimer complex with its DR3 target DNA. EMBO J 31:291–300
110. Pérez E, Bourguet W, Gronemeyer H, De Lera AR (2012) Modulation of RXR function through ligand design. Biochim Biophys Acta (BBA) Mol Cell Biol Lipids 1821:57–69
111. Perlmann T, Jansson L (1995) A novel pathway for vitamin A signaling mediated by RXR heterodimerization with NGFI-B and NURR1. Genes Dev 9:769–782
112. Perlmann T, Umesono K, Rangarajan PN, Forman BM, Evans RM (1996) Two distinct dimerization interfaces differentially modulate target gene specificity of nuclear hormone receptors. Mol Endocrinol 10:958–966
113. Pogenberg V, Guichou J-FO, Vivat-Hannah V, Kammerer S, Perez E, Germain P, De Lera AR, Gronemeyer H, Royer CA, Bourguet W (2005) Characterization of the Interaction between retinoic acid receptor/retinoid X receptor (RAR/RXR) heterodimers and transcriptional coactivators through structural and fluorescence anisotropy studies. J Biol Chem 280:1625–1633
114. Putcha B-DK, Fernandez EJ (2009) Direct interdomain interactions can mediate allosterism in the thyroid receptor. J Biol Chem 284:22517–22524
115. Ramlau R, Zatloukal P, Jassem J, Schwarzenberger P, Orlov SV, Gottfried M, Pereira JR, Temperley G, Negro-Vilar R, Rahal S, Zhang JK, Negro-Vilar A, Dziewanowska ZE (2008) Randomized phase III trial comparing bexarotene (L1069-49)/cisplatin/vinorelbine with cisplatin/vinorelbine in chemotherapy-naive patients with advanced or metastatic non-small-cell lung cancer: SPIRIT I. J Clin Oncol 26:1886–1892
116. Rizvi NA, Marshall JL, Dahut W, Ness E, Truglia JA, Loewen G, Gill GM, Ulm EH, Geiser R, Jaunakais D, Hawkins MJ (1999) A phase I study of LGD1069 in adults with advanced cancer. Clin Cancer Res 5:1658–1664
117. Rochel N, Ciesielski F, Godet J, Moman E, Roessle M, Peluso-Iltis C, Moulin M, Haertlein M, Callow P, Mély Y, Svergun DI, Moras D (2011) Common architecture of nuclear receptor heterodimers on DNA direct repeat elements with different spacings. Nat Struct Mol Biol 18:564–570
118. Saga Y, Kobayashi M, Ohta H, Murai N, Nakai N, Oshima M, Taketo MM (1999) Impaired extrapyramidal function caused by the targeted disruption of retinoid X receptor RXRγ1 isoform. Genes Cells 4:219–228
119. Sanguedolce MV, Leblanc BP, Betz JL, Stunnenberg HG (1997) The promoter context is a decisive factor in establishing selective responsiveness to class II nuclear receptors. EMBO J 15:2861–2873
120. Santos GM, Fairall L, Schwabe JWR (2011) Negative regulation by nuclear receptors: a plethora of mechanisms. Trends Endocrinol Metab 22:87–93
121. Shen Q, Bai Y, Chang KCN, Wang Y, Burris TP, Freedman LP, Thompson CC, Nagpal S (2011) Liver X receptor-retinoid X receptor (LXR-RXR) heterodimer cistrome reveals coordination of LXR and AP1 signaling in keratinocytes. J Biol Chem 286:14554–14563
122. Shen Q, Cline GW, Shulman GI, Leibowitz MD, Davies PJA (2004) Effects of rexinoids on glucose transport and insulin-mediated signaling in skeletal muscles of diabetic (db/db) mice. J Biol Chem 279:19721–19731
123. Shulman AI, Larson C, Mangelsdorf DJ, Ranganathan R (2004) Structural determinants of allosteric ligand activation in RXR heterodimers. Cell 116:417–429
124. Son YL, Lee YC (2009) Molecular determinants of the interactions between LXR/RXR heterodimers and TRAP2 20. Biochem Biophys Res Commun 384:389–393
125. Sucov HM, Dyson E, Gumeringer CL, Price J, Chien KR, Evans RM (1994) RXR alpha mutant mice establish a genetic basis for vitamin A signaling in heart morphogenesis. Genes Dev 8:1007–1018
126. Suino K, Peng L, Reynolds R, Li Y, Cha J-Y, Repa JJ, Kliewer SA, Xu HE (2004) The nuclear xenobiotic receptor CAR: structural determinants of constitutive activation and heterodimerization. Mol Cell 16:893–905
127. Svensson S, Ostberg T, Jacobsson M, Norstrom C, Stefansson K, Hallen D, Johansson IC, Zachrisson K, Ogg D, Jendeberg L (2003) Crystal structure of the heterodimeric complex

of LXR[alpha] and RXR[beta] ligand-binding domains in a fully agonistic conformation. EMBO J 22:4625–4633
128. Szeles L, Poliska S, Nagy G, Szatmari I, Szanto A, Pap A, Lindstedt M, Santegoets SJAM, Ruhl R, Dezso B, Nagy L (2010) Research resource: transcriptome profiling of genes regulated by RXR and its permissive and nonpermissive partners in differentiating monocyte-derived dendritic cells. Mol Endocrinol 24:2218–2231
129. Takiyama Y, Miyokawa N, Sugawara A, Kato S, Ito K, Sato K, Oikawa K, Kobayashi H, Kimura S, Tateno M (2004) Decreased expression of retinoid X receptor isoforms in human thyroid carcinomas. J Clin Endocrinol Metab 89:5851–5861
130. Tanaka T, Rodriguez De La Conception ML, De Luca LM (2001) Involvement of all-trans-retinoic acid in the breakdown of retinoic acid receptors alpha and gamma through proteasomes in mcf-7 human breast cancer cells. Biochem Pharmacol 61:1347–1355
131. Tudor C, Feige JRMN, Pingali H, Lohray VB, Wahli W, Desvergne BA, Engelborghs Y, Gelman L (2007) Association with coregulators is the major determinant governing peroxisome proliferator-activated receptor mobility in living cells. J Biol Chem 282:4417–4426
132. Urquiza AMD, Liu S, Sjoberg M, Zetterstrom RH, Griffiths W, Sjovall J, Perlmann T (2000) Docosahexaenoic acid, a ligand for the retinoid X receptor in mouse brain. Science 290:2140–2144
133. Wahlstrom GM, Sjoberg M, Andersson M, Nordstrom K, Vennstrom B (1992) Binding characteristics of the thyroid hormone receptor homo- and heterodimers to consensus AGGTCA repeat motifs. Mol Endocrinol 6
134. Wan YJ, An D, Cai Y, Repa JJ, Chen THP, Flores M, Postic C, Magnuson MA, Chen J, Chien KR, French S, Mangelsdorf DJ, Sucov HM (2000) Hepatocyte-specific mutation establishes retinoid X receptor a as a heterodimeric integrator of multiple physiological processes in the liver. Mol Cell Biol 20:4436–4444
135. Wang Z, Benoit G, Liu J, Prasad S, Aarnisalo P, Liu X, Xu HE, Walker NP, Perlmann T (2003) Structure and function of Nurr1 identifies a class of ligand-independent nuclear receptors. Nature 423:555–560
136. Weinberger C, Hollenberg SM, Ong ES, Harmon JM, Brower ST, Cidlowski J, Thompson EB, Rosenfeld MG, Evans RM (1985) Identification of human glucocorticoid receptor complementary DNA clones by epitope selection. Science 228:740–742
137. Westin S, Korokawa R, Nolte RT, Wisely GB, Mcinerney EM, Milburn MV, Rosenfeld MG, Glass CK (1998) Interactions controlling the assembly of nuclear-receptor heterodimers and coactivators. Nature 395:199–202
138. Willy P, Umesono K, Ong E, Evans R, Heyman R, Mangelsdorf D (1995) LXR, a nuclear receptor that defines a distinct retinoid response pathway. Genes Dev 9:1033–1045
139. Wu C, Orozco C, Boyer J, Leglise M, Goodale J, Batalov S, Hodge C, Haase J, Janes J, Huss J, Su A (2009) BioGPS: an extensible and customizable portal for querying and organizing gene annotation resources. Genome Biol 10:R130
140. Xie W, Barwick JL, Simon CM, Pierce AM, Safe S, Blumberg B, Guzelian PS, Evans RM (2000) Reciprocal activation of xenobiotic response genes by nuclear receptors SXR/PXR and CAR [In Process Citation]. Genes Dev 14:3014–3023
141. Xu HE, Lambert MH, Montana VG, Plunket KD, Moore LB, Collins JL, Oplinger JA, Kliewer SA, Gampe RT Jr, Mckee DD, Moore JT, Willson TM (2001) Structural determinants of ligand binding selectivity between the peroxisome proliferator-activated receptors. Proc Natl Acad Sci USA 98:13919–13924
142. Xu RX, Lambert MH, Wisely BB, Warren EN, Weinert EE, Waitt GM, Williams JD, Collins JL, Moore LB, Willson TM, Moore JT (2004) A structural basis for constitutive activity in the human CAR/RXRalpha heterodimer. Mol Cell 16:919–928
143. Yasmin R, Yeung KT, Chung RH, Gaczynska ME, Osmulski PA, Noy N (2004) DNA-looping by rxr tetramers permits transcriptional regulation "at a distance". J Mol Biol 343:327–338
144. Yu VC, Delsert C, Andersen B, Holloway JM, Devary OV, Naar AM, Kim SY, Boutin JM, Glass CK, Rosenfeld MG (1991) RXRbeta: a coregulator that enhances binding of retinoic

acid, thyroid hormone, and vitamin D receptors to their cognate response elements. Cell 67:1251–1266

145. Zeisig BB, Kwok C, Zelent A, Shankaranyanan H, Gronemeyer H, Dong S, So CW (2007) Recruitment of RXR by homotetrameric RARalpha fusion proteins is essential for transformation. Cancer Cell 12:36–51
146. Zhang J, Chalmers MJ, Stayrook KR, Burris LL, Wang Y, Busby SA, Pascal BD, Garcia-Ordonez RD, Bruning JB, Istrate MA, Kojetin DJ, Dodge JA, Burris TP, Griffin PR (2011) DNA binding alters coactivator interaction surfaces of the intact VDR-RXR complex. Nat Struct Mol Biol 18:556–563
147. Zhang XK, Hoffman BE, Tran PBV, Graupner G, Pfahl M (1992) Retinoid X receptor is an auxiliary protein for thyroid hormone and retinoic acid receptors. Nature 355:441–446
148. Zhou H, Liu W, Su Y, Wei Z, Liu J, Kolluri SK, Wu H, Cao Y, Chen J, Wu Y, Yan T, Cao X, Gao W, Molotkov A, Jiang F, Li W-G, Lin B, Zhang H-P, Yu J, Luo S-P, Zeng J-Z, Duester G, Huang P-Q, Zhang X-K (2010) NSAID sulindac and its analog bind RXRalpha and inhibit RXRalpha-dependent AKT signaling. Cancer Cell 17:560–573
149. Ziouzenkova O, Orasanu G, Sharlach M, Akiyama TE, Berger JP, Viereck J, Hamilton JA, Tang G, Dolnikowski GG, Vogel S, Duester G, Plutzky J (2007) Retinaldehyde represses adipogenesis and diet-induced obesity. Nat Med 13:695–702
150. Ziouzenkova O, Orasanu G, Sukhova G, Lau E, Berger JP, Tang G, Krinsky NI, Dolnikowski GG, Plutzky J (2007) Asymmetric cleavage of beta carotene yields a transcriptional repressor of retinoid X receptor and peroxisome proliferator-activated receptor responses. Mol Endocrinol 21:77–88

Chapter 6
Nuclear and Extra-Nuclear Effects of Retinoid Acid Receptors: How They Are Interconnected

Aleksandr Piskunov, Ziad Al Tanoury and Cécile Rochette-Egly

Abstract The nuclear retinoic acid receptors (RAR α, β and γ) and their isoforms are ligand-dependent regulators of transcription, which mediate the effects of all-*trans* retinoic acid (RA), the active endogenous metabolite of Vitamin A. They heterodimerize with Retinoid X Receptors (RXRs α, β and γ), and regulate the expression of a battery of target genes involved in cell growth and differentiation. During the two last decades, the description of the crystallographic structures of RARs, the characterization of the polymorphic response elements of their target genes, and the identification of the multiprotein complexes involved in their transcriptional activity have provided a wealth of information on their pleiotropic effects. However, the regulatory scenario became even more complicated once it was discovered that RARs are phosphoproteins and that RA can activate kinase signaling cascades via a pool of RARs present in membrane lipid rafts. Now it is known that these RA-activated kinases translocate to the nucleus where they phosphorylate RARs and other retinoid signaling factors. The phosphorylation state of the RARs dictates whether the transcriptional programs which are known to be induced by RA are facilitated and/or switched on. Thus, kinase signaling pathways appear to be crucial for fine-tuning the appropriate physiological activity of RARs.

A. Piskunov · Z. Al Tanoury · C. Rochette-Egly (✉)
IGBMC (Institut de Génétique et de Biologie Moléculaire et Cellulaire), INSERM, U964, CNRS, UMR7104, Université de Strasbourg, 1 rue Laurent Fries, BP 10142, 67404 Illkirch Cedex, France
e-mail: cegly@igbmc.fr

A. Piskunov
e-mail: piskunov@ibcgenerium.ru

Z. Al Tanoury
e-mail: tanouryz@igbmc.fr

M. A. Asson-Batres and C. Rochette-Egly (eds.), *The Biochemistry of Retinoic Acid Receptors I: Structure, Activation, and Function at the Molecular Level*, Subcellular Biochemistry 70,
DOI: 10.1007/978-94-017-9050-5_6, © Springer Science+Business Media Dordrecht 2014

Abbreviations

CAK	Cyclin-dependent kinase (CDK)-activating kinase
ChIP	Chromatin immunoprecipitation
ChIP-seq	ChIP coupled with deep sequencing
CRABP	Cellular retinoic acid binding protein
DBD	DNA binding domain
DR	Direct repeat
Erks	Extracellular-signal-regulated kinases
ES cells	Embryonic stem cells
FABP	Fatty acid binding protein
HAT	Histone acetyltransferase
HDAC	Histone deacetylase
HMT	Histone methyl transferase
iLBP	Intracellular lipid binding protein
IR	Inverted repeat
LBD	Ligand binding domain
LBP	Ligand binding pocket
MAPK	Mitogen activated protein kinase
MSK	Mitogen-and stress-activated protein kinase
N-CoR	Nuclear receptor corepressor
NMR	Nuclear magnetic resonance
NTD	N-terminal domain
PcG	Polycomb group proteins
PI3K	Phosphoinositide 3-kinase
Pin1	Protein interacting with NIMA (never in mitosis A)
PPAR	Peroxysome proliferator activated receptor
PRM	Proline rich motif
RA	Retinoic acid
RAR	Retinoic acid receptor
RARE	Retinoic acid response element
RNA-seq	high throughput qPCR sequencing
RXR	Retinoid X receptor
SAXS	Small angle X-ray
SH3	Src-homology-3
SRC	Steroid receptor coactivator
SMRT	Silencing mediator of retinoic acid receptor and thyroid hormone receptor
TBL1	Transducin beta like
TBLR1	TBL1-related protein 1
WW	Tryptophan-tryptophan

History: The Canonical Model for the Regulation of RAR-Target Gene Expression

The Functional Domains of RARs

Like most nuclear receptors (NRs) [49–51], RARs and RXRs exhibit a well-defined domain organization consisting of a central conserved DNA binding domain (DBD) linked to a variable N-terminal domain (NTD), and a C-terminal Ligand-Binding Domain (LBD) [10, 28, 73, 108] (Fig. 6.1a).

The structures of RAR and RXR LBDs are rather similar. This domain in each protein is composed of 12 conserved alpha helices and a beta-turn, separated by loops and folded into a three-layered, and parallel helical sandwich [20, 28, 103], with helices H4, H5, H8, H9 and H11 sandwiched between H1, H2 and H3 on one side, and H6, H7 and H10 on the other (Fig. 6.1a, b). In contrast, the C-terminal helix, H12, is more flexible and adopts conformations that differ from one RAR to the other. The conformation pf H12 also changes after RA binding.

The primary feature of the LBD is its functional complexity. It contains the Ligand-Binding Pocket (LBP) [19], the heterodimerization surface [21], and interaction surfaces involved in the binding of multiple coregulators (Fig. 6.1b). A well-described hydrophobic surface, generated by H3 and H4, is involved in the binding of corepressors/coactivators [57, 111] (see also Chap. 3 in this volume). The LBD also contains a recently described docking site for cyclin H, a subunit of the cyclin-dependent activating (CAK) sub complex of the general transcription factor, TFIIH, that is formed by loop L8-9 and the first amino acids of H9 [15] (Fig. 6.1b).

The DBD is composed of two zinc-nucleated modules and two alpha-helices [137, 138], which contribute to a second dimerization interface and define the contacts for specific DNA sequences, named RA response elements (RAREs). Classically, RAREs are composed of two direct repeats of a core hexameric motif (A/G) G (G/T) TCA separated by 1, 2 or 5 nucleotides and referred as DR1, DR2 and DR5 [8, 10, 48] (Fig. 6.1c). However, recent genome wide chromatin immunoprecipitation coupled with deep sequencing (ChIP-seq) technology has allowed identification of new RAR binding loci [85, 91] and revealed that RARs can occupy a larger repertoire of sites with an unexpected diversity in the spacing and the topology of the DNA binding elements, including DR0, DR8 and IR0 (inverted repeats) elements (Fig. 6.1c). Recent structural studies have also indicated that the architecture of DNA-bound heterodimers is dictated by the DNA sequence (Fig. 6.1d) [22, 104] (see also Chap. 2 in this volume).

In contrast to the DBD and the LBD, the NTD is not conserved between RARs and RXRs and even between the different subtypes and isoforms. As yet, high-resolution, three-dimensional structures of this region have not been produced [108]. Several biochemical and structural studies coupled to structure prediction algorithms suggest that the NTDs of RARs and RXRs, as well as of any member

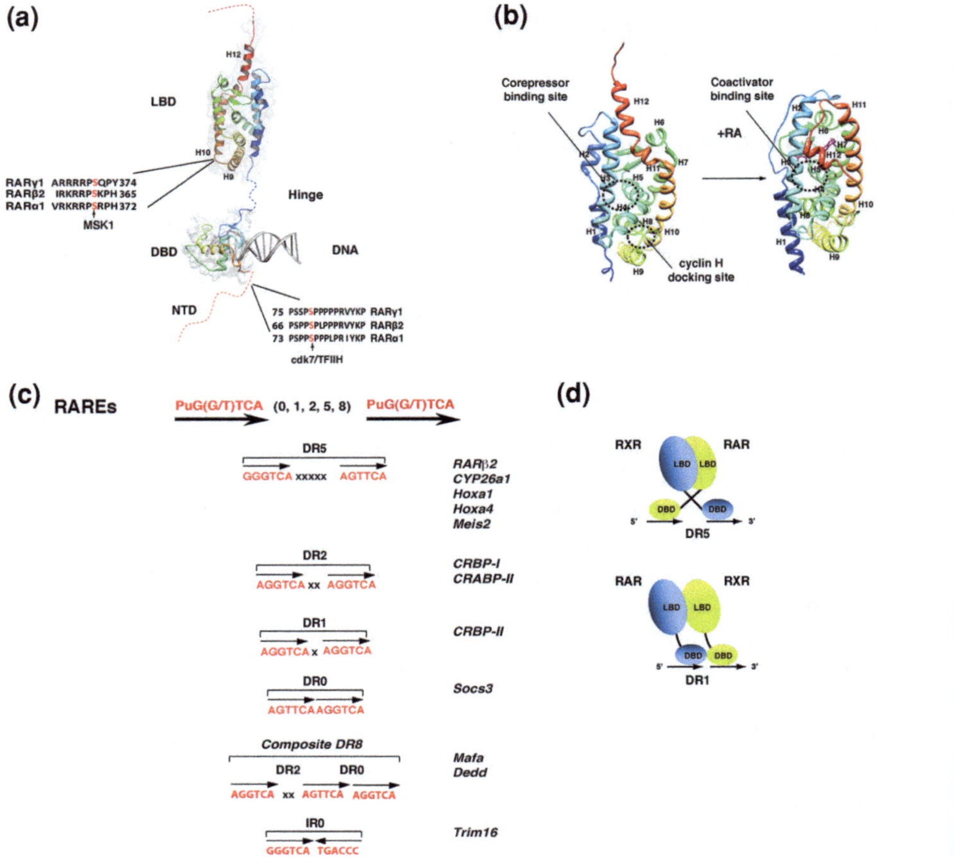

Fig. 6.1 Structure of RARs and of their DNA binding sites. **a** RARs depict a domain organization with an unstructured N-terminal domain (*NTD*), and two well structured domains: a central DNA binding domain (*DBD*) and a C-terminal ligand-binding domain (*LBD*). The phosphorylation sites located in the NTD and the LBD are shown. **b** Structural changes induced upon RA binding. The crystal structures of the unliganded RXRα and liganded RARγ LBDs are shown with the binding domains for corepressors, coactivators and cyclin H. Helices are represented as ribbons and labelled from H1 to H12. Adapted from Protein Data Bank 1lbd and 2lbd. **c** The retinoid response elements (*RAREs*) are composed of a direct repeat of the motif 5'- Pu G (G/T) TCA spaced by 0 (DR0), 1 (DR1), 2 (DR2), 5 (DR5) or 8 (DR8) base pairs. DR8 comprise three half sites with DR2 and DR0 spacing. Some RARE-associated genes are shown. **d** Binding of the RAR/RXR heterodimers to DR1 and DR5 RAREs (adapted from [104])

of the NR family, are naturally disordered [74, 131]. An interesting feature of these NTDs is that they contain phosphorylation sites [105], which are conserved between RARs (Fig. 6.1a) [112]. Moreover, they contain proline-rich motifs (PRMs) which are well known to bind proteins with Src-homology-3 (SH3) or tryptophan-tryptophan (WW) domains. Phosphorylation prevents or favors these interactions [7].

RAR-Mediated Gene Expression

RAR/RXR heterodimers control transcription *via* several distinct mechanisms, including both repression and activation. According to the canonical model, the transcriptional regulation of RA-target genes relies not only on the binding of RAR/RXR heterodimers to specific RAREs, but also on corepressors that dissociate and coactivators that associate with the LBD upon ligand-induced conformational changes [57, 81, 95, 133]. At the molecular level, the discrimination between corepressors and coactivators is governed by the ligand-induced, orientation of H12, which contributes in a critical manner to the generation or removal of cofactor interaction surfaces.

Repression of Transcription in the Absence of Ligand

In the absence of ligand and in a context of chromatin where the nucleosomes do not impede binding to RAREs, the RARα subtype is a strong repressor of target gene expression (Fig. 6.2a) [34]. In this unliganded state, H12 adopts an open conformation that unmasks a hydrophobic groove generated by H3 and H4 [76] (see also Chap. 3 in this volume). This interface specifically binds an LxxI/HIxxxI/L motif in the extended alpha helix box of the corepressors, NCoR or SMRT [98]. NCoR and SMRT are genetic paralogs with multiple protein variants, but SMRT is the favored corepressor for RARs [90]. According to recent studies, SMRT is recruited by the heterodimer only through the RAR partner [76] (see also Chap. 3 in this volume).

SMRT does not have intrinsic enzymatic activity, but serves as an adaptor to recruit other high molecular weight complexes that are endowed with histone deacetylase activity (HDACs) [95]. These complexes deacetylate lysine residues in the N-terminal tails of histones and maintain chromatin in a condensed and repressed state over the target promoter [34, 111] (Fig. 6.2a). The corepressor complexes also contain other components such as transducer β-like proteins (TBL1 and TBLR1), which serve as adaptors regulating corepressor assembly and function [94].

In contrast to RARα, the RARγ and RARβ subtypes poorly interact with corepressors [39, 60, 102], most probably due to the fact that, in these receptors, H12 interacts with H3 even in the absence of ligand, thus occluding the corepressor docking site.

Initiation of Transcription in Response to the Ligand: A Process Governed by the LBD via the Exchange of Coregulators

According to the canonical model, ligand binding to RARs must be understood in terms of structural features (Fig. 6.1b). When entering the cavity of the RARa LBP, the ligand induces a β-strand-to-α-helix secondary structure switch [76], which induces

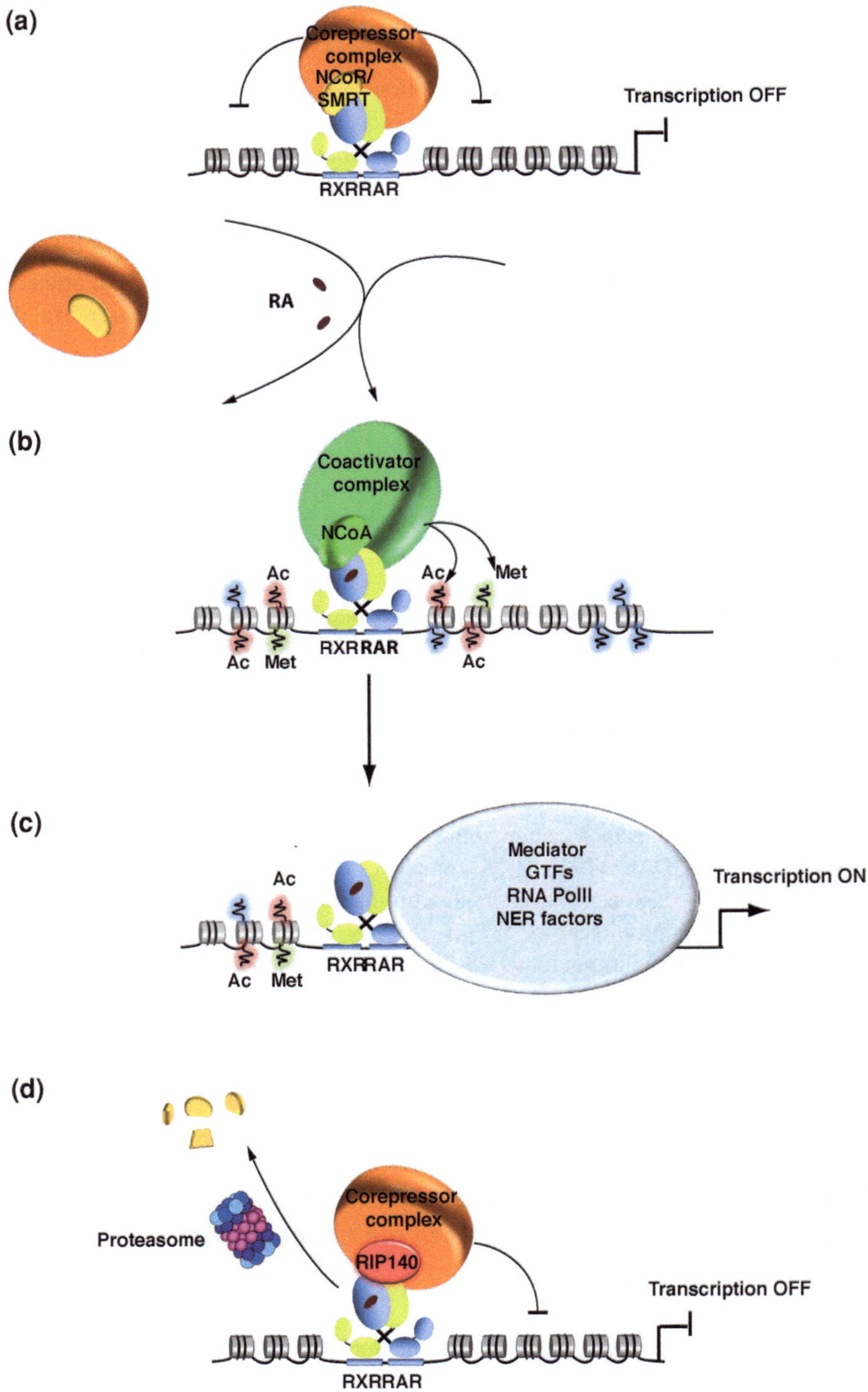
(a)
Corepressor complex NCoR/ SMRT
Transcription OFF
RXRRAR
RA
(b)
Coactivator complex
NCoA
Ac
Met
Ac
Ac Met
RXRRAR
Ac
(c)
Ac
Mediator
GTFs
RNA PolII
NER factors
Transcription ON
Ac Met
RXRRAR
(d)
Proteasome
Corepressor complex
RIP140
Transcription OFF
RXRRAR

Fig. 6.2 The classical model of activation of RA-target genes. **a** Repression in the absence of ligand. **b** Corepressors and coactivators exchange after ligand binding. **c** Recruitment of the transcription machinery and initiation of transcription. **d** End of the RA signal upon recruitment of non-conventional coativators such as RIP140, associated to large complexes with chromatin repressing activity. The end of the RA signal occurs also through the degradation of RARs by the ubiquitin proteasome system

repositioning of H11 relative to H10 and a concomitant swinging of H12 inward to pack against H3 and H4 in a mouse trap model that locks RA in the LBP. Consequently, corepressors are released and a new hydrophobic cleft is formed between H3, H4 and H12 [28], with a charge clamp between a conserved glutamate residue in H12 and a lysine in H3. This charge clamp specifically grips the ends of the helix specified by the LxxLL motif of the p160 subfamily of steroid receptor coactivators, SRC-1, SRC-2 and SRC-3 [81, 96]. The p160 coactivators have an intrinsic histone acetyl transferase (HAT) activity, and according to recent structural studies, only one coactivator molecule is recruited by the heterodimer through the RAR partner [22, 92].

When recruited to the LBD of the liganded RAR, the p160 coactivators initiate a dynamic, ordered and coordinated recruitment of other proteins with HAT activity [p300/CBP (CREB binding protein) and p/CAF (p300/CBP-associated factor)] or with histone methyl transferase (HMT) activities, such as Coactivator-associated arginine methyl-transferase 1 (CARM1) or Protein-arginine methyl transferase 1 (PRMT1) [57, 81, 111, 133] (Fig. 6.2b). Acetylation and methylation weaken histone DNA contacts and create marks forming an «*histone code*», which coordinates the recruitment of additional HATs or HMTs to histones for further chromatin decompaction. This code also orchestrates the recruitment of chromatin remodelers, which use the energy of ATP-hydrolysis to reposition nucleosomes by sliding them in *cis* or displacing them in *trans*, allowing the formation of nucleosome-free or nucleosome-spaced regions at the promoter.

The p160 coregulators also recruit large complexes which contain modules with other enzymatic activities, such as histone de-ubiquitinases [139] and histone lysine methyl transferases [42, 80]. Lysine methylation creates marks for the binding of other enzymes that erase repressive marks. According to a recent study, this step requires Poly (ADP-Ribose) glycohydrolase (PARG) which is corecruited at RAR-dependent promoters in response to RA [78]. Thus, due to their varied composition, the coactivator complexes provide an elegant mechanism to reorganize chromatin by writing activating histone marks, erasing repressive marks, and remodeling nucleosomes.

All of these steps pave the way for recruitment of the transcription machinery to the promoter region (Fig. 6.2c) [34, 111] including RNA Polymerase II, General Transcription Factors, and a specific subunit (DRIP205/TRAP220) of the multi-subunit Mediator complex. According to recent studies, the transcription machinery assembles sequentially with nucleotide excision repair (NER) factors, in order to achieve optimal histone modifications, and thus, efficient RNA synthesis [77].

According to recent chromatin conformation capture technologies, RARs that bind to different enhancer elements of a gene can form loops [25]. An emerging view is that RARs, similar to other TFs, promote the formation of long range chromatin loops, bridging genomic loci located even on different chromosomes, thus creating hot spots of transcription [115].

Turn "OFF" of Transcription

After the "ON" switch, transcription of the RA target genes has to be terminated. Several scenarios have been proposed, but it is still unclear whether termination activities are gene or cell-specific. One possibility is that liganded RARs recruit unconventional coregulators with LxxLL motifs which, in contrast to classical p160 SRCs, inhibit, rather than activate, the transcriptional activity of RARs (Fig. 6.2d). These coregulators include the receptor interacting protein of 140 kDa (RIP140/NRIP1) [62], the preferentially expressed antigen in melanoma (PRAME) [38], and the transcription Intermediary factor-1 alpha (TIF1α/Trim24) [75]. The mechanism of TIF1α-mediated repression has not been elucidated yet [67], but the repressive activities of RIP140 and PRAME have been attributed to the recruitment of HDACs and PcG proteins, respectively [38, 132].

According to another scenario, an efficient way to limit RAR function and/or to signal the end of the transcriptional process would be the degradation of RARs and RXRs by the ubiquitin proteasome system [126] (Fig. 6.2d). Supporting such a hypothesis, RARs have been shown to be ubiquitinated and degraded by the proteasome through the recruitment of TRIP1/SUG-1, which is a subunit of the 19S regulatory sub complex of the proteasome with an ATPase activity [41, 52].

Development of the Field: RARs also Have Non-canonical Extra-Nuclear Effects, Which Are Integrated in the Nucleus

RA Activates Kinase Signaling Pathways via a Pool of RARs Present in Membrane Lipid Rafts

It is now appreciated that RARs have additional extra-nuclear and non-transcriptional effects that activate kinase signaling pathways [4]. Studies from several laboratories have demonstrated that RA rapidly (within minutes after RA addition) and transiently activates several kinase cascades. RA activates p38 mitogen activated kinase (MAPK) in fibroblasts, mouse embryo carcinoma cells, mammary breast tumor cells, and leukemia cells [5, 25, 52, 100]. RA activates the p42/p44 MAPKs (also called Erks) in neurons, Sertoli cells, and embryonic stem cells [30, 33, 59, 87, 123].

Since the RA-induced activation of the MAPK pathways occurs after the activation of upstream cascades involving RhoGTPases [5, 33, 100], PI3 kinase and/or protein kinase B (PKB)/Akt [31, 87, 93] in the cytosol, it has been suggested that the RA-induced cytosolic activities must involve an atypical, non-genomic event similar to that described for steroid NRs [82, 99]. In line with this concept, RARα

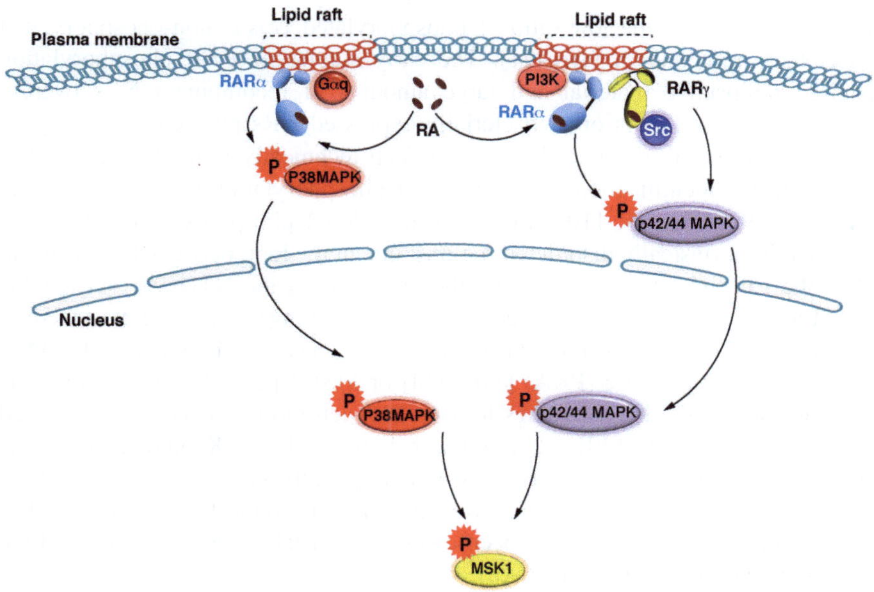

Fig. 6.3 Activation of kinase cascades by an extra nuclear pool of RARs. A subpopulation of RARα is present in membrane lipid rafts. Depending on the cell type, in response to RA, this pool of RARα can either interact with Gαq proteins or with PI3K to activate the p38 or p42/p44 MAPK pathways respectively. In other cell types, RARγ can also activate the p42/p44MAPK pathway via Src. Then the activated MAPKs translocate into the nucleus where they phosphorylate and activate MSK1

proteins have recently been found in lipid rafts isolated from the membranes of several cell types [87, 100] (Fig. 6.3). Moreover, the activation of p38MAPK has been shown to involve the interaction of RARα present in lipid rafts with Gαq proteins [100] (Fig. 6.3). However, the activation of Erks by RA did not involve Gαq proteins (Piskunov et al., unpublished results), but rather PI3K [31, 87, 93] or the *Src* kinase [33] (Fig. 6.3). Thus depending on the cell type, the extra nuclear effects of RA appear to involve different mechanisms and kinase cascades.

Once activated by RARα, p38MAPK and Erks translocate to the nucleus where they phosphorylate MSK1 (Fig. 6.3) [99]. Finally MAPKs and MSK1 phosphorylate several nuclear factors involved in the expression of RA-target genes, including RARs themselves and their coregulators.

In the Nucleus, RARs Are Rapidly Phosphorylated by a Cascade of Kinases

A few years after the cloning of RARs, it emerged that these receptors are also phosphoproteins [45, 106, 107, 109, 110]. However, at that time, the analysis of RAR phosphorylation was a challenging task because of its highly dynamic nature and also because of the low ratio of phosphorylated *versus* non-phosphorylated RARs that are found in vivo [71]. The early studies of RAR phosphorylation required radioactive material and large amounts of recombinant NRs overexpressed in cultured cells or of bacterially expressed NRs purified and phosphorylated in vitro with different kinases. Though technically limited, these studies resulted in the identification of a number of phosphorylation sites in RARs and RXRs [1, 79, 105, 106, 110, 121]. Two main RAR phosphorylation sites were identified in intrinsically disordered regions; one in the loop between helices 9 and 10 of the LBD (S369 in RARα) and the other one in the NTD (S77 in RARα) [25] (Fig. 6.1a). The serine located in the LBD belongs to an Arginine-Lysine-rich motif and can be phosphorylated by several kinases such as the cyclic AMP dependent Protein Kinase (PKA) [110, 114] or MSK1 [25, 112]. In contrast, the serine located in the NTD belongs to a proline-rich motif and is phosphorylated by cdk7/cyclin H [106, 130], a kinase that belongs to the CAK subcomplex of the general transcription factor TFIIH. Most importantly, the correct positioning of the cdk7 kinase and thereby the efficiency of the NTD phosphorylation by cdk7 relies on the docking of cyclin H at a specific site in the LBD located in L8-9 and the N-terminal part of H9 [15] (Fig. 6.1b).

More recently, the emergence of new methods for enrichment of phosphopeptide samples, as well as the availability of phosphospecific antibodies, made it possible to analyze endogenous RAR phosphorylation. These studies revealed that the serines located in the LBD and in the NTD are both rapidly phosphorylated in response to RA via a cascade of kinases (Fig. 6.4). First, RA-activated MSK1 phosphorylates RARα at the serine located in the LBD [25]. Then, phosphorylation of this residue promotes phosphorylation of the serine located in the NTD through subtle conformational changes [25, 43]. Molecular dynamic simulations of the RARα LBD showed that phosphorylation of S369 (located in loop L9-10) leads to changes in the structural dynamics of the cyclin H binding site (composed of loop L8-9) situated at a 30 Å distance. This change in dynamics has been correlated with an increase in cyclin H binding and phosphorylation of the NTD at S77 by cdk7 [29, 112]. Thus, the coordinated phosphorylation of RARα results from a coordinated cascade that can be explained by changes in the structural features of the molecule.

This phosphorylation cascade has been described in cells that respond to RA via the activation of p38MAPK [25]. Whether it also occurs in cells that respond via the activation of Erks requires further investigations. Remarkably, the two phosphorylation sites are conserved between the mammalian RAR subtypes (α, β and γ) [112] (Fig. 6.1a) and the RARγ subtype is also phosphorylated at the same residues [9, 70, 110] through a similar cascade (our unpublished results). It is interesting to note that the serine residue located in the NTD has been conserved during evolution of chordates, indicating that the phosphorylation of this residue is likely important for RARs activity. In contrast, the serine residue located in the LBD is not present in non-mammalian RARs, suggesting that in other vertebrates, the phosphorylation cascade described above does not occur. Consequently, in other vertebrates, the phosphorylation of the NTD would be controlled by different regulatory circuits [112].

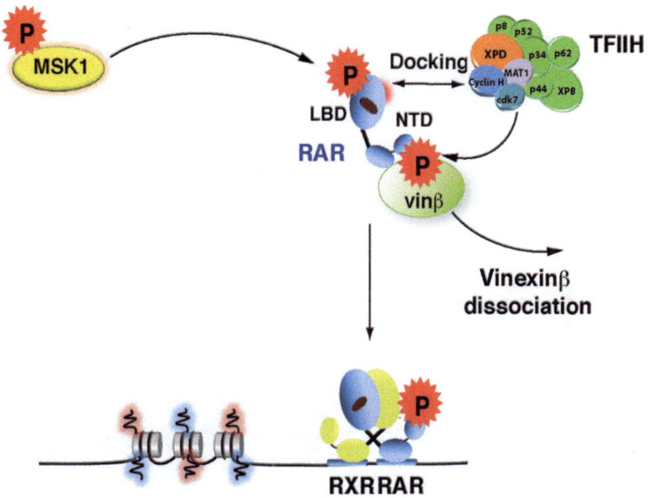

Fig. 6.4 RARs are phosphorylated by a cascade of cascades. Activated MSK1 phosphorylates RARs at a serine residue located in the LBD (loop L9-10). Phosphorylation of this residue induces conformational changes in loop L8-9, which promote the binding of the cyclin H subunit of the CAK subcomplex of TFIIH. Consequently the cdk7 kinase can phosphorylate the serine residue located in the NTD. In the case of RARγ, phosphorylation of the NTD induces the dissociation of vinexin β. Finally the phosphorylated RAR can be recruited to response elements located in the promoters of target genes

Consequences of RARs Phosphorylation at the NTD: Dissociation of Coregulators and Degradation by the Proteasome

The NTD is an intrinsically disordered region [74, 131], but the serine residue of this domain belongs to a PRM (Fig. 6.1a), which can form polyproline helices and bind proteins with SH3 or WW domains [65, 84, 136]. Recently, our laboratory identified vinexinβ as a new binding partner for the RARγ PRM [17]. Vinexinβ is an adaptor protein characterized by the presence of three SH3 domains, the third C-terminal one interacting with the PRM of RARγ. Recently, the combination of nuclear magnetic resonance (NMR), Circular dichroïsm, small angle X-ray scattering (SAXS) and molecular dynamics simulations revealed that phosphorylation of the serine residue located in the PRM of RARγ changes the global hydrodynamic behavior of the polyproline helix and decreases the propensity of the PRM to bind SH3 domains (Kieffer et al., unpublished results). Consequently, vinexinβ dissociates from RARγ [70] (Fig. 6.4).

In addition, at the end of the transcriptional process, phosphorylation of the N-terminal serine residue of RARγ and of an additional one located at position -2 has been shown to promote the ubiquitination of the receptor and its subsequent degradation by the proteasome [52, 68]. This is a typical example of interplay between different posttranslational modifications [118].

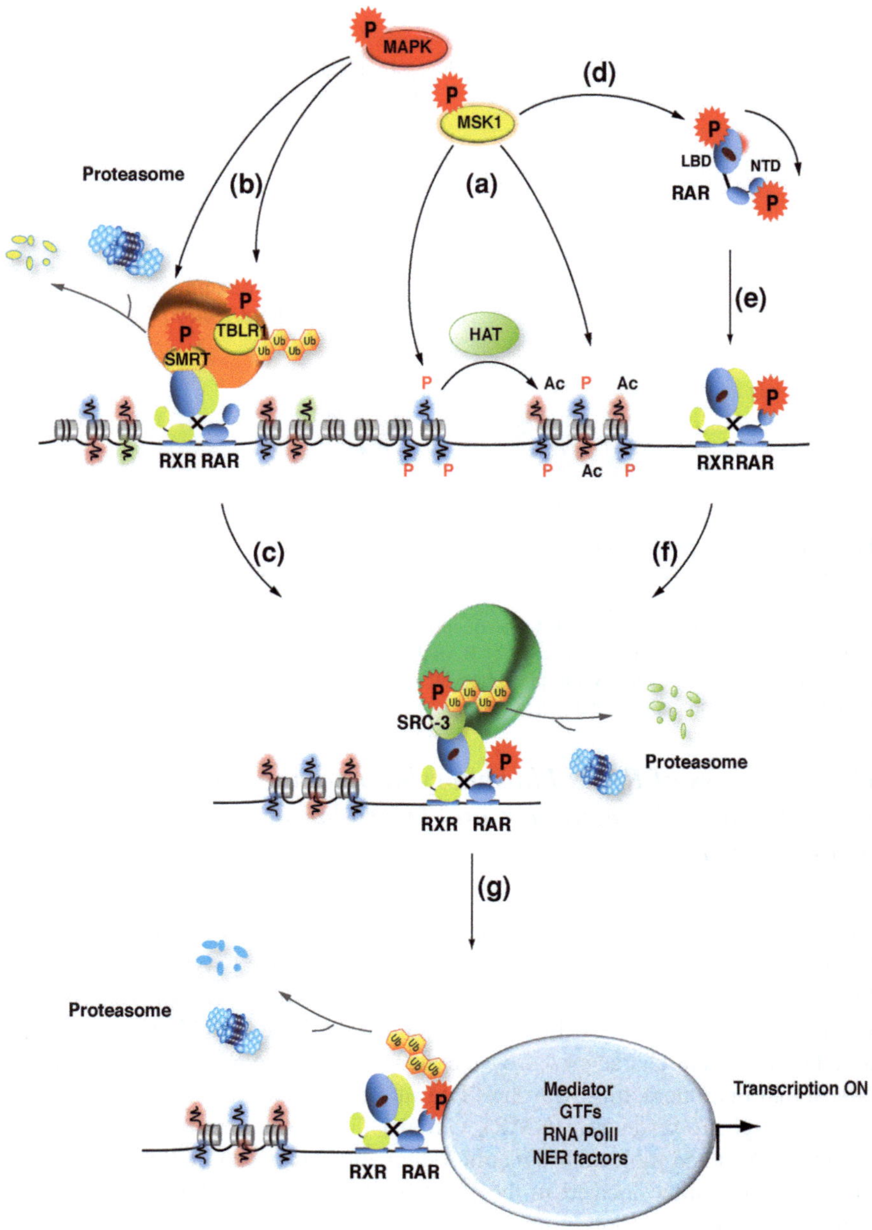
MAPK
MSK1
(d)
(a)
(b)
Proteasome
LBD
NTD
RAR
TBLR1
SMRT
Ub
HAT
(e)
Ac
Ac
Ac
RXR RAR
RXRRAR
(c)
(f)
SRC-3
Proteasome
RXR RAR
(g)
Proteasome
Mediator
GTFs
RNA PolII
NER factors
Transcription ON
RXR RAR

Fig. 6.5 Working models for the role of phosphorylations in the activation of RAR-target genes. When RARs already occupy RAREs in the absence of ligand, the RA-activated MAPKs phosphorylate components of the corepressor complexes such as SMRT and TBLR1. (**b**) Phosphorylation promotes their dissociation and their degradation by the proteasome, thus facilitating their exchange for coactivators. Then the coactivators such as SRC3 become also phosphorylated (**c**). Subsequently, they dissociate from RARs and are degraded, allowing the recruitment of other coregulators. Histones are also phosphorylated (**a**) and their phosphorylation induces the recruitment of HATs and remodeling complexes. Altogether these events cooperate to decompact chromatin at the promoters and pave the way for the recruitment of the transcription machinery (**g**). When RAREs are not occupied, phosphorylation of histones (**a**) and RARs (**d**) cooperate for the recruitment of RARs at their response elements (**e**). Next, liganded and DNA bound RARs recruit coactivator complexes, which decompact further chromatin. As above, coactivators phosphorylation (**f**) leads to their dissociation and degradation by the proteasome, thus facilitating the dynamics of coregulators exchange and the recruitment of the transcription machinery (**g**). Finally, RARs are degraded by the ubiquitin-proteasome system

The N-terminal PRM of RARα has been shown to bind the proline isomerase, Pin1, in a phospho-dependent manner [23, 53]. Pin1 is a WW domain-containing protein that is well known to induce *cis-trans* isomerization of proline residues that follow phosphorylated serines, and in so doing, to create new specific recognition sites for other interacting factors [135]. Pin1 interaction has been correlated with the degradation of RARα by the proteasome and the inhibition of RARα activity [53]. However the mechanism of the Pin1-mediated degradation of RARα remains to be defined.

Not only RARs but also Several Other Proteins Integrate MAPK Signaling and Become Phosphorylated in Response to RA

RA-activated MAPKs and MSK1 also phosphorylate other factors involved in RA target gene transcription, their phosphorylation favoring programs induced by the ligand. MSK1 is recruited at RAR-target promoters where it phosphorylates histone H3 tails at serines S10 and S28 (Fig. 6.5a). H3S10 phosphorylation has been correlated with the recruitment of HATs and the SWI/SNF ATP-dependent chromatin-remodeling complex [25, 99], while H3S28 phosphorylation induces the displacement of PcG complexes that maintain chromatin in a repressive state [47].

In addition, MSK1 and the upstream kinases, p38MAPK and Erks, phosphorylate several other actors in RA signaling, including RXRs, corepressors and coactivators. The important point is that phosphorylations alter protein structure, protein-protein interactions and protein activity, thus constituting an important cellular integration mechanism.

As an example, RA-activated p38MAPK rapidly phosphorylates rapidly RXRα at three residues located in the NTD [24, 128] by an as yet, unknown mechanism. MAPKs also phosphorylate components of corepressor complexes, such as SMRT and TBLR1 (Fig. 6.5b). Phosphorylation of SMRT induces its release from RARα [64] and disrupts its interaction with HDACs and other proteins in the corepressor complexes [129]. Consequently, the architecture, composition, and function of the corepressor complexes are disrupted. A current model of TBLR1phosphorylation is that this adaptor mediates recruitment of the ubiquitin proteasome system to ubiquitinate and degrade NCoR, SMRT and HDACs [94, 97]. It has been proposed that the phosphorylation-dependent dissociation and degradation of components of the corepressor complexes mediates the exchange of corepressors with coactivators.

Coactivators are also phosphorylated in response to RA [54]. The p160 coactivator, SRC-3, is phosphorylated by p38MAPK at a serine residue located in the vicinity of the RAR binding domain. Phosphorylation of this residue results in dissociation of SRC3 from RARα. It also marks SRC-3 for ubiquitination and degradation by the proteasome [40] (Fig. 6.5c). This phosphorylation-ubiquitination-degradation process facilitates the dynamics of RARs-mediated transcription by allowing other coregulators to bind. The other components of the coactivator complexes, such as p300/CBP, can be also phosphorylated by several kinases in response to several signaling pathways [94, 96, 97], but whether they are phosphorylated in response to RA requires further investigation. Overall, coregulators respond to several signals that fine-tune their functional interactions with RARs and thus, their ability to modulate RAR transcriptional activity.

Current State of the Field

The RA-Induced Kinases and RAR Recruitment to DNA

Recent ChIP-seq profiles have confirmed that the occupancy of many RAREs is increased in response to RA [25, 70, 85, 89], and that there is a significant correlation between transcription activation and the binding of RAR/RXR heterodimers to DNA [89]. Though the mechanism of RAR/RXR recruitment to DNA in response to RA is still ill-defined, one cannot exclude a role for phosphorylation processes.

In the absence of RA, many RAREs are inaccessible due to a compact epigenetic landscape of chromatin. This implies that RAR binding requires an initial rapid modification of the chromatin environment in order to alleviate compaction and make the RAREs accessible. Among the candidates for chromatin reorganization, there are the RA-activated kinases. Indeed, RA-activated MSK1 is rapidly recruited to RAR-target genes promoters and phosphorylates histones H3 at serines S10 and S28 [25, 47]. Once phosphorylated, these serine residues are marks that induce the recruitment of remodeling complexes and the displacement of repressive complexes [47, 55, 56].

Another possibility is that the phosphorylation state of RARs themselves control their recruitment to DNA. To validate such a hypothesis, RNA-seq and ChIP experiments were performed with mutant mouse embryonic stem cell (mESC) lines expressing RAR phosphomutants in a RAR null background. Such experiments highlighted direct target genes whose expression is controlled by phosphorylation of the N-terminal serine residue of RARγ [2]. These studies also revealed that only RAREs with specific spacings recruit the phosphorylated form of RARγ in response to RA [2].

Then, the question was how phosphorylation of the N-terminal serine could promote recruitment of RARγ to specific RAREs in response to RA. It must be noted that this serine belongs to a PRM that is located in the vicinity of the DBD and which interacts with vinexin β, a repressor of RARγ-mediated transcription (see discussion, above). Remarkably, vinexin β bound to the nonphosphorylated form of RARγ prevents DNA binding, while vinexin β dissociation upon RARγ phosphorylation allows the binding of the receptor to DNA [70] (Fig. 6.4). Thus phosphorylation of the NTD that occurs in response to RA would promote DNA binding via the dissociation of proteins that occlude the DBD.

A Working Model for the Role of Phosphorylation in the Activation of RAR-Target Genes

RA-induced phosphorylation processes play an important role in the expression of RAR-target genes through modulating RAR recruitment to DNA, the sequential recruitment of the different classes of coregulators, and also the stability of the target proteins, thus constituting an important cellular integration mechanism. Two models can be proposed for the role of phosphorylation in the activation of RAR target genes. One for RARs already bound to DNA and another for RARs that are recruited to RAREs in response to RA (Fig. 6.5).

When RARs are constitutively bound to their DNA targets, ligand binding is the crucial molecular event that switches transcription from repression to activation via coregulator exchanges and chromatin reorganization. However it is now evident that RA-activated kinases provide additional layers of regulation for this switch through a phosphorylation code examplified by the phosphorylation of the corepressors and coactivators that promotes their dissociation from RARs and their degradation by the proteasome (Fig. 6.5b, c). Such a code facilitates the exchange of coregulator complexes and the reorganization of the epigenetic landscape. The kinases also phosphorylate histones, introducing additional marks for the recruitment of activating complexes and dissociation of repressive cofactors (Fig. 6.5a).

Not all RAREs are occupied in the absence of RA due to a compact genetic landscape. In this case, it is hypothesized that RA-activated kinases promote recruitment of RARs to DNA by first, phosphorylating histones, a process that alleviates chromatin compaction (Fig. 6.5a). The RA-activated kinases also phosphorylate RARs, but this process controls the recruitment of the receptors only to a subset of RAREs with a specific spacing (Fig. 6.5d, e). Additional approaches are required to investigate why phosphorylation controls RARs recruitment only to certain RAREs and not to the others. When RARs are recruited to DNA, it is evident that phosphorylation of the different coactivators also facilitates the dynamics of their association-dissociation for further chromatin decompaction and recruitment of transcription machinery (Fig. 6.5f, g). Finally, the phosphorylation of RARs signals their degradation by the ubiquitin proteasome system in order to stop the transcriptional process [16] (Fig. 6.5g).

Relevance: In vivo Relevance of the Cross Talk Between the Nuclear and Extra-Nuclear Effects of RARs

Embryonic Development and cell differentiation

Several genetic approaches performed in animals demonstrate that RARs and RXRs are the conductors of RA signaling during development [86, 113]. However, without further development of appropriate technology, current animal models cannot be used to study the role of RAR and RXR phosphorylation due to the complexity of the processes and signaling pathways. Instead, cell differentiation models have provided interesting tools to study the influence of RAR phosphorylation. These have included mouse embryo carcinoma cells (F9 cell line) which markedly resemble embryonic cells from the blastocyst and differentiate into primitive, parietal or visceral endoderm-like cells [18] after RA addition. They also included mouse embryonic stem cells that are pluripotent cells which self-renew indefinitely and have the propensity to differentiate in vitro into a larger variety of cell types [58, 134], such as neurons in response to RA [13].

Experiments from our laboratory revealed that differentiation of F9 cells into primitive endoderm [125] and ES cells into neurons [2] involves the RARγ2 subtype. Most interestingly, the generation of stable rescue cell lines expressing RARγ2 phosphomutants in a RARγ null background indicated that phosphorylation of the RARγ2 NTD is critically required for the RA-induced differentiation of these cells [2, 125]. Moreover, recent genome wide RNA-seq analysis experiments highlighted a subset of genes belonging to early phosphoRAR-regulated gene programs that are critical for triggering the effects of RA [2, 3]. These data suggest an important role for RAR phosphorylation in RA signaling, and pave the way for further investigations during embryonic and tissue development.

Cancer and Diseases

Available evidence suggests that the integrity of signaling pathways is required for the proper activity of RARs. Consequently, one can speculate that deregulation of the "*kinome*" would have deleterious downstream effects. Accordingly, in Xeroderma Pigmentosum patients, who are characterized by mutations affecting subunits of the general transcription factor, TFIIH, cdk7 does not efficiently phosphorylate RARα. This deficient phosphorylation has characteristic downstream consequences on the expression of RAR target genes [66] and has been correlated at least in part, to clinical abnormalities observed in patients.

In addition, in several cancers characterized by amplified or deregulated cytosolic kinase cascades [14], ending at Akt or MAPKs [119, 127] RARα has been shown to be aberrantly phosphorylated [121, 122]. Moreover, the RA-induced activation of the MAPK pathway is abrogated [100]. Subsequently, RARα is degraded and/or its transcriptional activity is suppressed. Similarly, in hepatocellular carcinoma, RXRα is aberrantly phosphorylated in its LBD, with characteristic inhibition of its transcriptional activity [88]. Thus, one can postulate that aberrant kinome signaling and RAR/RXR phosphorylation and activity may correlate with tumoral growth and/or RA resistance [37].

There is an increasingly body of evidence indicating that RA signaling plays an important role in brain function, such as synaptic plasticity and learning and memory via transcriptional effects [72], as well as through non genomic effects involving a pool of RARα present in dendrites that becomes phosphorylated in response to RA [6, 30, 101]. An interesting observation is that RA signaling is also involved in the pathophysiology of Alzheimer's disease [72]. Alzheimer's disease is complex, but its striking and increasingly important characteristic is the aberrant expression and activity of several protein kinases [35, 61]. Therefore, one can suggest that in this disease, aberrant phosphorylation of RARα could have consequences on both its transcriptional activity and its extra nuclear effects on synaptic plasticity.

In conclusion, RA signaling and RAR phosphorylation represent potentially exploitable pathways for devising novel therapies in several diseases, including the Alzheimer disease [120].

Relevance in the Biology of Other Nuclear Receptors

According to the classical model, all-*trans* RA is channeled to RARs in the nucleus via the cellular RA-binding protein CRABPII, which is a small cytosolic protein belonging to the family of intracellular lipid binding proteins (iLBP) [26, 32, 36]. Such a process markedly facilitates formation of the liganded receptor. However, recent studies revealed that in certain CRABPII-deficient, cell types, such as brain, adipose tissue, skeletal muscle and skin, RA binds another fatty acid-binding protein of the iLBP family, FABP5. The interesting point is that upon RA binding, FABP5 does not deliver the ligand to RARs, but to the Peroxisome Proliferator-Activated Receptor β/δ (PPARβ/δ), another nuclear receptor, resulting in the regulation of genes that are not direct RAR targets [11, 116, 117]. Consequently, new functions of RA in the regulation of energy homeostasis and insulin responses were revealed [12]. PPARs are also known to be phosphoproteins [27] and to have nongenomic effects [83], but the β/δ subtype is the least studied in terms of phosphorylation. Whether it becomes phosphorylated in response to RA requires further investigation.

Some in vitro studies suggest that RA signaling could be mediated by other nuclear factors such as the Retinoic acid receptor related Orphan Receptor beta (RORβ) [124], the Chicken Ovalbumin Upstream Promoter Transcription Factor (COUP-TFII) [69] or the Testicular Receptors (TR2/4) [140], which are also able to integrate several signaling pathways [46, 63]. However, the in vivo relevance of such observations remains to be determined as well as whether these receptors can be also phosphorylated in response to RA.

Future Directions: What Is Still Left to do

The regulation of RAR-target gene activation by RA is controlled not only by simple on/off conformational switches of RARs, but also by kinase signaling pathways. Importantly, these signaling pathways target several actors in retinoid regulatory processes through phosphorylations that fine-tune the RA response via rapid changes in chromatin organization, RAR dynamics, coregulator interactions, and structural and functional shifts in protein-DNA interactions. The future challenges are to connect these data directly with new highly sensitive, real-time or large-scale technologies in order to get novel, critical information about the influence of phosphorylations on the regulation of RARs and RXRs activity. The last-generation dual linear ion trap mass spectrometers coupled with the Orbitrap technology should allow the identification of new phosphorylation sites in endogenous RARs, RXRs and their coregulators, and should provide information about their regulation by RA. Biophysical approaches, such as NMR, coupled to molecular dynamics simulations are other promising tools to investigate how phosphorylations fine-tune the structure of RARs and RXRs to control their recruitment to RAREs with specific spacings, but not to others with different spacings and/or sequences.

Finally, the recent TALEN (Transcription Activator-like Effector Nucleases) or CRISPR (Clustered Regularly Interspaced Short Palindromic Repeats)-based technologies [44] should make possible the generation of point mutations at phosphorylation sites in vivo, providing a more powerful tool than the classical re-expression of a mutant in a null background. These tools coupled to RNA-seq, ChIP-seq and quantitative proteomics should provide interesting information about downstream gene expression and changes in protein complexes induced by RAR phosphorylation. Large-scale and quantitative phosphorylation screens of RARs, combined with other large-scale data sets, should pave the way to breakthroughs in disease-related research. In conclusion, further insights into the effects of RA will likely continue to reveal new targets and mechanisms that will help explain their pleiotropic effects and how these features might be manipulated in the treatment of metabolic disorders.

Acknowledgments Funds from the Agence Nationale pour la Recherche (ANR-05-BLAN-0390-02 and ANR-09-BLAN-0297-01), the Foundation pour la Recherche Medicale (FRM, DEQ20090515423), the Institut National du Cancer (INCa-PLO7-96099 and PL09-194) and the Association pour la Recherche sur le Cancer (ARC 3169, ARC SL220110603474) supported this work. This study was also supported by the grant ANR-10-LABX-0030-INRT, a French State fund managed by the ANR under the frame programme Investissements d'Avenir labelled ANR-10-IDEX-0002-02. INCA supported ZA while FRM and the lady TATA Memorial Trust supported AP.

References

1. Adam-Stitah S, Penna L, Chambon P, Rochette-Egly C (1999) Hyperphosphorylation of the retinoid X receptor alpha (RXRa) by activated c-Jun N-terminal Kinases (JNKs). J Biol Chem 274:18932–18941
2. Al Tanoury Z, Gaouar S, Piskunov A, Ye T, Urban S, Jost B, Keime C, Davidson I, Dierich A, Rochette-Egly C (2014) Phosphorylation of the retinoic acid receptor RARγ2 is crucial for the neuronal differentiation of mouse embryonic stem cells. J Cell Sci 127:2095–2105

3. Al Tanoury Z, Piskunov A, Andriamoratsiresy D, Gaouar S, Lutzing R, Ye T, Jost B, Keime C, Rochette-Egly C (2013a) Genes involved in cell adhesion and signaling: a new repertoire of retinoic acid receptors target genes in mouse embryonic fibroblasts. J Cell Sci 127:521–533
4. Al Tanoury Z, Piskunov A, Rochette-Egly C (2013b) Vitamin A and retinoid signaling: genomic and nongenomic effects: thematic review series: fat-soluble vitamins: vitamin A. J Lipid Res 54:1761–1775
5. Alsayed Y, Uddin S, Mahmud N, Lekmine F, Kalvakolanu DV, Minucci S, Bokoch G, Platanias LC (2001) Activation of Rac1 and the p38 mitogen-activated protein kinase pathway in response to all-trans-retinoic acid. J Biol Chem 276:4012–4019
6. Aoto J, Nam CI, Poon MM, Ting P, Chen L (2008) Synaptic signaling by all-trans retinoic acid in homeostatic synaptic plasticity. Neuron 60:308–320
7. Ball LJ, Kuhne R, Schneider-Mergener J, Oschkinat H (2005) Recognition of proline-rich motifs by protein-protein-interaction domains. Angew Chem Int Ed Engl 44:2852–2869
8. Balmer JE, Blomhoff R (2005) A robust characterization of retinoic acid response elements based on a comparison of sites in three species. J Steroid Biochem Mol Biol 96:347–354
9. Bastien J, Adam-Stitah S, Riedl T, Egly JM, Chambon P, Rochette-Egly C (2000) TFIIH interacts with the retinoic acid receptor gamma and phosphorylates its AF-1-activating domain through cdk7. J Biol Chem 275:21896–21904
10. Bastien J, Rochette-Egly C (2004) Nuclear retinoid receptors and the transcription of retinoid-target genes. Gene 328:1–16
11. Berry DC, Noy N (2007) Is PPARbeta/delta a retinoid receptor? PPAR Res 2007:73256
12. Berry DC, Noy N (2009) All-trans-retinoic acid represses obesity and insulin resistance by activating both peroxisome proliferation-activated receptor beta/delta and retinoic acid receptor. Mol Cell Biol 29:3286–3296
13. Bibel M, Richter J, Lacroix E, Barde YA (2007) Generation of a defined and uniform population of CNS progenitors and neurons from mouse embryonic stem cells. Nat Protoc 2:1034–1043
14. Blume-Jensen P, Hunter T (2001) Oncogenic kinase signalling. Nature 411:355–365
15. Bour G, Gaillard E, Bruck N, Lalevee S, Plassat JL, Busso D, Samama JP, Rochette-Egly C (2005) Cyclin H binding to the RARalpha activation function (AF)-2 domain directs phosphorylation of the AF-1 domain by cyclin-dependent kinase 7. Proc Natl Acad Sci USA 102:16608–16613
16. Bour G, Lalevee S, Rochette-Egly C (2007) Protein kinases and the proteasome join in the combinatorial control of transcription by nuclear retinoic acid receptors. Trends Cell Biol 17:302–309
17. Bour G, Plassat JL, Bauer A, Lalevee S, Rochette-Egly C (2005) Vinexin beta Interacts with the non-phosphorylated AF-1 Domain of retinoid receptor gamma (RARg) and represses RARg-mediated transcription. J Biol Chem 280:17027–17037
18. Bour G, Taneja R, Rochette-Egly C (2006) Mouse embryocarcinoma F9 cells and retinoic acid. A model to study the molecular mechanisms of endodermal differentiation. In: Taneja R (ed) Nuclear Receptors in development, vol 16. Elsevier Press Inc, Netherlands, pp 211–253
19. Bourguet W, Germain P, Gronemeyer H (2000) Nuclear receptor ligand-binding domains: three-dimensional structures, molecular interactions and pharmacological implications. Trends Pharmacol Sci 21:381–388
20. Bourguet W, Ruff M, Chambon P, Gronemeyer H, Moras D (1995) Crystal structure of the ligand-binding domain of the human nuclear receptor RXR-alpha. Nature 375:377–382
21. Bourguet W, Vivat V, Wurtz JM, Chambon P, Gronemeyer H, Moras D (2000) Crystal structure of a heterodimeric complex of RAR and RXR ligand-binding domains. Mol Cell 5:289–298
22. Brelivet Y, Rochel N, Moras D (2012) Structural analysis of nuclear receptors: from isolated domains to integral proteins. Mol Cell Endocrinol 348:466–473
23. Brondani V, Schefer Q, Hamy F, Klimkait T (2005) The peptidyl-prolyl isomerase Pin1 regulates phospho-Ser77 retinoic acid receptor alpha stability. Biochem Biophys Res Commun 328:6–13

24. Bruck N, Bastien J, Bour G, Tarrade A, Plassat JL, Bauer A, Adam-Stitah S, Rochette-Egly C (2005) Phosphorylation of the retinoid x receptor at the omega loop, modulates the expression of retinoic-acid-target genes with a promoter context specificity. Cell Signal 17:1229–1239
25. Bruck N, Vitoux D, Ferry C, Duong V, Bauer A, de The H, Rochette-Egly C (2009) A coordinated phosphorylation cascade initiated by p38MAPK/MSK1 directs RARalpha to target promoters. EMBO J 28:34–47
26. Budhu AS, Noy N (2002) Direct channeling of retinoic acid between cellular retinoic acid-binding protein II and retinoic acid receptor sensitizes mammary carcinoma cells to retinoic acid-induced growth arrest. Mol Cell Biol 22:2632–2641
27. Burns KA, Vanden Heuvel JP (2007) Modulation of PPAR activity via phosphorylation. Biochim Biophys Acta 1771:952–960
28. Chambon P (1996) A decade of molecular biology of retinoic acid receptors. FASEB J 10:940–954
29. Chebaro Y, Amal I, Rochel N, Rochette-Egly C, Stote R, Dejaegere A (2013) Phosphorylation of the retinoic acid receptor alpha induces a mechanical allosteric regulation and changes in internal dynamics. Plos Comp Biol 9:e1003012
30. Chen N, Napoli JL (2008) All-trans-retinoic acid stimulates translation and induces spine formation in hippocampal neurons through a membrane-associated RARalpha. FASEB J 22:236–245
31. Cheung YT, Lau WK, Yu MS, Lai CS, Yeung SC, So KF, Chang RC (2009) Effects of all-trans-retinoic acid on human SH-SY5Y neuroblastoma as in vitro model in neurotoxicity research. Neurotoxicology 30:127–135
32. Delva L, Bastie JN, Rochette-Egly C, Kraiba R, Balitrand N, Despouy G, Chambon P, Chomienne C (1999) Physical and functional interactions between cellular retinoic acid binding protein II and the retinoic acid-dependent nuclear complex. Mol Cell Biol 19:7158–7167
33. Dey N, De PK, Wang M, Zhang H, Dobrota EA, Robertson KA, Durden DL (2007) CSK controls retinoic acid receptor (RAR) signaling: a RAR-c-SRC signaling axis is required for neuritogenic differentiation. Mol Cell Biol 27:4179–4197
34. Dilworth FJ, Chambon P (2001) Nuclear receptors coordinate the activities of chromatin remodeling complexes and coactivators to facilitate initiation of transcription. Oncogene 20:3047–3054
35. Ding Y, Qiao A, Wang Z, Goodwin JS, Lee ES, Block ML, Allsbrook M, McDonald MP, Fan GH (2008) Retinoic acid attenuates beta-amyloid deposition and rescues memory deficits in an Alzheimer's disease transgenic mouse model. J Neurosci 28:11622–11634
36. Dong D, Ruuska SE, Levinthal DJ, Noy N (1999) Distinct roles for cellular retinoic acid-binding proteins I and II in regulating signaling by retinoic acid. J Biol Chem 274:23695–23698
37. Duong V, Rochette-Egly C (2011) The molecular physiology of nuclear retinoic acid receptors. From health to disease. Biochim Biophys Acta 1812:1023–1031
38. Epping MT, Wang L, Edel MJ, Carlee L, Hernandez M, Bernards R (2005) The human tumor antigen PRAME is a dominant repressor of retinoic acid receptor signaling. Cell 122:835–847
39. Farboud B, Hauksdottir H, Wu Y, Privalsky ML (2003) Isotype-restricted corepressor recruitment: a constitutively closed helix 12 conformation in retinoic acid receptors beta and gamma interferes with corepressor recruitment and prevents transcriptional repression. Mol Cell Biol 23:2844–2858
40. Ferry C, Gaouar S, Fischer B, Boeglin M, Paul N, Samarut E, Piskunov A, Pankotai-Bodo G, Brino L, Rochette-Egly C (2011) Cullin 3 mediates SRC-3 ubiquitination and degradation to control the retinoic acid response. Proc Natl Acad Sci USA 108:20603–20608
41. Ferry C, Gianni M, Lalevee S, Bruck N, Plassat JL, Raska I Jr, Garattini E, Rochette-Egly C (2009) SUG-1 plays proteolytic and non-proteolytic roles in the control of retinoic acid target genes via its interaction with SRC-3. J Biol Chem 284:8127–8135
42. Fujiki R, Chikanishi T, Hashiba W, Ito H, Takada I, Roeder RG, Kitagawa H, Kato S (2009) GlcNAcylation of a histone methyltransferase in retinoic-acid-induced granulopoiesis. Nature 459:455–459

43. Gaillard E, Bruck N, Brelivet Y, Bour G, Lalevee S, Bauer A, Poch O, Moras D, Rochette-Egly C (2006) Phosphorylation by PKA potentiates retinoic acid receptor alpha activity by means of increasing interaction with and phosphorylation by cyclin H/cdk7. Proc Natl Acad Sci USA 103:9548–9553
44. Gaj T, Gersbach CA, Barbas CF 3rd (2013) ZFN, TALEN, and CRISPR/Cas-based methods for genome engineering. Trends Biotechnol 31:397–405
45. Gaub MP, Rochette-Egly C, Lutz Y, Ali S, Matthes H, Scheuer I, Chambon P (1992) Immunodetection of multiple species of retinoic acid receptor alpha: evidence for phosphorylation. Exp Cell Res 201:335–346
46. Gay F, Barath P, Desbois-Le Peron C, Metivier R, Le Guevel R, Birse D, Salbert G (2002) Multiple phosphorylation events control chicken ovalbumin upstream promoter transcription factor I orphan nuclear receptor activity. Mol Endocrinol 16:1332–1351
47. Gehani SS, Agrawal-Singh S, Dietrich N, Christophersen NS, Helin K, Hansen K (2010) Polycomb group protein displacement and gene activation through MSK-dependent H3K27me3S28 phosphorylation. Mol Cell 39:886–900
48. Germain P, Altucci L, Bourguet W, Rochette-Egly C, Gronemeyer H (2003) Nuclear receptor superfamily: principles of signaling. Pure Appl Chem IUPAC 75:1619–1664
49. Germain P, Chambon P, Eichele G, Evans RM, Lazar MA, Leid M, De Lera AR, Lotan R, Mangelsdorf DJ, Gronemeyer H (2006a) International union of pharmacology. LX. Retinoic acid receptors. Pharmacol Rev 58:712–725
50. Germain P, Chambon P, Eichele G, Evans RM, Lazar MA, Leid M, De Lera AR, Lotan R, Mangelsdorf DJ, Gronemeyer H (2006b) International union of pharmacology. LXIII. Retinoid X receptors. Pharmacol Rev 58:760–772
51. Germain P, Staels B, Dacquet C, Spedding M, Laudet V (2006) Overview of nomenclature of nuclear receptors. Pharmacol Rev 58:685–704
52. Gianni M, Bauer A, Garattini E, Chambon P, Rochette-Egly C (2002) Phosphorylation by p38MAPK and recruitment of SUG-1 are required for RA-induced RARγ degradation and transactivation. EMBO J 21:3760–3769
53. Gianni M, Boldetti A, Guarnaccia V, Rambaldi A, Parrella E, Raska I Jr, Rochette-Egly C, Del Sal G, Rustighi A, Terao M, Garattini E (2009) Inhibition of the peptidyl-prolyl-isomerase Pin1 enhances the responses of acute myeloid leukemia cells to retinoic acid via stabilization of RARalpha and PML-RARalpha. Cancer Res 69:1016–1026
54. Gianni M, Parrella E, Raska I, Gaillard E, Nigro EA, Gaudon C, Garattini E, Rochette-Egly C (2006) P38MAPK-dependent phosphorylation and degradation of SRC-3/AIB1 and RARalpha-mediated transcription. EMBO J 25:739–751
55. Gillespie RF, Gudas LJ (2007) Retinoic acid receptor isotype specificity in F9 teratocarcinoma stem cells results from the differential recruitment of coregulators to retinoic response elements. J Biol Chem 282:33421–33434
56. Gillespie RF, Gudas LJ (2007) Retinoid regulated association of transcriptional co-regulators and the polycomb group protein SUZ12 with the retinoic acid response elements of Hoxa1, RARbeta(2), and Cyp26A1 in F9 embryonal carcinoma cells. J Mol Biol 372:298–316
57. Glass CK, Rosenfeld MG (2000) The coregulator exchange in transcriptional functions of nuclear receptors. Genes Dev 14:121–141
58. Gudas LJ, Wagner JA (2011) Retinoids regulate stem cell differentiation. J Cell Physiol 226:322–330
59. Gupta P, Ho PC, Huq MM, Ha SG, Park SW, Khan AA, Tsai NP, Wei LN (2008) Retinoic acid-stimulated sequential phosphorylation, PML recruitment, and SUMOylation of nuclear receptor TR2 to suppress Oct4 expression. Proc Natl Acad Sci USA 105:11424–11429
60. Hauksdottir H, Farboud B, Privalsky ML (2003) Retinoic acid receptors beta and gamma do not repress, but instead activate target gene transcription in both the absence and presence of hormone ligand. Mol Endocrinol 17:373–385
61. Hilgeroth A, Tell V, Kramer S, Totzke F, Schachtele C (2013) Approaches to a multitargeting drug development: first profiled 3-Ethoxycarbonyl-1-aza-9-oxafluorenes representing a perspective compound class targeting alzheimer disease relevant kinases CDK1, CDK5 and GSK-3beta. Med Chem

62. Hu X, Chen Y, Farooqui M, Thomas MC, Chiang CM, Wei LN (2004) Suppressive effect of receptor-interacting protein 140 on coregulator binding to retinoic acid receptor complexes, histone-modifying enzyme activity, and gene activation. J Biol Chem 279:319–325
63. Jetten AM, Kurebayashi S, Ueda E (2001) The ROR nuclear orphan receptor subfamily: critical regulators of multiple biological processes. Prog Nucleic Acid Res Mol Biol 69:205–247
64. Jonas BA, Varlakhanova N, Hayakawa F, Goodson M, Privalsky ML (2007) Response of SMRT (silencing mediator of retinoic acid and thyroid hormone receptor) and N-CoR (nuclear receptor corepressor) corepressors to mitogen-activated protein kinase kinase kinase cascades is determined by alternative mRNA splicing. Mol Endocrinol 21:1924–1939
65. Kay BK, Williamson MP, Sudol M (2000) The importance of being proline: the interaction of proline-rich motifs in signaling proteins with their cognate domains. FASEB J 14:231–241
66. Keriel A, Stary A, Sarasin A, Rochette-Egly C, Egly JM (2002) XPD mutations prevent TFIIH-dependent transactivation by nuclear receptors and phosphorylation of RARalpha. Cell 109:125–135
67. Khetchoumian K, Teletin M, Tisserand J, Mark M, Herquel B, Ignat M, Zucman-Rossi J, Cammas F, Lerouge T, Thibault C, Metzger D, Chambon P, Losson R (2007) Loss of Trim24 (Tif1alpha) gene function confers oncogenic activity to retinoic acid receptor alpha. Nat Genet 39:1500–1506
68. Kopf E, Plassat JL, Vivat V, de The H, Chambon P, Rochette-Egly C (2000) Dimerization with retinoid X receptors and phosphorylation modulate the retinoic acid-induced degradation of retinoic acid receptors alpha and gamma through the ubiquitin-proteasome pathway. J Biol Chem 275:33280–33288
69. Kruse SW, Suino-Powell K, Zhou XE, Kretschman JE, Reynolds R, Vonrhein C, Xu Y, Wang L, Tsai SY, Tsai MJ, Xu HE (2008) Identification of COUP-TFII orphan nuclear receptor as a retinoic acid-activated receptor. PLoS Biol 6:e227
70. Lalevee S, Bour G, Quinternet M, Samarut E, Kessler P, Vitorino M, Bruck N, Delsuc MA, Vonesch JL, Kieffer B, Rochette-Egly C (2010) Vinexinss, an atypical "sensor" of retinoic acid receptor gamma signaling: union and sequestration, separation, and phosphorylation. FASEB J 24:4523–4534
71. Lalevee S, Ferry C, Rochette-Egly C (2010) Phosphorylation control of nuclear receptors. Methods Mol Biol 647:251–266
72. Lane MA, Bailey SJ (2005) Role of retinoid signalling in the adult brain. Prog Neurobiol 75:275–293
73. Laudet V, Gronemeyer H (2001) Nuclear receptor factsbook. Academic Press, London
74. Lavery DN, McEwan IJ (2005) Structure and function of steroid receptor AF1 transactivation domains: induction of active conformations. Biochem J 391:449–464
75. Le Douarin B, Zechel C, Garnier JM, Lutz Y, Tora L, Pierrat P, Heery D, Gronemeyer H, Chambon P, Losson R (1995) The N-terminal part of TIF1, a putative mediator of the ligand-dependent activation function (AF-2) of nuclear receptors, is fused to B-raf in the oncogenic protein T18. EMBO J 14:2020–2033
76. le Maire A, Teyssier C, Erb C, Grimaldi M, Alvarez S, de Lera AR, Balaguer P, Gronemeyer H, Royer CA, Germain P, Bourguet W (2010) A unique secondary-structure switch controls constitutive gene repression by retinoic acid receptor. Nat Struct Mol Biol 17:801–807
77. Le May N, Mota-Fernandes D, Velez-Cruz R, Iltis I, Biard D, Egly JM (2010) NER factors are recruited to active promoters and facilitate chromatin modification for transcription in the absence of exogenous genotoxic attack. Mol Cell 38:54–66
78. Le May N, Iltis I, Amé J, Zhovner A, Biard D, Egly JM, Schreiber V, Coin F (2012) Poly (ADP-ribose) glycohydrolase regulates retinoic acid receptor-mediated gene expression. Mol Cell 48:785–798
79. Lee HY, Suh YA, Robinson MJ, Clifford JL, Hong WK, Woodgett JR, Cobb MH, Mangelsdorf DJ, Kurie JM (2000) Stress pathway activation induces phosphorylation of retinoid X receptor. J Biol Chem 275:32193–32199
80. Lee MG, Villa R, Trojer P, Norman J, Yan KP, Reinberg D, Di Croce L, Shiekhattar R (2007) Demethylation of H3K27 regulates polycomb recruitment and H2A ubiquitination. Science 318:447–450

81. Lefebvre P, Martin PJ, Flajollet S, Dedieu S, Billaut X, Lefebvre B (2005) Transcriptional activities of retinoic acid receptors. Vitam Horm 70:199–264
82. Losel R, Wehling M (2003) Nongenomic actions of steroid hormones. Nat Rev Mol Cell Biol 4:46–56
83. Luconi M, Cantini G, Serio M (2010) Peroxisome proliferator-activated receptor gamma (PPARgamma): is the genomic activity the only answer? Steroids 75:585–594
84. Macias MJ, Wiesner S, Sudol M (2002) WW and SH3 domains, two different scaffolds to recognize proline-rich ligands. FEBS Lett 513:30–37
85. Mahony S, Mazzoni EO, McCuine S, Young RA, Wichterle H, Gifford DK (2011) Ligand-dependent dynamics of retinoic acid receptor binding during early neurogenesis. Genome Biol 12:R2
86. Mark M, Ghyselinck NB, Chambon P (2009) Function of retinoic acid receptors during embryonic development. Nucl Recept Signal 7:e002
87. Masia S, Alvarez S, de Lera AR, Barettino D (2007) Rapid, nongenomic actions of retinoic acid on phosphatidylinositol-3-kinase signaling pathway mediated by the retinoic acid receptor. Mol Endocrinol 21:2391–2402
88. Matsushima-Nishiwaki R, Okuno M, Adachi S, Sano T, Akita K, Moriwaki H, Friedman SL, Kojima S (2001) Phosphorylation of retinoid X receptor alpha at serine 260 impairs its metabolism and function in human hepatocellular carcinoma. Cancer Res 61:7675–7682
89. Mendoza-Parra MA, Walia M, Sankar M, Gronemeyer H (2011) Dissecting the retinoid-induced differentiation of F9 embryonal stem cells by integrative genomics. Mol Syst Biol 7:538
90. Mengeling BJ, Goodson ML, Bourguet W, Privalsky ML (2012) SMRTepsilon, a corepressor variant, interacts with a restricted subset of nuclear receptors, including the retinoic acid receptors alpha and beta. Mol Cell Endocrinol 351:306–316
91. Moutier E, Ye T, Choukrallah MA, Urban S, Osz J, Chatagnon A, Delacroix L, Langer D, Rochel N, Moras D, Benoit G, Davidson I (2012) Retinoic acid receptors recognise the mouse genome through binding elements with diverse spacing and topology. J Biol Chem 287:26328–26341
92. Osz J, Brelivet Y, Peluso-Iltis C, Cura V, Eiler S, Ruff M, Bourguet W, Rochel N, Moras D (2012) Structural basis for a molecular allosteric control mechanism of cofactor binding to nuclear receptors. Proc Natl Acad Sci USA 109:E588–E594
93. Pan J, Kao YL, Joshi S, Jeetendran S, Dipette D, Singh US (2005) Activation of Rac1 by phosphatidylinositol 3-kinase in vivo: role in activation of mitogen-activated protein kinase (MAPK) pathways and retinoic acid-induced neuronal differentiation of SH-SY5Y cells. J Neurochem 93:571–583
94. Perissi V, Aggarwal A, Glass CK, Rose DW, Rosenfeld MG (2004) A corepressor/coactivator exchange complex required for transcriptional activation by nuclear receptors and other regulated transcription factors. Cell 116:511–526
95. Perissi V, Jepsen K, Glass CK, Rosenfeld MG (2010) Deconstructing repression: evolving models of co-repressor action. Nat Rev Genet 11:109–123
96. Perissi V, Rosenfeld MG (2005) Controlling nuclear receptors: the circular logic of cofactor cycles. Nat Rev Mol Cell Biol 6:542–554
97. Perissi V, Scafoglio C, Zhang J, Ohgi KA, Rose DW, Glass CK, Rosenfeld MG (2008) TBL1 and TBLR1 phosphorylation on regulated gene promoters overcomes dual CtBP and NCoR/SMRT transcriptional repression checkpoints. Mol Cell 29:755–766
98. Perissi V, Staszewski LM, McInerney EM, Kurokawa R, Krones A, Rose DW, Lambert MH, Milburn MV, Glass CK, Rosenfeld MG (1999) Molecular determinants of nuclear receptor-corepressor interaction. Genes Dev 13:3198–3208
99. Piskunov A, Rochette-Egly C (2011) MSK1 and nuclear receptor signaling. In: MSKs, Arthur S, Vermeulen L (eds) Landes Biosciences, Austin
100. Piskunov A, Rochette-Egly C (2012) A retinoic acid receptor RARalpha pool present in membrane lipid rafts forms complexes with G protein alphaQ to activate p38MAPK. Oncogene 31:3333–3345
101. Poon MM, Chen L (2008) Retinoic acid-gated sequence-specific translational control by RARalpha. Proc Natl Acad Sci USA 105:20303–20308

102. Privalsky ML (2004) The role of corepressors in transcriptional regulation by nuclear hormone receptors. Annu Rev Physiol 66:315–360
103. Renaud JP, Rochel N, Ruff M, Vivat V, Chambon P, Gronemeyer H, Moras D (1995) Crystal structure of the RAR-gamma ligand-binding domain bound to all-trans retinoic acid. Nature 378:681–689
104. Rochel N, Ciesielski F, Godet J, Moman E, Roessle M, Peluso-Iltis C, Moulin M, Haertlein M, Callow P, Mely Y, Svergun DI, Moras D (2011) Common architecture of nuclear receptor heterodimers on DNA direct repeat elements with different spacings. Nat Struct Mol Biol 18:564–570
105. Rochette-Egly C (2003) Nuclear receptors: integration of multiple signalling pathways through phosphorylation. Cell Signal 15:355–366
106. Rochette-Egly C, Adam S, Rossignol M, Egly JM, Chambon P (1997) Stimulation of RAR alpha activation function AF-1 through binding to the general transcription factor TFIIH and phosphorylation by CDK7. Cell 90:97–107
107. Rochette-Egly C, Gaub MP, Lutz Y, Ali S, Scheuer I, Chambon P (1992) Retinoic acid receptor-beta: immunodetection and phosphorylation on tyrosine residues. Mol Endocrinol 6:2197–2209
108. Rochette-Egly C, Germain P (2009) Dynamic and combinatorial control of gene expression by nuclear retinoic acid receptors. Nucl Recept Signal 7:e005
109. Rochette-Egly C, Lutz Y, Saunders M, Scheuer I, Gaub MP, Chambon P (1991) Retinoic acid receptor gamma: specific immunodetection and phosphorylation. J Cell Biol 115:535–545
110. Rochette-Egly C, Oulad-Abdelghani M, Staub A, Pfister V, Scheuer I, Chambon P, Gaub MP (1995) Phosphorylation of the retinoic acid receptor-alpha by protein kinase A. Mol Endocrinol 9:860–871
111. Rosenfeld MG, Lunyak VV, Glass CK (2006) Sensors and signals: a coactivator/corepressor/epigenetic code for integrating signal-dependent programs of transcriptional response. Genes Dev 20:1405–1428
112. Samarut E, Amal I, Markov G, Stote R, Dejaegere A, Laudet V, Rochette-Egly C (2011) Evolution of nuclear retinoic acid receptors alpha (RARa) phosphorylation sites. Serine gain provides fine-tuned regulation. Mol Biol Evol 28:2125–2137
113. Samarut E, Rochette-Egly C (2012) Nuclear retinoic acid receptors: conductors of the retinoic acid symphony during development. Mol Cell Endocrinol 348:348–360
114. Santos NC, Kim KH (2010) Activity of retinoic acid receptor-alpha is directly regulated at its protein kinase A sites in response to follicle-stimulating hormone signaling. Endocrinology 151:2361–2372
115. Schoenfelder S, Sexton T, Chakalova L, Cope NF, Horton A, Andrews S, Kurukuti S, Mitchell JA, Umlauf D, Dimitrova DS, Eskiw CH, Luo Y, Wei CL, Ruan Y, Bieker JJ, Fraser P (2010) Preferential associations between co-regulated genes reveal a transcriptional interactome in erythroid cells. Nat Genet 42:53–61
116. Schug TT, Berry DC, Shaw NS, Travis SN, Noy N (2007) Opposing effects of retinoic acid on cell growth result from alternate activation of two different nuclear receptors. Cell 129:723–733
117. Schug TT, Berry DC, Toshkov IA, Cheng L, Nikitin AY, Noy N (2008) Overcoming retinoic acid-resistance of mammary carcinomas by diverting retinoic acid from PPARbeta/delta to RAR. Proc Natl Acad Sci USA 105:7546–7551
118. Sims RJ 3rd, Reinberg D (2008) Is there a code embedded in proteins that is based on post-translational modifications? Nat Rev Mol Cell Biol 9:815–820
119. Slamon DJ, Clark GM, Wong SG, Levin WJ, Ullrich A, McGuire WL (1987) Human breast cancer: correlation of relapse and survival with amplification of the HER-2/neu oncogene. Science 235:177–182
120. Sodhi RK, Singh N (2013) All-trans retinoic acid rescues memory deficits and neuropathological changes in mouse model of streptozotocin-induced dementia of Alzheimer's type. Prog Neuropsychopharmacol Biol Psychiatry 40:38–46

121. Srinivas H, Juroske DM, Kalyankrishna S, Cody DD, Price RE, Xu XC, Narayanan R, Weigel NL, Kurie JM (2005) c-Jun N-terminal kinase contributes to aberrant retinoid signaling in lung cancer cells by phosphorylating and inducing proteasomal degradation of retinoic acid receptor alpha. Mol Cell Biol 25:1054–1069
122. Srinivas H, Xia D, Moore NL, Uray IP, Kim H, Ma L, Weigel NL, Brown PH, Kurie JM (2006) Akt phosphorylates and suppresses the transactivation of retinoic acid receptor alpha. Biochem J 395:653–662
123. Stavridis MP, Collins BJ, Storey KG (2010) Retinoic acid orchestrates fibroblast growth factor signalling to drive embryonic stem cell differentiation. Development 137:881–890
124. Stehlin-Gaon C, Willmann D, Zeyer D, Sanglier S, Van Dorsselaer A, Renaud JP, Moras D, Schule R (2003) All-trans retinoic acid is a ligand for the orphan nuclear receptor ROR beta. Nat Struct Biol 10:820–825
125. Taneja R, Rochette-Egly C, Plassat JL, Penna L, Gaub MP, Chambon P (1997) Phosphorylation of activation functions AF-1 and AF-2 of RAR alpha and RAR gamma is indispensable for differentiation of F9 cells upon retinoic acid and cAMP treatment. EMBO J 16:6452–6465
126. Tansey WP (2001) Transcriptional activation: risky business. Genes Dev 15:1045–1050
127. Tari AM, Lim SJ, Hung MC, Esteva FJ, Lopez-Berestein G (2002) Her2/neu induces all-trans retinoic acid (ATRA) resistance in breast cancer cells. Oncogene 21:5224–5232
128. Tarrade A, Bastien J, Bruck N, Bauer A, Gianni M, Rochette-Egly C (2005) Retinoic acid and arsenic trioxide cooperate for apoptosis through phosphorylated RXR alpha. Oncogene 24:2277–2288
129. Varlakhanova N, Hahm JB, Privalsky ML (2011) Regulation of SMRT corepressor dimerization and composition by MAP kinase phosphorylation. Mol Cell Endocrinol 332:180–188
130. Wang A, Alimova IN, Luo P, Jong A, Triche TJ, Wu L (2010) Loss of CAK phosphorylation of RAR{alpha} mediates transcriptional control of retinoid-induced cancer cell differentiation. FASEB J: Official Publ Fed Am Soc Exp Biol 24:833–843
131. Warnmark A, Treuter E, Wright AP, Gustafsson JA (2003) Activation functions 1 and 2 of nuclear receptors: molecular strategies for transcriptional activation. Mol Endocrinol 17:1901–1909
132. Wei LN, Hu X, Chandra D, Seto E, Farooqui M (2000) Receptor-interacting protein 140 directly recruits histone deacetylases for gene silencing. J Biol Chem 275:40782–40787
133. Westin S, Rosenfeld MG, Glass CK (2000) Nuclear receptor coactivators. Adv Pharmacol 47:89–112
134. Wilson V, Olivera-Martinez I, Storey KG (2009) Stem cells, signals and vertebrate body axis extension. Development 136:1591–1604
135. Wulf G, Finn G, Suizu F, Lu KP (2005) Phosphorylation-specific prolyl isomerization: is there an underlying theme? Nat Cell Biol 7:435–441
136. Zarrinpar A, Lim WA (2000) Converging on proline: the mechanism of WW domain peptide recognition. Nat Struct Biol 7:611–613
137. Zechel C, Shen XQ, Chambon P, Gronemeyer H (1994) Dimerization interfaces formed between the DNA binding domains determine the cooperative binding of RXR/RAR and RXR/TR heterodimers to DR5 and DR4 elements. EMBO J 13:1414–1424
138. Zechel C, Shen XQ, Chen JY, Chen ZP, Chambon P, Gronemeyer H (1994) The dimerization interfaces formed between the DNA binding domains of RXR, RAR and TR determine the binding specificity and polarity of the full-length receptors to direct repeats. EMBO J 13:1425–1433
139. Zhao Y, Lang G, Ito S, Bonnet J, Metzger E, Sawatsubashi S, Suzuki E, Le Guezennec X, Stunnenberg HG, Krasnov A, Georgieva SG, Schule R, Takeyama K, Kato S, Tora L, Devys D (2008) A TFTC/STAGA module mediates histone H2A and H2B deubiquitination, coactivates nuclear receptors, and counteracts heterochromatin silencing. Mol Cell 29:92–101
140. Zhou XE, Suino-Powell KM, Xu Y, Chan CW, Tanabe O, Kruse SW, Reynolds R, Engel JD, Xu HE (2011) The orphan nuclear receptor TR4 is a vitamin A-activated nuclear receptor. J Biol Chem 286:2877–2885

Chapter 7
The Roles of Retinoic Acid and Retinoic Acid Receptors in Inducing Epigenetic Changes

Alison Urvalek, Kristian Bruun Laursen and Lorraine J. Gudas

Abstract Epigenetics is "the branch of biology which studies the causal interactions between genes and their products which bring the phenotype into being" as defined by Conrad Waddington in 1942 in a discussion of the mechanisms of cell differentiation. More than seven decades later we know that these mechanisms include histone tail post-translational modifications, DNA methylation, ATP-dependent chromatin remodeling, and non-coding RNA pathways. Epigenetic modifications are powerful drugs targets, and combined targeting of multiple pathways is expected to significantly advance cancer therapy.

Abbreviations

SAH	S-adenosylhomocysteine
SAHA	Suberoylanilidehydroxamic acid
KMT/PRMT	Lysine/arginine methyltransferase
KDM/PRDM	Lysine/arginine demethylase
ES	Embryonic stem
HDAC	Histone deacetylase
CpG	Phosphodiester-bonded cytosine–guanine dinucleotide
MECP	Methyl-CpG-binding domain proteins
DNMT	DNA methyltransferase

A. Urvalek (✉) · K. B. Laursen (✉) · L. J. Gudas
Department of Pharmacology, Weill Cornell Medical College,
1300 York Avenue, New York, NY 10065, USA
e-mail: alu2004@med.cornell.edu

K. B. Laursen
e-mail: krl2004@med.cornell.edu

L. J. Gudas
e-mail: ljgudas@med.cornell.edu

M. A. Asson-Batres and C. Rochette-Egly (eds.), *The Biochemistry of Retinoic Acid Receptors I: Structure, Activation, and Function at the Molecular Level*, Subcellular Biochemistry 70,
DOI: 10.1007/978-94-017-9050-5_7,

RAR	Retinoic acid receptor
RA	retinoic acid
Hox	Homeobox
ChIP	Chromatin immunoprecipitation
RXR	Retinoid X receptor
RARE	Retinoic acid responsive DNA element
PRC	Polycomb repressive complex
5hmC	5-hydroxymethylcytosine
5-Aza	5-Aza-2′-deoxycytidine
DZNep	3-Deazaneplanocin A
HSCs	Hematopoietic stem cells
TRAIL	TNF-related apoptosis-inducing ligand
TDG	thymine DNA glycosylase
PP	proximal promoter

Standardized Gene Names/Nomenclature

KDM1	LSD1/2
KDM4A	JMJD2A
KDM5A	Jarid1A/B/C/D
KDM6	JMJD3/UTX/(UTY)
KAT3A/B	CBP/p300
KAT6A	MOZ
KAT6B	MORF
MT2A	MLL1
KMT2B/C	MLL2/3

Introduction

> What determines whether a given piece of DNA along the chromosome is functioning, since it's covered with the histones? You can inherit something beyond the DNA sequence. That's where the real excitement of genetics is now.
> —James Watson, 2003

The term "epigenetics" was coined by Conrad Waddington in 1942 in a discussion of the mechanisms of cell differentiation. Waddington defined epigenetics as "the branch of biology which studies the causal interactions between genes and their products which bring the phenotype into being" [97]. The specific epigenetic mechanisms that regulate genetic programming were not discovered until decades after Waddington first coined the term [30, 73]. These mechanisms are now known to include histone tail post-translational modifications, DNA methylation, ATP-dependent chromatin remodeling, and non-coding RNA pathways [88]. With these discoveries, Waddington's original definition of "epigenetics" has changed and evolved to the currently accepted view that "epigenetics (epi—being a Greek prefix for "on top of") refers to "the study of heritable changes in genes that are not the result of changes in the DNA sequence" [74].

Dr. James Watson won the Nobel Prize for his seminal role in discovering the structure of the DNA double helix structure in 1953, but 50 years later he acknowledged that DNA is not the sole regulator of gene inheritance and expression. Instead, epigenetic changes that occur "above" the DNA may be just as or more important than genetics in terms of their effects on development and disease state. Retinoic acid (RA), a vitamin A derivative that functions as the active metabolite in cellular signaling, induces cell differentiation in stem cells and some cancer cells. Along with the more well known effects of RA signaling on cell lineage specification through transcriptional activation of retinoic acid receptor (RAR)-regulated genes, recent studies are demonstrating that RA also mediates cell differentiation via rapid, profound effects on the epigenome. This observation is opening up a new area of fundamental research into transcriptional regulation as well as pointing the way to new clinical applications of RA. The use of RA in combination with drugs that modify the epigenome is showing promise in the treatment and/or prevention of several types of cancer. This type of combination therapy is increasingly relevant, as many types of cancer exhibit aberrant levels of or mutations in epigenetic regulatory proteins.

History: Epigenetic Regulation Is Achieved by a Number of Different Mechanisms

Waddington nicely illustrated the idea of genotype-to-phenotype changes along cell development pathways by his drawing of an "epigenetic landscape" [96]. In this model specific epigenetic modifications are acquired as progenitor cells, depicted as marbles, differentiate and commit to a specific cell fate, conceptualized as marbles rolling down into one of several valleys. This idea has been substantiated by experimental findings where it has been demonstrated that commitment of cells into specific differentiation pathways is associated with progressive epigenetic modifications [34].

Histone Protein Tail Modifications and Transcriptional Regulation

Cellular chromatin is composed of DNA-wrapped nucleosomes packed into regions of either compacted or loose nucleosomal structure, referred to as hetero- and euchromatin, respectively. In general, genes residing in heterochromatic regions are silenced, whereas genes located in euchromatic regions are actively transcribed.

The nucleosome is a histone octamer composed of two of each of the core histones H2A, H2B, H3, and H4, and one auxiliary H1 linker histone [22]. The histone proteins are each composed of a globular domain with an extended, positively charged N-terminal tail that interacts with the phosphodiester backbone of the DNA. Histone proteins were discovered in 1884, but it was not until 1963 that histone tails were shown to be post-transcriptionally modified [66]. Subsequently, the effects of these histone modifications on gene regulation began to be elucidated [2]. Importantly, the histones, in particular the lysine/arginine rich tails, were shown to be targets for extensive post-transcriptional modifications. Recently, Yuan et al. [101] have shown that RA-mediated transcriptional activation of the Cytochrome P450 26a1 gene is associated with a loosening of the chromatin structure, which is required for transcriptional activation.

Histone protein tail regulation is highly complex, and numerous post-translational modifications can regulate different aspects of gene transcription; these include phosphorylation, sumoylation, ubiquitination, ribosylation, neddylation, ADP-ribosylation, citrullination, and others [82, 100]. The enzymes that regulate these modifications can be divided into three groups of epigenetic regulators: "writers", "readers", and "erasers".

Histone modifications are generated by "writer" enzymes, which include families of lysine/arginine methyltransferases (KMTs/PRMTs), histone lysine acetyltransferases (KATs), and serine/threonine kinases. Methylation of histone tails, including trimethylation of histone 3 lysine 9 (H3K9me3) and 27 (H3K27me3), is generally associated with gene repression, whereas the acetylation of the same residues (H3K9 and H3K27) correlates strongly with gene activation [31, 93]. There are also instances in which methylation of certain residues is associated with gene activation, such as methylation of H3K4 and H3K36 [79, 94]. Acetylation by KAT proteins, such as KAT3A (p300), KAT3B (CBP), PCAF, or KAT13B (pCIP), promotes the activation of gene expression by neutralizing the positive charges of histone tails [93], leading to loosening of the negatively charged chromatin and subsequent binding of DNA binding factors that promote gene transcription.

The actions of "writer" enzymes are countered by a group of enzymes known as "erasers". "Erasers" are responsible for the removal of specific histone modifications. This group includes lysine/arginine demethylases (KDMs/PRDMs), histone deacetylases (HDACs), and serine/threonine phosphatases (summarized in Table 7.1). Typically, these "erasers" counteract the actions of "writer" enzymes, having direct effects on gene transcription. For example, "writer" enzymes of the KAT family mediate the deposition of acetylation marks onto lysine residues of histone tails, thereby neutralizing the attraction of positively charged histones with negatively charged DNA. This allows for the unraveling of DNA, thereby allowing general transcriptional machinery and other proteins to bind that mediate gene activation. Conversely, "eraser" proteins of the HDAC family remove the acetylation mark. This allows the DNA to again wrap around histones, preventing the binding of general transcription machinery, thereby leading to gene repression.

"Reader" proteins specifically bind to post-translationally modified chromatin, and recognize these specific histone modifications to alter chromatin structure and dynamics. Often, "reader" proteins are part of larger protein complexes that contain "reader" and/or "eraser" proteins. Without "reader" proteins, posttranslational modifications would not be recognized, and the protein complex or specific "writers" or "erasers" would not be recruited. Alternatively, "writer" or "eraser" proteins themselves can also serve as "readers" proteins. For example, KAT

Table 7.1 Groups of Epigenetic Modifiers and their functions

–	Writer	Eraser	Reader
Histone marks			
Lysine methylation	KMT	KDM	CBX proteins
Arginine methylation	PRMT		14-3-3-proteins
Lysine acetylation	KAT	HDAC	
Serine/threonine phosphorylation	S/T Kinase	Phosphatase	
DNA methylation			
CpG (5meC)	DNMT	DNA demethylase	MeCP MBD1-4

"Writers" are a group of enzymes that mediate the addition of epigenetic modifications (marks). "Erasers" are proteins with enzymatic activity that mediate the removal of these marks. "Readers" are proteins, generally with no enzymatic activity, that recognize and bind to posttranslational modifications to mediate downstream effects

Key Lysine methyltransferase (*KMT*); protein arginine methyltransferase *(PRMT*); lysine acetyltransferase (*KAT*); serine/threonine kinases (*S/T kinase*); lysine demethylase (*KDM*); histone deacetylase (*HDAC*); DNA methyltransferase (*DNMT*); methyl-CpG-binding domain proteins (*MeCP*); chromobox homolog (*CBX*)

proteins possess a bromodomain that recognizes and binds acetylated lysine residues on histone tails. This allows for KAT proteins to further mediate acetylation at these specific DNA regions. Because of this, these "reader" proteins also mediate changes in transcription or DNA replication [14].

The distinction between "writers", "readers", and "erasers" is complicated by the fact that protein complexes that add marks ("writer" complexes) are frequently composed of several subunits with different enzymatic properties. For example, the polycomb repressive complex 2 (PRC2) is comprised of at least four subunits which include the Suz12 (zinc finger), Eed, Ezh2 (SET domain with histone methyltransferase activity) and RbAp48 (histone binding domain) proteins. Importantly, the Ezh2 protein has enzymatic activity and can add methyl groups specifically to the H3K27 resulting in trimethylation of this histone residue. This posttranslational modification is deposited onto histone tails at lysine 27 by the PRC2 complex—a "writer", but is recognized by the polycomb repressive complex 1 (PRC1)—a "reader" (Min et al. [59]. However, PRC1 mediates the deposition of ubiquitin, another histone modification, onto histone 2A lysine 119. PRC1 can in other words "read" the H3K27me3 and "write" the H2A. K119Ub, and can thus be considered both a "reader" and a "writer" enzyme [80]. Additional modifications of the histone tail (e.g. H3S28ph) proximal to the site of the initial modification (H3K27me3) add another layer complexity. In this example the recognition of H3K27me3 by the INHAT "reader" protein is prevented by phosphorylation of Serine 28 (H3S28ph) [44]. This illustrates how modification of nearby residues can interfere with the recognition of specific histone marks by "reader" proteins. The effect of combinatorial histone modifications is commonly referred to as the histone code, a term coined by Charles D. Allis in 2001 [38]. As exemplified above by the context dependent recognition of H3K27me3, the emerging view is that the recognition by "reader" proteins is not dictated only by specific histone modifications, but rather by an interplay between different histone modifications.

DNA Methylation and Gene Silencing

In contrast to histone modifications, which are relatively transient in nature, DNA methylation provides a more persistent, long-term gene silencing. DNA methylation occurs when a methyl group is deposited on the cytosine of a phosphodiester-bonded cytosine-guanine dinucleotide (CpG) sequence. DNA methylation, e.g. the formation of 5-methylcytosine (5mC), was first proposed as a mechanism for changing gene expression in 1975 [35, 73]. CpG sequences are typically concentrated in large clusters called CpG islands, predominantly located at or near gene promoters, but CpG islands are also found in intergenic regions. Members of a family of DNA methyltransferase (DNMT) enzymes transfer methyl groups to DNA and this engenders stable, long term gene silencing [18]. DNA methylation is introduced by the recruitment of DNMT3a and 3b by sequence specific repressors that silence gene transcription [25]. Newly replicated DNA is transiently hemi-methylated until DNMT1 uses the methylated parent strand to direct deposition of corresponding methylation on the daughter strand, thus maintaining the overall pattern of DNA methylation [48]. In the context of DNA methylation, the DNMTs function as "writers", whereas methyl-CpG-binding domain proteins (MECP), which recognize methylated CpGs, function as "readers" [76]. Recently, researchers have determined that DNA methylation is reversible [47, 81], which suggests that DNA methylation is a dynamic process rather than a one-way mechanism of gene silencing, as was previously thought to be the case.

Selected groups of epigenetic regulators are listed in Table 7.1, where families of "writers", "erasers", and "readers" are listed for each type of epigenetic modification (individual rows). In Table 7.2 are listed a number of commonly investigated epigenetic modifications (individual rows), and their effects on transcription.

Other Epigenetic Regulators of Gene Expression

ATP-dependent remodeling of chromatin structure and long intergenic non-coding RNAs (lincs) are other major epigenetic regulators of gene expression, but to date, little is known about their roles, if any, in RA regulated gene transcription. Here, we will focus on what is known about RA involvement in histone modifications and DNA methylation.

Development of the Field: Retinoids and RARs Mediate Histone Modifications

RA functions as the ligand for retinoic acid receptors (RARs), and can regulate several developmentally important genes, including the Hox (homeobox) gene clusters [41, 42, 50]. At these gene clusters as well as at other RA regulated genes,

Table 7.2 Selected histone modifications and their enzymatic regulators

–	Histone mark	Writer	Eraser
Repression	H3K27me3	EZH2, NSD3	KDM6A/B (JMJD3)
	H3K9me3	SETDB1/2 SUV39H1/2	Lysine specific demethylase 4A/B/C/D
Activation	H3K4me1	SETD7	KDM1A KDM5B
	H3K4me2	NSD3	KDM5A/D KDM1A KDM5B
	H3K4me3	MLL MLL3/4 PRDM9 SETD1A/B SET AND MYND domain-containing protein 3	Lysine specific demethylase 4A/B/C/D KDM5B
	H3K36me3	SETD2 NSD2	Lysine specific demethylase 4A Lysine specific demethylase NO66
	H3K14Ac	PCAF MYST3	HDAC3
	H3K9Ac	PCAF KAT13B (pCIP) KAT6A (Moz)	SIRT1 SIRT6
	H3K27Ac	KAT3A/B (P300/CBP)	–
	H3S28Ph	MAPKKK-MLT MSK1/2 STK5	–

Specific histone marks involved in transcriptional regulation, and the enzymes that modify these marks. Activating marks are modifications that generally favor transcription (Activation), whereas repressive marks are modifications that favor transcriptional silencing (Repression). Examples are given of specific histone modifications (Histone marks), and of the specific enzymes depositing ("writers") and removing ("erasers") these marks. This is a not a comprehensive list, but rather a list of the most well understood regulators of epigenetic changes

Key enhancer of zeste homolog 2 (*EZH2*); nuclear SET domain-containing protein (*NSD3*); lysine (K)-specific demethylase (*KDM*); SET domain, bifurcated (*SETDB*); methyltransferase variant (*SUV39H12*); SET domain (*SETD*); mixed-lineage leukemia (*MLL*); positive regulatory domain (*PRDM*); p300/CBP associated factor (*PCAF*); histone deacetylase (*HDAC*); lysine acetyltransferase (*KAT*); MAP-kinase-kinase-kinase; mitogen- and stress-activated protein kinase (*MSK*); aurora kinase (*STK*)

heterodimers of RARγ and retinoid X receptor α (RXRα) recognize and bind to RA responsive DNA elements (RAREs), then inducing epigenetic changes and transcriptional induction in response to RA (Fig. 7.1) [27, 28]. The transcriptional induction by RA is associated with increased levels of the co-activator proteins KAT3A (p300), KAT13B (pCIP), and of RNA polymerase II at target RAREs. Conversely, co-repressor proteins such as SUZ12, a key protein component of PRC2, are associated with specific RAREs, but dissociate in response to RA

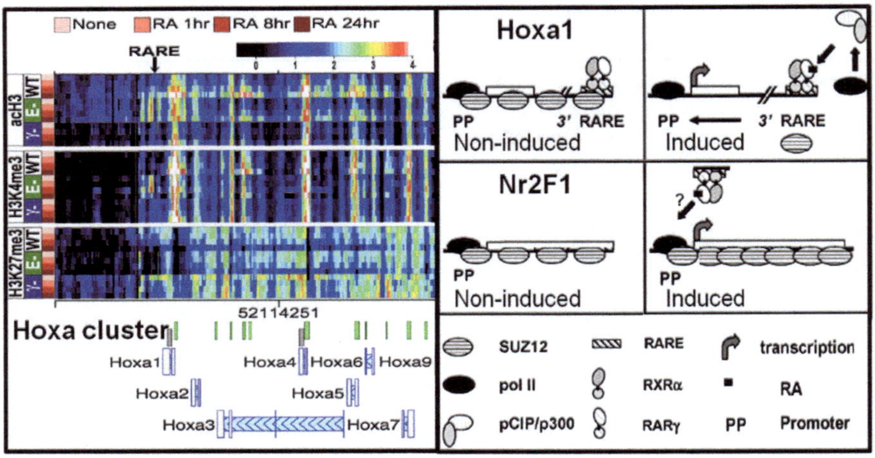

Fig. 7.1 Epigenetic changes induced along the Hoxa cluster in response to RA. The epigenetic changes of the RA responsive Hoxa gene cluster are shown, with the locations of the Hoxa1 proximal promoter (*PP*) and RA responsive element (*RARE*) indicated by arrows. The levels of acH3, H3K4me3, and H3K27me3 determined by ChIP-chip are presented as heat maps, with rows representing individual timepoints for each genotype, and columns indicating specific genomic regions. The genotypes of the stem cell lines are as follows: Wild type (WT), RARE knockout (E-), and RARγ knockout (γ-). The cells were cultured in RA for 1, 8, and 24 h, as indicated. The color scale representing log2-transformed ChIP enrichment is indicated at the top of the figure. Note the reduced levels of acH3 and H3K4me3 at Hoxa1 PP and RARE in the RARγ–knockout cell line. Models for RA mediated transcription of RA target genes Hoxa1 and Nr2F1. In the absence of RA, RARγ–RXRα heterodimers associated with Hoxa1 RAREs presumably associate with co-repressors, thereby generating a SUZ12-rich environment which represses transcription. Binding of the RA ligand causes a conformational change in the RARγ–RXRα heterodimer bound to the Hoxa1 RARE. This results in the recruitment of pCIP/p300, which generates an euchromatic environment, presumably by acetylating the histone tails. This allows pol II to initiate transcription of Hoxa1. The Nr2F1 promoter region (PP) is bound by SUZ12 in the absence of RA. Upon exposure to RA the increase in activating marks is initially counteracted by a concomitant increase in SUZ12, which attenuates the transcription of Nr2F1. Eventually, the SUZ12 levels decline, allowing the increased transcriptional activation of Nr2F1 (modified from Kashyap et al. [41] and Gillespie and Gudas [28])

(Figs. 7.1 and 7.2) [41, 51]. Furthermore, the re-association of SUZ12 with RAREs upon RA removal [27, 28] exemplifies the highly reversible nature of cofactor association. Extensive changes in histone marks can be observed in response to RA, as illustrated by a heat-map showing RA-associated changes in H3K27me3, H3K4me3, and H3 acetylation (acH3) levels at the Hoxa cluster (Fig. 7.1). Importantly, for the Hoxa cluster the levels of activating marks (H3K4me3 and acH3) increase, whereas the levels of repressive marks (H3K27me3) decrease in response to RA (Fig. 7.1) [41]. This is not the case for all RA inducible genes; for CoupTF1 (Nr2F1) the levels of both activating (H3K4me3) and repressive (H3K27me3) marks show an initial increase in response to RA (Fig. 7.2). However, the H3K27me3 levels then start to decline, thereby increasing the extent of the induction [51]. The simultaneous presence of active H3K4me3 marks and

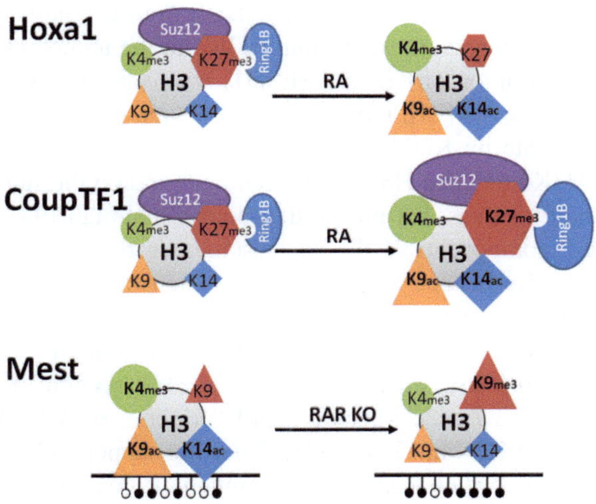

Fig. 7.2 Epigenetic Signatures Associated with RA and RAR regulated transcription. Hoxa1 represents a group of direct target genes induced by RA (*upper panel*). The induction is characterized by dissociation of PRCs (*ovals*) and depletion of the H3K27me3 repressive mark (*hexagons*), and by increased levels of transcriptionally permissive marks, H3K4me3 (*circle*), H3K9ac (*triangle*), and H3K14ac (*diamond*). CoupTF1 represents a group of target genes with delayed transcriptional induction by RA (*middle panel*). The induction is characterized by an initial increase of PRCs (*ovals*) and of the H3K27me3 repressive mark (*hexagons*), concurrent with increased levels of transcriptionally permissive marks; H3K4me3, H3K9ac, and H3K14ac. The imprinted gene Mest is transcribed in the presence of RARα, but is silenced by DNA methylation upon knockout of RARα (*lower panel*). The transcriptional silencing of Mest is associated with increased DNA methylation, increased levels of the H3K9me3 repressive mark, and with decreased levels of transcriptionally permissive marks; H3K4me3, H3K9ac, and H3K14ac. Note that for Hoxa1 and CoupTF1 (Nr2F1) the transcriptionally active state is shown to the *right*, whereas for Mest the transcriptionally active state is shown to the *left* (modified from Laursen et al. [51])

repressive H3K27me3 marks is referred to as a bivalent domain, and this chromatin structure can often be found at promoters of RA inducible genes [41, 51, 56, 63]. The bivalent chromatin structure signifies that these genes are in a poised state in which changes in the H3K4me3/H3K27me3 ratio are associated with transcriptional induction (presence of RA; H3K4me3 ↑, H3K27me3 ↓) or silencing (absence or removal of RA; H3K27me3 ↑, H3K4me3 ↓), as illustrated in Fig. 7.2 [7, 41, 42, 50, 51].

Now the question is; what regulates the levels of histone marks? Since histone marks are actually covalent modifications of histones, the levels need to be regulated by enzymes located in proximity to the histone. This brings us back to the "writers", "readers", and "erasers" mentioned in the beginning of the chapter. The levels of the active H3K4me3 mark are regulated by lysine methyl transferases (KMT) "writers" and lysine demethylases (KDM) "erasers". Specifically, MLL proteins, which are KMT "writers" of the trithorax family, trimethylate H3K4 [64], and KDM5 proteins, which are H3K4me3-specific "erasers," can subsequently convert the H3K4me3 mark into H3K4me2 [37, 46, 64]. Once H3K4me3 has been converted to H3K4me2, the KDM1A/B "eraser" proteins can remove the remaining methyl marks, thus returning H3K4 to the unmethylated state [12]. Pharmacological inhibition of KDM1A (LSD1) reactivates the RA differentiation pathway in leukemia cells [78], indicating that enzymatic conversion of H3K4 to the unmethylated state by KDM1A plays a key role in antagonizing RA signaling. Curiously, H3K4me3 and H3K27me3 can be found on the same nucleosome, composed of eight histones, but not on the same histone tail [94]. Thus, each of the two histone 3 components of the nucleosome can be differentially modified. This specification of a histone as simultaneously activating and repressive is the core of the bivalent domains.

Conversely, the levels of the repressive H3K27me3 mark are regulated by Polycomb group proteins, which are H3K27me1/2 specific KMT "writers" [10, 49], and the H3K27me2/3-specific KDM "erasers" KDM6A/B [1, 53]. The antagonistic effects of H3K4me3 and H3K27me3 are supported by the observation that RA-induced transcription leads to a concomitant decrease in H3K27me3 levels as well as an increase in H3K4me3 levels along the Hoxa cluster (Fig. 7.2). In this respect, it is interesting that the MLL2 complex contains KDM6A, an H3K27 demethylase [1]. Consequently, the MLL complex combines methyltransferase activity targeting H3K4 with demethylase activity targeting the opposing H3K27me3 mark. A similar "push-pull" effect is observed with the PRC2 complex which contains EZH2, an H3K27 methyltransferase, and KDM5B, an H3K4 demethylase [104]. The depletion of H3K27me3 through knockdown of the EZH2 methyltransferase failed to induce Hoxa1 expression [53]. This suggests that without increased H3K4 methylation the loss of H3K27 methylation is insufficient to induce transcription of Hoxa1. Consequently, the combined actions of H3K4 methyltransferases and H3K27 demethylases may be required for gene transcriptional activation of at least some genes. Also, the transcriptional activation of Hoxa1 precedes the removal of the H3K27me3 mark by many hours, showing that the removal of the H3K27me3 mark is not required for Hoxa1 transcriptional activation by RA [41].

A different scenario of the "push-pull" effect is observed when RA induces the CoupTF1 (Nr2F1) gene. In this case, activating H3K4me3 and repressive H3K27me3 marks are simultaneously recruited to the CoupTF1 promoter (Fig. 7.2), initiating a repressed or dampened induction characteristic of several late RA target genes [51]. While the functional depletion of PRC2 did not enhance RA induction of Hoxa5 and Hoxa1 (early genes), the depletion potently enhanced RA mediated induction of CoupTF1 and CoupTF2 (late genes) [51]. This finding is important since it provides a mechanistic rationale for distinguishing between early and late targets of RA induction. It has been shown that PRC2 can sense chromatin density, and thereby distinguish active chromatin (marked by H3K4me3 and H3K36me2/3) from inactive chromatin, on which PRC2 will target H3K27 for methylation. This helps to explain how PRC2 maintains target genes in an inactive, compacted chromatin state for long periods [101]. Taken together, these data further point to the presence of a combined "push-pull" effect, wherein the effects of specific KMTs are supported by the effects of specific KDMs, which together place and remove specific lysine methylation marks in a coordinated manner.

RA induced transcription of the Hox genes increases not only histone H3K4 methylation, but also histone acetylation [41] (Figs. 7.1 and 7.2). H3K27 for example can be modified by either acetylation or methylation, with opposite effects on the chromatin environment, and thus on the transcriptional activity. Acetylation [20, 70] and methylation of H3K27 are mutually exclusive marks positioned by KAT3A/B (CBP/p300) and PRC2 (EZH2), respectively [65, 87]. H3K27 thus provides an example in which the enzymatic activities of KATs/KMTs and HDACs/KDMs converge in regulating gene activity. However, H3K27 is not the only target of acetylation; H3K9 and H3K14 are acetylated concurrently with RA induced transcriptional activation (Figs. 7.1 and 7.2) [41, 42]. The RA-dependent recruitment of the acetyltransferases KAT3B (p300) and KAT13B (NCoA3, Actr, pCIP, Src3) to the RAREs of Hoxa1 (Fig. 7.1) and Cyp26a1 in F9 teratocarcinoma stem cells and embryonic stem (ES) cells suggests that these KATs also play key roles in RA-induced transcription [27, 28, 40]. Finally, KAT6A (Moz) is involved in H3K9 acetylation of the Hox gene loci, yet RA can activate the Hox loci independently of KAT6A [95]. The plethora of coregulators involved in RA induced transcription allows for fine-tuning of a highly gene specific response (Fig. 7.1).

Current State of the Field: DNA Demethylation Is Involved in the RA Transcriptional Response

Passive DNA demethylation takes place when maintenance methylation is inhibited during DNA replication, while active DNA demethylation requires specific enzymes and can occur without DNA replication [105]. Activation induced cytidine deaminase (AICDA, AID) is an active, reprogramming DNA demethylase expressed in ES cells and other cell types [61]. A second, more recently discovered DNA demethylase family, Tet 1, 2 and 3, removes DNA methylation through oxidative demethylation, a mechanism also employed by JmjC proteins to demethylate histones [36, 81, 91]. Tet1 mediated hydroxylation of 5mC to 5-hydroxylmethylcytosine (5hmC) is enhanced by AICDA, which generates 5hmC as a step towards the demethylation of 5mC. This requires thymine DNA glycosylase (TDG), a base excision repair enzyme, which excises the 5hmC [32]. Through the active prevention of DNA methylation, TDG maintains bivalent chromatin domains in ES cells [16]. Considering that several RA primary target genes reside in bivalent domains, it is worth noting that Um et al. [92] identified interactions between TDG and the RARs/RXRs which may link RA to active demethylation of DNA. TDG forms a complex with AICDA and GADD45a, and is required for the recruitment of the coactivator protein KAT3B (p300) to the promoters of RA-inducible genes [17]. Thus, a loss of TDG activity could result in a decrease in RAR/RA-associated gene transcription and a resultant block in cell differentiation, which would be consistent with the observed increase in DNA methylation of the Mest promoter region in response to knockout of RARα [52]. This indicates that RARα (and possibly other RARs) plays a direct role in maintaining gene expression by keeping specific promoters in a hypomethylated state, and conversely, underscores the fact that reduced expression of RARα can have adverse consequences, such as leukemogenesis [29]. A reduction in RARα signaling also impairs the survival of tumor reactive CD8(+) T-cells within the tumor microenvironment [33]. Whether this is related to RARα's ability to control the methylation state of certain genes has not yet been elucidated.

During gametogenesis, maternal or paternal genomes can be modified so that one parental allele is expressed, whereas the other is transcriptionally silenced. This genomic imprinting typically occurs through DNA methylation of CpG islands [68]. An exciting, recent, finding suggests that RARα, independently of RA, maintains the DNA methylation status of specific imprinted genes [52]. This was highlighted by the identification of several aberrantly expressed, imprinted, genes in RARα knockout F9 stem cells [52]. Under normal conditions RARα associates with the promoter region of the paternally expressed gene, Mest; upon RARα knockout, resulting in the absence of RARα, the levels of H3K9me3 and the DNA methylation of the Mest promoter region significantly increase [52] (Fig. 7.2). Several of the changes in gene expression associated with the RARα knockout are similar to those observed during the differentiation of stem-like progenitors to hypertrophic chondrocytes in the developing growth plate [15]. This similarity between the in vivo and in vitro data supports the idea that in vivo imprinting may be regulated by RARα, and highlights the important roles of specific RARs in regulating epigenetic changes during development. Further exploration of this topic is expected to deepen our understanding of genomic imprinting and to expand the realm of RAR regulated transcription beyond the well-known ligand-induced regulation of gene activity.

Relevance: RA Regulated Epigenetic Changes in Carcinogenesis

Retinoid signaling is often disrupted during carcinogenesis, suggesting that restoration of retinoid signaling may be a viable option for cancer prevention and/or treatment [60, 85]. Synthetic retinoids modify the levels of the various RARs during breast carcinogenesis [8], and RA inhibits the growth of human osteosarcoma by promoting cell differentiation [99]. In a glioma animal model, RA also promoted the differentiation of cancer stem cells [9]. As a result, retinoids are currently being tested and/or used for treatment of many different cancers, including breast, ovarian, renal, head and neck, melanoma, leukemias, and prostate cancers. However, epigenetic changes, such as histone modifications and DNA methylation, and subsequent changes in gene expression are also thought to play major roles in cancer initiation and progression. Therefore, in line with the aforementioned "push-pull" model, combination cancer therapies that include retinoids together with epigenetic therapeutic agents are believed to be more effective in treating different cancers.

Histone deacetylase inhibitors, such as suberoylanilidehydroxamic acid (SAHA) have been extensively studied as potential cancer therapies, and are currently being used to treat multiple cancers, including cutaneous T-cell lymphoma and non-small cell lung cancer [3]. Chemoproteomics profiling of HDAC inhibitors revealed selective targeting of histone deacetylase (HDAC) complexes as promising cancer therapies [3].

It is believed that treatment with HDAC inhibitors together with retinoid therapies may be an even more effective treatment regimen for certain cancers. When combined with HDAC inhibitors such as Trichostatin A and valproic acid, RA can re-induce RARβ expression in kidney [89] or breast [60] cancers, and inhibit cell proliferation in many types of cancers [13, 23, 43, 67, 69, 77, 86, 90, 98]. Furthermore, RA synergizes with valproic acid to promote the degradation of the PML-RARα oncoprotein, destroying the leukemia initiating cells in vivo [54]. Recently a phase I trial using valproic acid and liposomal RA for patients with solid tumors yielded positive results, suggesting that this therapy may be used for various solid tumors [21].

Another promising treatment approach is the co-administration of retinoids with DNA methyltransferase inhibitors. Mice treated with a combination of RA and the DNA methyltransferase inhibitor 5-Aza-2′-deoxycytidine (5-Aza) exhibit a decreased incidence of oral cancer after carcinogen treatment [84], and valproic acid, 5-Aza, and RA promoted growth arrest and cell differentiation of cultured human head and neck squamous cell carcinoma cells [26]. Additionally, a phase II clinical trial for patients with acute myeloid leukemia combining 5-Aza with RA was just completed with promising results [55].

Finally, other studies have examined the potential efficacy of treatments with RA, HDAC inhibitors, and DNA methyltransferases together. RA treatment in the presence of both valproic acid and 5-Aza promotes the re-expression of RARβ and inhibits cell growth in breast cancer cell lines [60]. Additionally, promyelocytic leukemia cells exhibit cell growth inhibition and increased granulocyte differentiation after treatment with all three drugs [77]. Overall, these studies indicate that combinations of retinoids and epigenetic modulating drugs are promising treatment options for multiple types of cancer, in part because of their actions in promoting cell differentiation and the inhibition of cell proliferation. Various epigenetic machinery inhibitors are being intensely studied as possible cancer treatments [75], and these could potentially be even more effective in combination with RA.

The Future: RA Action and Epigenetics, Cell Differentiation and Cancer

Further studies are needed to determine the roles and specificities of various KATs and KDMs with respect to RA transcriptional activation and to develop a better understanding of how RARα (and possibly other RARs) plays a direct role in maintaining gene expression by keeping specific promoters in a hypomethylated state. Many different epigenetic changes must take place for stem cells to differentiate

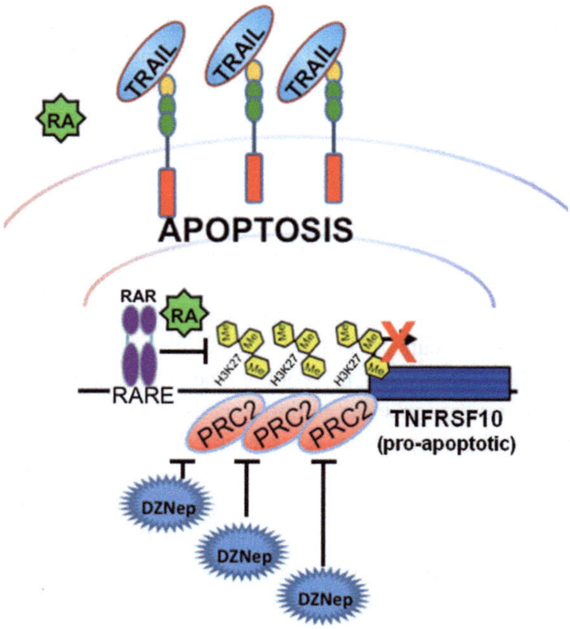

Fig. 7.3 Representation of RA and DZNep effects on apoptosis regulation in human colon cancer cells. Retinoic acid treatment promotes TRAIL-related apoptosis in RARβ / RARγ-positive HT29 cells, but not in SW480 cells which express only low levels of RARβ / RARγ. The functional depletion of PRC2 by either inhibition with DZNep or by knockdown of SUZ12 increases TRAIL-mediated apoptosis in both HT29 and SW480 cell lines. In this scenario the PRC2-mediated repression is alleviated, thereby activating TNFRSF10 even in the absence of RARβ /RARγ

properly, and when these changes do not proceed normally, increased tumorigenesis can result. Aberrant expression of the polycomb protein EZH2, a core component of PRC2, has been found in human breast, prostate, bladder, and colon cancers, and this overexpression is correlated with a poor prognosis [58, 72]. Overexpression of EZH2 in hematopoietic stem cells (HSCs) eliminates the exhaustion of the long-term repopulation potential of these stem cells during multiple, sequential transplantations [39]. EZH2 also enhances leukemogenesis by enhancing the differentiation block in acute myeloid leukemia [62, 83]. Thus, these epigenetic modifications by EZH2 have profound consequences in terms of reducing the ability of HSCs to differentiate and enhancing tumorigenesis. Likewise, in prostate cancer EZH2 can block differentiation by affecting transcriptional regulation by the androgen receptor [19]. The recent development of EZH2 inhibitors for treatment of lymphomas shows the power of manipulating epigenetic modifications for cancer treatment [4].

DZNep, an S-adenosylhomocysteine (SAH) hydrolase inhibitor, can eradicate tumor initiating cells in hepatocellular carcinoma cells and induce apoptosis in acute myeloid leukemia [11, 24, 103]. DZNep can also inhibit tumorigenicity and progression in prostate cancer [19]. The inhibition of SAH hydrolase causes an increase in SAH, resulting in inhibition of S-adenosyl-L-methionine dependent methyltransferases such as EZH2. We recently showed that human colon cancer cells, when exposed to RA, DZNep, or to a genetic knockdown of the PRC2 core protein SUZ12, exhibited enhanced PTEN mediated apoptosis, whereas the survival of ES cells was unaffected [6]. The apoptotic effects of RA, DZNep, or SUZ12 depletion were further enhanced by combination with the TNF-related apoptosis-inducing ligand (TRAIL) death receptor [5]. The synergy between TRAIL and RA was confirmed by another report in which the authors demonstrate that treatment with retinyl acetate (another vitamin A metabolite) in combination with TRAIL not only induced apoptosis specifically in intestinal polyps, but also inhibited tumor growth and prolonged survival in a murine model of human colon cancer [102]. These results suggest that one mechanism by which RA enhances TRAIL associated apoptosis is via removing PRC2 complexes from various genes involved in differentiation and/or apoptosis (Fig. 7.3).

Research in this field will be enhanced by the recent development of more specific EZH2 inhibitors [45, 57], and by the evaluation of new drug combinations that more efficiently target specific epigenetic regulators. The fact that so many different types of cancer exhibit altered epigenetic profiles and/or mutations in proteins that modify the epigenome indicates that this will be a fruitful area of research that will provide major benefits to cancer patients in terms of new combination therapies.

Acknowledgements We would like to thank Weill Cornell and the NIH (NCI R01-CA043796 to LJG, NIDCR R01-DE010389 to LJG), and (NIAAA F32-AA021045 to AU) for support for this chapter. We also thank members of the Gudas lab for suggestions and comments, and in particular, Dr. Yannick Benoit for critically reading this chapter. We thank Tamara Weissman for editing this chapter.

References

1. Agger K, Cloos PA, Christensen J, Pasini D, Rose S, Rappsilber J, Issaeva I, Canaani E, Salcini AE, Helin K (2007) UTX and JMJD3 are histone H3K27 demethylases involved in HOX gene regulation and development. Nature 449:731–734
2. Allfrey VG, Faulkner R, Mirsky AE (1964) Acetylation and methylation of histones and their possible role in the regulation of RNA synthesis. Proc Natl Acad Sci USA 51:786–794
3. Bantscheff M, Hopf C, Savitski MM, Dittmann A, Grandi P, Michon AM, Schlegl J, Abraham Y, Becher I, Bergamini G, Boesche M, Delling M, Dumpelfeld B, Eberhard D, Huthmacher C, Mathieson T, Poeckel D, Reader V, Strunk K, Sweetman G, Kruse U, Neubauer G, Ramsden NG, Drewes G (2011) Chemoproteomics profiling of HDAC inhibitors reveals selective targeting of HDAC complexes. Nat Biotechnol 29:255–265
4. Béguelin W, Popovic R, Teater M, Jiang Y, Bunting KL, Rosen M, Shen H, Yang SN, Wang L, Ezponda T, Martinez-Garcia E, Zhang H, Zheng Y, Verma SK, McCabe MT, Ott HM, Van Aller GS, Kruger RG, Liu Y, McHugh CF, Scott DW, Chung YR, Kelleher N, Shaknovich R, Creasy CL, Gascoyne RD, Wong KK, Cerchietti L, Levine RL, Abdel-Wahab O, Licht JD, Elemento O, Melnick AM (2013) EZH2 is required for germinal center formation and somatic EZH2 mutations promote lymphoid transformation. Cancer Cell 23:677–692
5. Benoit YD, Laursen KB, Witherspoon MS, Lipkin SM, Gudas LJ (2013) Inhibition of PRC2 histone methyltransferase activity increases TRAIL-mediated apoptosis sensitivity in human colon cancer cells. J Cell Physiol 228:764–772

6. Benoit YD, Witherspoon MS, Laursen KB, Guezguez A, Beauséjour M, Beaulieu JF, Lipkin SM, Gudas LJ (2013) Pharmacological inhibition of polycomb repressive complex-2 activity induces apoptosis in human colon cancer stem cells. Exp Cell Res 319:1463
7. Bernstein BE, Mikkelsen TS, Xie X, Kamal M, Huebert DJ, Cuff J, Fry B, Meissner A, Wernig M, Plath K, Jaenisch R, Wagschal A, Feil R, Schreiber SL, Lander ES (2006) A bivalent chromatin structure marks key developmental genes in embryonic stem cells. Cell 125:315–326
8. Bosch A, Bertran SP, Lu Y, Garcia A, Jones AM, Dawson MI, Farias EF (2012) Reversal by RARalpha agonist Am 580 of c-Myc-induced imbalance in RARalpha/RARgamma expression during MMTV-Myc tumorigenesis. Breast Cancer Res 14:R121
9. Campos B, Wan F, Farhadi M, Ernst A, Zeppernick F, Tagscherer KE, Ahmadi R, Lohr J, Dictus C, Gdynia G, Combs SE, Goidts V, Helmke BM, Eckstein V, Roth W, Beckhove P, Lichter P, Unterberg A, Radlwimmer B, Herold-Mende C (2010) Differentiation therapy exerts antitumor effects on stem-like glioma cells. Clin Cancer Res 16:2715–2728
10. Cao R, Wang L, Wang H, Xia L, Erdjument-Bromage H, Tempst P, Jones RS, Zhang Y (2002) Role of histone H3 lysine 27 methylation in Polycomb-group silencing. Science 298:1039–1043
11. Chiba T, Suzuki E, Negishi M, Saraya A, Miyagi S, Konuma T, Tanaka S, Tada M, Kanai F, Imazeki F, Iwama A, Yokosuka O (2012) 3-Deazaneplanocin A is a promising therapeutic agent for the eradication of tumor-initiating hepatocellular carcinoma cells. Int J Cancer 130:2557–2567
12. Ciccone DN, Su H, Hevi S, Gay F, Lei H, Bajko J, Xu G, Li E, Chen T (2009) KDM1B is a histone H3K4 demethylase required to establish maternal genomic imprints. Nature 461:415–418
13. Cimino G, Lo-Coco F, Fenu S, Travaglini L, Finolezzi E, Mancini M, Nanni M, Careddu A, Fazi F, Padula F, Fiorini R, Spiriti MA, Petti MC, Venditti A, Amadori S, Mandelli F, Pelicci PG, Nervi C (2006) Sequential valproic acid/all-trans retinoic acid treatment reprograms differentiation in refractory and high-risk acute myeloid leukemia. Cancer Res 66:8903–8911
14. Cohen I, Poręba E, Kamieniarz K, Schneider R (2011) Histone modifiers in cancer: friends or foes? Genes Cancer 2:631–647
15. Constância M, Angiolini E, Sandovici I, Smith P, Smith R, Kelsey G, Dean W, Ferguson-Smith A, Sibley CP, Reik W, Fowden A (2005) Adaptation of nutrient supply to fetal demand in the mouse involves interaction between the Igf2 gene and placental transporter systems. Proc Natl Acad Sci USA 102:19219–19224
16. Cortázar D, Kunz C, Selfridge J, Lettieri T, Saito Y, MacDougall E, Wirz A, Schuermann D, Jacobs AL, Siegrist F, Steinacher R, Jiricny J, Bird A, Schär P (2011) Embryonic lethal phenotype reveals a function of TDG in maintaining epigenetic stability. Nature 470:419–423
17. Cortellino S, Xu J, Sannai M, Moore R, Caretti E, Cigliano A, Le Coz M, Devarajan K, Wessels A, Soprano D, Abramowitz LK, Bartolomei MS, Rambow F, Bassi MR, Bruno T, Fanciulli M, Renner C, Klein-Szanto AJ, Matsumoto Y, Kobi D, Davidson I, Alberti C, Larue L, Bellacosa A (2011) Thymine DNA glycosylase is essential for active DNA demethylation by linked deamination-base excision repair. Cell 146:67–79
18. Coskun V, Tsoa R, Sun YE (2012) Epigenetic regulation of stem cells differentiating along the neural lineage. Curr Opin Neurobiol 22:762–767
19. Crea F, Hurt EM, Mathews LA, Cabarcas SM, Sun L, Marquez VE, Danesi R, Farrar WL (2011) Pharmacologic disruption of polycomb repressive complex 2 inhibits tumorigenicity and tumor progression in prostate cancer. Mol Cancer 10:40
20. Creyghton MP, Cheng AW, Welstead GG, Kooistra T, Carey BW, Steine EJ, Hanna J, Lodato MA, Frampton GM, Sharp PA, Boyer LA, Young RA, Jaenisch R (2010) Histone H3 K27ac separates active from poised enhancers and predicts developmental state. Proc Natl Acad Sci USA 107:21931–21936
21. David KA, Mongan NP, Smith C, Gudas LJ, Nanus DM (2010) Phase I trial of ATRA-IV and depakote in patients with advanced solid tumor malignancies. Cancer Biol Ther 9:678–684

22. Dawson MA, Kouzarides T, Huntly BJ (2012) Targeting epigenetic readers in cancer. N Engl J Med 367:647–657
23. Feng D, Cao Z, Li C, Zhang L, Zhou Y, Ma J, Liu R, Zhou H, Zhao W, Wei H, Ling B (2012) Combination of valproic acid and ATRA restores RARβ2 expression and induces differentiation in cervical cancer through the PI3K/Akt pathway. Curr Mol Med 12:342–354
24. Fiskus W, Buckley K, Rao R, Mandawat A, Yang Y, Joshi R, Wang Y, Balusu R, Chen J, Koul S, Joshi A, Upadhyay S, Atadja P, Bhalla KN (2009) Panobinostat treatment depletes EZH2 and DNMT1 levels and enhances decitabine mediated de-repression of JunB and loss of survival of human acute leukemia cells. Cancer Biol Ther 8:939–950
25. Fuks F, Burgers WA, Godin N, Kasai M, Kouzarides T (2001) Dnmt3a binds deacetylases and is recruited by a sequence-specific repressor to silence transcription. EMBO J 20:2536–2544
26. Gan CP, Hamid S, Hor SY, Zain RB, Ismail SM, Wan Mustafa WM, Teo SH, Saunders N, Cheong SC (2012) Valproic acid: growth inhibition of head and neck cancer by induction of terminal differentiation and senescence. Head Neck 34:344–353
27. Gillespie RF, Gudas LJ (2007) Retinoic acid receptor isotype specificity in F9 teratocarcinoma stem cells results from the differential recruitment of coregulators to retinoic response elements. J Biol Chem 282:33421–33434
28. Gillespie RF, Gudas LJ (2007) Retinoid regulated association of transcriptional co-regulators and the polycomb group protein SUZ12 with the retinoic acid response elements of Hoxa1, RARbeta(2), and Cyp26A1 in F9 embryonal carcinoma cells. J Mol Biol 372:298–316
29. Glasow A, Barrett A, Petrie K, Gupta R, Boix-Chornet M, Zhou DC, Grimwade D, Gallagher R, von Lindern M, Waxman S, Enver T, Hildebrandt G, Zelent A (2008) DNA methylation-independent loss of RARA gene expression in acute myeloid leukemia. Blood 111:2374–2377
30. Griffith JS, Mahler HR (1969) DNA ticketing theory of memory. Nature 223:580–582
31. Guillemette B, Drogaris P, Lin HH, Armstrong H, Hiragami-Hamada K, Imhof A, Bonneil E, Thibault P, Verreault A, Festenstein RJ (2011) H3 lysine 4 is acetylated at active gene promoters and is regulated by H3 lysine 4 methylation. PLoS Genet 7:e1001354
32. Guo JU, Su Y, Zhong C, Ming GL, Song H (2011) Hydroxylation of 5-methylcytosine by TET1 promotes active DNA demethylation in the adult brain. Cell 145:423–434
33. Guo Y, Pino-Lagos K, Ahonen CA, Bennett KA, Wang J, Napoli JL, Blomhoff R, Sockanathan S, Chandraratna RA, Dmitrovsky E, Turk MJ, Noelle RJ (2012) A retinoic acid-rich tumor microenvironment provides clonal survival cues for tumor-specific CD8+ T cells. Cancer Res 72:5230–5239
34. Hochedlinger K, Plath K (2009) Epigenetic reprogramming and induced pluripotency. Development 136:509–523
35. Holliday R, Pugh JE (1975) DNA modification mechanisms and gene activity during development. Science 187:226–232
36. Ito S, D'Alessio AC, Taranova OV, Hong K, Sowers LC, Zhang Y (2010) Role of Tet proteins in 5mC to 5hmC conversion, ES-cell self-renewal and inner cell mass specification. Nature 466:1129–1133
37. Iwase S, Lan F, Bayliss P, de la Torre-Ubieta L, Huarte M, Qi HH, Whetstine JR, Bonni A, Roberts TM, Shi Y (2007) The X-linked mental retardation gene SMCX/JARID1C defines a family of histone H3 lysine 4 demethylases. Cell 128:1077–1088
38. Jenuwein T, Allis CD (2001) Translating the histone code. Science 293:1074–1080
39. Kamminga LM, Bystrykh LV, de Boer A, Houwer S, Douma J, Weersing E, Dontje B, de Haan G (2006) The Polycomb group gene Ezh2 prevents hematopoietic stem cell exhaustion. Blood 107:2170–2179
40. Kashyap V, Gudas LJ (2010) Epigenetic regulatory mechanisms distinguish retinoic acid-mediated transcriptional responses in stem cells and fibroblasts. J Biol Chem 285:14534–14548

41. Kashyap V, Gudas LJ, Brenet F, Funk P, Viale A, Scandura JM (2011) Epigenomic reorganization of the clustered Hox genes in embryonic stem cells induced by retinoic acid. J Biol Chem 286:3250–3260
42. Kashyap V, Laursen KB, Brenet F, Viale AJ, Scandura JM, Gudas LJ (2013) RARγ is essential for retinoic acid induced chromatin remodeling and transcriptional activation in embryonic stem cells. J Cell Sci 126:999–1008
43. Kato Y, Salumbides BC, Wang XF, Qian DZ, Williams S, Wei Y, Sanni TB, Atadja P, Pili R (2007) Antitumor effect of the histone deacetylase inhibitor LAQ824 in combination with 13-cis-retinoic acid in human malignant melanoma. Mol Cancer Ther 6:70–81
44. Kim JY, Kim KB, Son HJ, Chae YC, Oh ST, Kim DW, Pak JH, Seo SB (2012) H3K27 methylation and H3S28 phosphorylation-dependent transcriptional regulation by INHAT subunit SET/TAF-Iβ. FEBS Lett 586:3159–3165
45. Knutson SK, Wigle TJ, Warholic NM, Sneeringer CJ, Allain CJ, Klaus CR, Sacks JD, Raimondi A, Majer CR, Song J, Scott MP, Jin L, Smith JJ, Olhava EJ, Chesworth R, Moyer MP, Richon VM, Copeland RA, Keilhack H, Pollock RM, Kuntz KW (2012) A selective inhibitor of EZH2 blocks H3K27 methylation and kills mutant lymphoma cells. Nat Chem Biol 8:890–896
46. Kouzarides T (2007) Chromatin modifications and their function. Cell 128:693–705
47. Kriaucionis S, Heintz N (2009) The nuclear DNA base 5-hydroxymethylcytosine is present in Purkinje neurons and the brain. Science 324:929–930
48. Kulis M, Esteller M (2010) DNA methylation and cancer. Adv Genet 70:27–56
49. Kuzmichev A, Nishioka K, Erdjument-Bromage H, Tempst P, Reinberg D (2002) Histone methyltransferase activity associated with a human multiprotein complex containing the enhancer of zeste protein. Genes Dev 16:2893–2905
50. Langston AW, Thompson JR, Gudas LJ (1997) Retinoic acid-responsive enhancers located 3' of the Hox A and Hox B homeobox gene clusters. Functional analysis. J Biol Chem 272:2167–2175
51. Laursen KB, Mongan NP, Zhuang Y, Ng MM, Benoit YD, Gudas LJ (2013) Polycomb recruitment attenuates retinoic acid-induced transcription of the bivalent NR2F1 gene. Nucleic Acids Res 41:6430–6443
52. Laursen KB, Wong PM, Gudas LJ (2012) Epigenetic regulation by RARalpha maintains ligand-independent transcriptional activity. Nucleic Acids Res 40:102–115
53. Lee MG, Villa R, Trojer P, Norman J, Yan KP, Reinberg D, Di Croce L, Shiekhattar R (2007) Demethylation of H3K27 regulates polycomb recruitment and H2A ubiquitination. Science 318:447–450
54. Leiva M, Moretti S, Soilihi H, Pallavicini I, Peres L, Mercurio C, Dal Zuffo R, Minucci S, de Thé H (2012) Valproic acid induces differentiation and transient tumor regression, but spares leukemia-initiating activity in mouse models of APL. Leukemia 26:1630–1637
55. Lübbert M, Rüter BH, Claus R, Schmoor C, Schmid M, Germing U, Kuendgen A, Rethwisch V, Ganser A, Platzbecker U, Galm O, Brugger W, Heil G, Hackanson B, Deschler B, Döhner K, Hagemeijer A, Wijermans PW, Döhner H (2012) A multicenter phase II trial of decitabine as first-line treatment for older patients with acute myeloid leukemia judged unfit for induction chemotherapy. Haematologica 97:393–401
56. Mark M, Ghyselinck NB, Chambon P (2006) Function of retinoid nuclear receptors: lessons from genetic and pharmacological dissections of the retinoic acid signaling pathway during mouse embryogenesis. Annu Rev Pharmacol Toxicol 46:451–480
57. McCabe MT, Ott HM, Ganji G, Korenchuk S, Thompson C, Van Aller GS, Liu Y, Graves AP, Della Pietra A, Diaz E, LaFrance LV, Mellinger M, Duquenne C, Tian X, Kruger RG, McHugh CF, Brandt M, Miller WH, Dhanak D, Verma SK, Tummino PJ, Creasy CL (2012) EZH2 inhibition as a therapeutic strategy for lymphoma with EZH2-activating mutations. Nature 492:108–112
58. Mills AA (2010) Throwing the cancer switch: reciprocal roles of polycomb and trithorax proteins. Nat Rev Cancer 10:669–682
59. Min J, Zhang Y, Xu RM (2003) Structural basis for specific binding of Polycomb chromodomain to histone H3 methylated at Lys 27. Genes Dev 17:1823–1828

60. Mongan NP, Gudas LJ (2005) Valproic acid, in combination with all-trans retinoic acid and 5-aza-2′-deoxycytidine, restores expression of silenced RARbeta2 in breast cancer cells. Mol Cancer Ther 4:477–486
61. Morgan HD, Dean W, Coker HA, Reik W, Petersen-Mahrt SK (2004) Activation-induced cytidine deaminase deaminates 5-methylcytosine in DNA and is expressed in pluripotent tissues: implications for epigenetic reprogramming. J Biol Chem 279:52353–52360
62. Neff T, Sinha AU, Kluk MJ, Zhu N, Khattab MH, Stein L, Xie H, Orkin SH, Armstrong SA (2012) Polycomb repressive complex 2 is required for MLL-AF9 leukemia. Proc Natl Acad Sci USA 109:5028–5033
63. Pan G, Tian S, Nie J, Yang C, Ruotti V, Wei H, Jonsdottir GA, Stewart R, Thomson JA (2007) Whole-genome analysis of histone H3 lysine 4 and lysine 27 methylation in human embryonic stem cells. Cell Stem Cell 1:299–312
64. Pasini D, Hansen KH, Christensen J, Agger K, Cloos PA, Helin K (2008) Coordinated regulation of transcriptional repression by the RBP2 H3K4 demethylase and polycomb-repressive complex 2. Genes Dev 22:1345–1355
65. Pasini D, Malatesta M, Jung HR, Walfridsson J, Willer A, Olsson L, Skotte J, Wutz A, Porse B, Jensen ON, Helin K (2010) Characterization of an antagonistic switch between histone H3 lysine 27 methylation and acetylation in the transcriptional regulation of Polycomb group target genes. Nucleic Acids Res 38:4958–4969
66. Phillips DM (1963) The presence of acetyl groups of histones. Biochem J 87:258–263
67. Pili R, Salumbides B, Zhao M, Altiok S, Qian D, Zwiebel J, Carducci MA, Rudek MA (2012) Phase I study of the histone deacetylase inhibitor entinostat in combination with 13-cis retinoic acid in patients with solid tumours. Br J Cancer 106:77–84
68. Prickett AR, Oakey RJ (2012) A survey of tissue-specific genomic imprinting in mammals. Mol Genet Genomics 287:621–630
69. Qi H, Ratnam M (2006) Synergistic induction of folate receptor beta by all-trans retinoic acid and histone deacetylase inhibitors in acute myelogenous leukemia cells: mechanism and utility in enhancing selective growth inhibition by antifolates. Cancer Res 66:5875–5882
70. Rada-Iglesias A, Bajpai R, Swigut T, Brugmann SA, Flynn RA, Wysocka J (2011) A unique chromatin signature uncovers early developmental enhancers in humans. Nature 470:279–283
71. Rai K, Sarkar S, Broadbent TJ, Voas M, Grossmann KF, Nadauld LD, Dehghanizadeh S, Hagos FT, Li Y, Toth RK, Chidester S, Bahr TM, Johnson WE, Sklow B, Burt R, Cairns BR, Jones DA (2010) DNA demethylase activity maintains intestinal cells in an undifferentiated state following loss of APC. Cell 142:930–942
72. Raman JD, Mongan NP, Tickoo SK, Boorjian SA, Scherr DS, Gudas LJ (2005) Increased expression of the polycomb group gene, EZH2, in transitional cell carcinoma of the bladder. Clin Cancer Res 11:8570–8576
73. Riggs AD (1975) X inactivation, differentiation, and DNA methylation. Cytogenet Cell Genet 14:9–25
74. Riggs AD, Matienssen RA, Russo VEA (1996) Epigenetic mechanisms of gene regulation. Cold Spring Harbor Laboratory Press, Cold Spring Harbor
75. Rodríguez-Paredes M, Esteller M (2011) Cancer epigenetics reaches mainstream oncology. Nat Med 17:330–339
76. Sasai N, Defossez PA (2009) Many paths to one goal? The proteins that recognize methylated DNA in eukaryotes. Int J Dev Biol 53:323–334
77. Savickiene J, Treigyte G, Jazdauskaite A, Borutinskaite VV, Navakauskiene R (2012) DNA methyltransferase inhibitor RG108 and histone deacetylase inhibitors cooperate to enhance NB4 cell differentiation and E-cadherin re-expression by chromatin remodelling. Cell Biol Int 36:1067–1078
78. Schenk T, Chen WC, Göllner S, Howell L, Jin L, Hebestreit K, Klein HU, Popescu AC, Burnett A, Mills K, Casero RA, Marton L, Woster P, Minden MD, Dugas M, Wang JC, Dick JE, Müller-Tidow C, Petrie K, Zelent A (2012) Inhibition of the LSD1 (KDM1A) demethylase reactivates the all-trans-retinoic acid differentiation pathway in acute myeloid leukemia. Nat Med 18:605–611

79. Schneider R, Bannister AJ, Myers FA, Thorne AW, Crane-Robinson C, Kouzarides T (2004) Histone H3 lysine 4 methylation patterns in higher eukaryotic genes. Nat Cell Biol 6:73–77
80. Stock JK, Giadrossi S, Casanova M, Brookes E, Vidal M, Koseki H, Brockdorff N, Fisher AG, Pombo A (2007) Ring1-mediated ubiquitination of H2A restrains poised RNA polymerase II at bivalent genes in mouse ES cells. Nat Cell Biol 9:1428–1435
81. Tahiliani M, Koh KP, Shen Y, Pastor WA, Bandukwala H, Brudno Y, Agarwal S, Iyer LM, Liu DR, Aravind L, Rao A (2009) Conversion of 5-methylcytosine to 5-hydroxymethylcytosine in mammalian DNA by MLL partner TET1. Science 324:930–935
82. Tan M, Luo H, Lee S, Jin F, Yang JS, Montellier E, Buchou T, Cheng Z, Rousseaux S, Rajagopal N, Lu Z, Ye Z, Zhu Q, Wysocka J, Ye Y, Khochbin S, Ren B, Zhao Y (2011) Identification of 67 histone marks and histone lysine crotonylation as a new type of histone modification. Cell 146:1016–1028
83. Tanaka S, Miyagi S, Sashida G, Chiba T, Yuan J, Mochizuki-Kashio M, Suzuki Y, Sugano S, Nakaseko C, Yokote K, Koseki H, Iwama A (2012) Ezh2 augments leukemogenicity by reinforcing differentiation blockage in acute myeloid leukemia. Blood 120:1107–1117
84. Tang XH, Albert M, Scognamiglio T, Gudas LJ (2009) A DNA methyltransferase inhibitor and all-trans retinoic acid reduce oral cavity carcinogenesis induced by the carcinogen 4-nitroquinoline 1-oxide. Cancer Prev Res (Phila) 2:1100–1110
85. Tang XH, Gudas LJ (2011) Retinoids, retinoic acid receptors, and cancer. Annu Rev Pathol 6:345–364
86. Tavares TS, Nanus D, Yang XJ, Gudas LJ (2008) Gene microarray analysis of human renal cell carcinoma: the effects of HDAC inhibition and retinoid treatment. Cancer Biol Ther 7:1607–1618
87. Tie F, Banerjee R, Conrad PA, Scacheri PC, Harte PJ (2012) Histone demethylase UTX and chromatin remodeler BRM bind directly to CBP and modulate acetylation of histone H3 lysine 27. Mol Cell Biol 32:2323–2334
88. Tollervey J, Lunyak VV (2012) Epigenetics: Judge, jury and executioner of stem cell fate. Epigenetics 7:823–840
89. Touma SE, Goldberg JS, Moench P, Guo X, Tickoo SK, Gudas LJ, Nanus DM (2005) Retinoic acid and the histone deacetylase inhibitor trichostatin a inhibit the proliferation of human renal cell carcinoma in a xenograft tumor model. Clin Cancer Res 11:3558–3566
90. Trus MR, Yang L, Suarez Saiz F, Bordeleau L, Jurisica I, Minden MD (2005) The histone deacetylase inhibitor valproic acid alters sensitivity towards all trans retinoic acid in acute myeloblastic leukemia cells. Leukemia 19:1161–1168
91. Tsukada Y, Fang J, Erdjument-Bromage H, Warren ME, Borchers CH, Tempst P, Zhang Y (2006) Histone demethylation by a family of JmjC domain-containing proteins. Nature 439:811–816
92. Um S, Harbers M, Benecke A, Pierrat B, Losson R, Chambon P (1998) Retinoic acid receptors interact physically and functionally with the T: G mismatch-specific thymine-DNA glycosylase. J Biol Chem 273:20728–20736
93. Verdone L, Agricola E, Caserta M, Di Mauro E (2006) Histone acetylation in gene regulation. Brief Funct Genomic Proteomic 5:209–221
94. Voigt P, Leroy G, Drury WJ, Zee BM, Son J, Beck DB, Young NL, Garcia BA, Reinberg D (2012) Asymmetrically modified nucleosomes. Cell 151:181–193
95. Voss AK, Collin C, Dixon MP, Thomas T (2009) Moz and retinoic acid coordinately regulate H3K9 acetylation, Hox gene expression, and segment identity. Dev Cell 17:674–686
96. Waddington CH (1957) The strategy of genes. Allen & Unwin, London
97. Waddington CH (1968) Towards a Theoretical Biology. Edinburgh University Press, Edinburgh, Scotland
98. Wang XF, Qian DZ, Ren M, Kato Y, Wei Y, Zhang L, Fansler Z, Clark D, Nakanishi O, Pili R (2005) Epigenetic modulation of retinoic acid receptor beta2 by the histone deacetylase inhibitor MS-275 in human renal cell carcinoma. Clin Cancer Res 11:3535–3542
99. Yang QJ, Zhou LY, Mu YQ, Zhou QX, Luo JY, Cheng L, Deng ZL, He TC, Haydon RC, He BC (2012) All-trans retinoic acid inhibits tumor growth of human osteosarcoma by activating Smad signaling-induced osteogenic differentiation. Int J Oncol 41:153–160

100. Yang XJ (2005) Multisite protein modification and intramolecular signaling. Oncogene 24:1653–1662
101. Yuan W, Wu T, Fu H, Dai C, Wu H, Liu N, Li X, Xu M, Zhang Z, Niu T, Han Z, Chai J, Zhou XJ, Gao S, Zhu B (2012) Dense chromatin activates Polycomb repressive complex 2 to regulate H3 lysine 27 methylation. Science 337:971–975
102. Zhang L, Ren X, Alt E, Bai X, Huang S, Xu Z, Lynch PM, Moyer MP, Wen XF, Wu X (2010) Chemoprevention of colorectal cancer by targeting APC-deficient cells for apoptosis. Nature 464:1058–1061
103. Zhou J, Bi C, Cheong LL, Mahara S, Liu SC, Tay KG, Koh TL, Yu Q, Chng WJ (2011) The histone methyltransferase inhibitor, DZNep, up-regulates TXNIP, increases ROS production, and targets leukemia cells in AML. Blood 118:2830–2839
104. Zhou W, Chen H, Zhang L (2009) The PcG protein hPc2 interacts with the N-terminus of histone demethylase JARID1B and acts as a transcriptional co-repressor. BMB Rep 42:154–159
105. Zhu JK (2009) Active DNA demethylation mediated by DNA glycosylases. Annu Rev Genet 43:143–166

Chapter 8
RARs and MicroRNAs

Clara Nervi and Francesco Grignani

Abstract MicroRNAs (miRNAs) are small noncoding RNAs acting as endogenous regulators of gene expression. Their discovery is one of the major recent breakthroughs in molecular biology. miRNAs establish a multiplicity of relationships with target mRNAs and exert pleiotropic biological effects in many cell physiological pathways during development and adult life. The dynamic nature of gene expression regulation by Retinoic Acid (RA) is consistent with an extensive functional interplay with miRNA activities. In fact, RA regulates the expression of many different miRNAs, thus suggesting a relevant function of miRNAs in RA-controlled gene expression programmes. miRNAs have been extensively studied as targets and mediators of the biological activity of RA during embryonic development as well as in normal and neoplastic cells. However, relatively few studies have experimentally explored the direct contribution of miRNA function to the RA signalling pathway. Here, we provide an overview of the mechanistic aspects that allow miRNA biogenesis, functional activation and regulation, focusing on recent evidence that highlights a functional interplay between miRNAs and RA-regulated molecular networks. We report examples of tissue-specific roles of miRNAs modulated by RA in stem cell pluripotency maintenance and regeneration, embryonic development, hematopoietic and neural differentiation, and other biological model systems, underlining their role in disease pathogenesis. We also address novel areas of research linking the RA signalling pathway to the nuclear activity of miRNAs.

C. Nervi (✉)
Department of Medical-Surgical Sciences and Biotechnologies, University "La Sapienza", Rome, Italy
e-mail: clara.nervi@uniroma1.it

F. Grignani
Department of Clinical and Experimental Medicine, General Pathology, University of Perugia, Perugia, Italy
e-mail: francesco.grignani@unipg.it

M. A. Asson-Batres and C. Rochette-Egly (eds.), *The Biochemistry of Retinoic Acid Receptors I: Structure, Activation, and Function at the Molecular Level*, Subcellular Biochemistry 70,
DOI: 10.1007/978-94-017-9050-5_8, © Springer Science+Business Media Dordrecht 2014

Abbreviations

Ago	Argonaute
ALDH1A2	Aldehyde dehydrogenase-1a2
AML	Acute myeloid leukemia
APL	Acute promyelocytic leukemia
C/EBPα	CCAAT/enhancer-binding protein-alpha
ceRNAs	Competing endogenous RNAs
DGCR8	DiGeorge syndrome critical region gene 8
ESC	Embryonic stem cells
HDAC	Histone deacetylase
Hox	Homeobox
HPCs	Hematopoietic progenitor cells
HSC	Hematopoietic stem cells
lncRNA	Long non-coding RNA
miRNAs	MicroRNAs
NFI-A	Nuclear factor I-A
PACT	Protein kinase R (PKR) activator
P-bodies	Processing bodies
PcG	Polycomb group proteins
Pitx3	Paired-like homeodomain transcription factor 3
pre-miRNA	MicroRNA precursor
pri-miRNA	Primary microRNA
PTB	Polypyrimidine tract binding protein
RA	Retinoic Acid
RALDH2	Retinaldehyde dehydrogenase 2
RAR	Retinoic acid receptor
RISC	RNA-induced silencing complex
RNA polymerase II	RNA polII
RXR	Retinoid X receptor
TF	Transcription factor
TRBP	Trans-activation response (TAR) RNA-binding protein
UTR	Untranslated region

It is now clear an extensive miRNA world was flying almost unseen by our genetic radar. As much as geneticists like to think that nothing can escape genetic analysis, the miRNA genes are so small that they almost escaped our notice [124].

History

The hypothesis of genomically-encoded regulators of protein expression acting in the cytoplasm as intermediate molecules between mRNA and proteins was formulated more than 50 years ago by Jacob and Monod. In a seminal paper they affirm that:

> ...the most firmly grounded of these conclusions is the existence of **regulator** genes, which control the rate of information-transfer from **structural** genes to proteins...the regulator gene acts via a specific cytoplasmic substance whose effect is to **inhibit** the expression of the structural genes ...the chemical identification of the **repressor as an RNA fraction** is a logical assumption. [62]

Thus, Jacob and Monod had already envisaged the existence of non-coding regulatory RNAs. More than 30 years later, in 1993, Victor Ambros' laboratory cloned the first microRNA (miRNA), *lin-4,* in *Caenorhabditis elegans*, and identified *Lin14* as one of its target mRNAs [81]. Seven years later, Gary Ruvkun's group cloned the second miRNA, *let-7*, and reported the conservation of miRNAs across species [117, 122].

The canonical miRNA genes, *lin-4* and *let-7*, were identified by genetic screening of mutant phenotypes that caused developmental timing defects in stage specific lineages of worm larvae [81, 122]. Since then, genetic cloning, bioinformatics, gene expression analysis and computational algorithms defining phylogenetic conservation and structural characteristics of miRNA precursors have allowed large-scale identification of miRNAs from various life forms, such as plants, *Drosophila melanogaster,* and vertebrates, including mammals. Animal viruses also use miRNA [2, 9, 77].

Global gene expression analyses have shown that miRNAs and their mRNA targets often have mutually exclusive expression in contiguous developmental stages or across tissues [38, 134, 137]. miRNAs participate in transcriptional programs that control development, cellular pluripotency, and differentiation. They affect expression levels of lineage-specific TFs in integrated transcriptional regulatory circuits, and, in turn, are regulated by the activities of these factors [39, 114, 151].

miRNAs are well-preserved in blood plasma or serum, urine, and formalin-fixed tissue blocks, and can be measured with much greater sensitivity than proteins [79, 103]. miRNA expression profiling is now considered to be a relevant approach in a broad range of biological and medical studies and may have many molecular diagnostic and therapeutic applications.

Development of the field

At present, the miRBase release 20 (June 2013) (http://www.mirbase.org/) reports 24,521 entries of hairpin precursor miRNAs which give rise to 30,424 mature miRNA products. MiRBase estimates 1872 precursors and 2578 mature miRNAs in humans. Well over half of the human transcriptome is computationally predicted to be subjected to miRNA regulation [43]. Expression analysis of miRNAs by commercially available oligonucleotide microarrays or quantitative polymerase chain reaction (qPCR) arrays are now powerful approaches for monitoring tissue specific, developmental, physiological and disease state miRNA expression across the whole genome.

The mechanisms related to miRNA biogenesis have been revealed in detail. Genes encoding miRNAs are an integral component of a cell's genomic program, and are generally conserved through evolution [1, 38]. They can be encoded in intergenic transcription units, in polycistronic clusters, within intronic sequences of protein-coding genes, or within the introns and exons of non-coding RNAs [1, 72, 85, 123]. For genes sharing the same promoters, the "host" transcript and miRNAs usually have similar expression profiles [5, 123]. Moreover, sequence variations in miRNA genes potentially influence the processing and/or target recognition of miRNAs [101]. miRNA gene mutation or mis-expression have been observed in various human cancers and indicate that miRNAs can function as tumor suppressors and oncogenes [35].

Most miRNA genes are initially transcribed by RNA polymerase II (Pol II) as long primary transcripts (pri-miRNA) that contain one or more hairpin-shaped structures (Fig. 8.1). The primary transcripts contain a 5′ cap and a poly(A) tail similar to that of mRNAs [15, 86]. RNA polymerase III can drive miRNA transcription from dense human clusters interspersed among repetitive Alu elements [13]. In the nucleus, the pri-miRNA is cleaved by a microprocessor complex consisting of the RNase III enzyme, Drosha, and the double-stranded RNA binding protein "DiGeorge syndrome critical region gene 8″ (DGCR8, also known as Pasha). Drosha and DGCR8 release an imperfect hairpin-structured precursor miRNA (pre-miRNA) of about 70 nucleotides [28, 55, 83, 85].

The Drosha complex includes several auxiliary proteins, such as RNA binding protein fused in sarcoma (FUS), Ewing sarcoma breakpoint region 1 (EWSR1), heterogeneous nuclear ribonucleoproteins (hnRNP), and DEAD (Asp-Glu-Ala-Asp) box helicases p68 (DDX5) and p72 (DDX17) [49]. Whereas some hnRNPs and p68/72 promote the fidelity and activity of Drosha processing, the exact role of these proteins in fine regulation of miRNAs expression is still largely unknown [50].

The Drosha-processed pre-miRNA is exported from the nucleus to the cytoplasm by the Exportin-5 [12, 93, 166]. The Exportin-5 recognizes the 2-nucleotide 3′ overhang structure and the double-stranded stem of the pre-miRNA [109]. In the cytoplasm, the pre-miRNA is captured by a second RNase III enzyme, Dicer. Dicer and its related double-stranded RNA binding partners, the transactivating response RNA-binding protein and protein kinase R activator, cleave the pre-miRNA at the stem-loop junction [20, 52, 61, 84, 95, 125]. This generates a miRNA:miRNA duplex of approximately 22 nucleotides with overhangs of 2-nucleotides at their 3′ ends. Generally, the strand whose 5′ end has a less tight base pairing (also known as the "*guide*" strand) is favored for incorporation into a functional protein complex called "*RNA-induced silencing complex*" (RISC), which is active in the repression of mRNA function. Within this riboprotein complex, miRNAs are driven to the target mRNA [129, 133]. The other miRNA strand [also known as "*passenger*" strand or star miRNA (miRNA*)] is usually degraded [100], although in some physiological conditions both strands of the miRNA duplex can be detectable in the RISC [111].

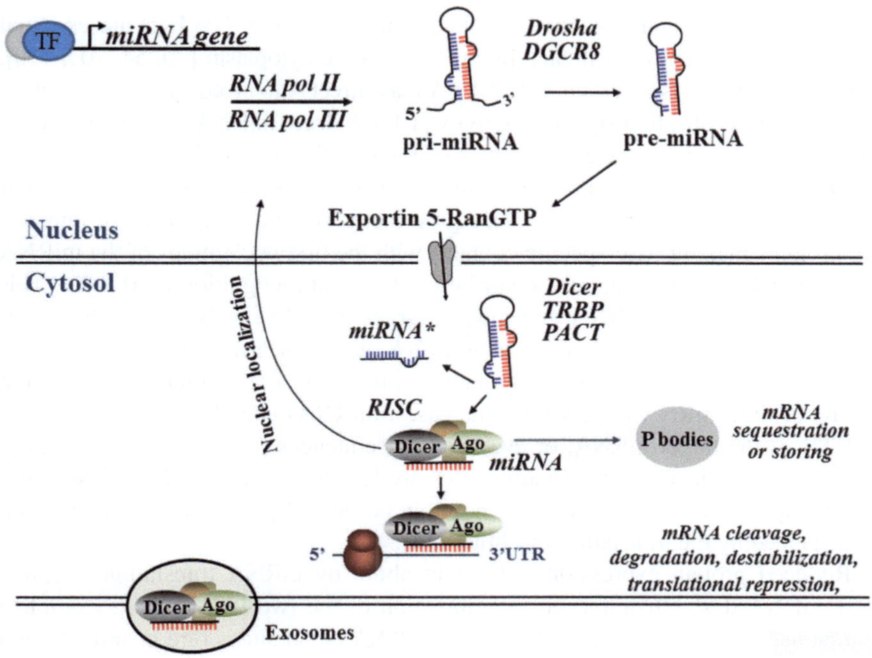

Fig. 8.1 miRNA biogenesis and function. miRNA genes are regulated by transcription factors (*TF*) and transcribed by RNA pol II and III into primary miRNA transcripts (*pri-miRNAs*). The pri-miRNA is processed into a precursor miRNA (*pre-miRNA*) stem-loop of ~70 nucleotides in length by the nuclear RNase III enzyme Drosha and its partner DiGeorge syndrome critical region gene 8 (*DGCR8*). Exportin-5 actively transports the pre-miRNA into the cytosol, where it is processed by the Dicer RNaseIII enzyme and its partner TRBP, into a double strand miRNA. The miRNA strand (in *red*) is recruited as a single-stranded molecule into the RNA-induced silencing (*RISC*) effector complex and assembled through processes that are dependent on Dicer and other double strand RNA binding domain proteins (*dsRBD*), as well as on members of the Argonaute (*AGO*) family. The other strand (miRNA*) is usually degraded. Mature miRNA guides the RISC complex to the 3′UTR of the complementary mRNA targets and repress their expression by several mechanisms: repression of mRNA translation, destabilization of mRNA transcripts through cleavage, de-adenylation, and localization in the processing body (*P-body*), where the miRNA-targeted mRNA can be sequestered from the translational machinery and degraded or stored for subsequent use. Nuclear localization of mature miRNAs has been described as a novel mechanism of action for miRNAs

The RISC contains a number of different proteins, including Argonaute (Ago) and Dicer proteins (Fig. 8.1). In humans, there is only one Dicer and four Ago proteins (Ago 1, 2, 3, and 4) [102]. Human Ago2 is the only Ago protein embodying the endonucleolytic "*slicer*" activity of RISC which catalyzes the cleavage of target mRNA [132]. The structure of human Ago2 in complex with miRNA-20A was recently determined. The RNA confers remarkable stability to the enzyme, most likely as a consequence of multiple interactions that spread along the entire protein [31]. All four Ago proteins and other proteins, such as a trinucleotide repeat containing 6 A, B and C subunits (TNRC6, also known as GW182) are essential in the silencing process facilitated by miRNAs in the cytoplasm [20, 54, 102, 110]. Some evidence also suggests that TNRC6A may direct Ago proteins into the nucleus via a nuclear localization signaling activity [107]. Indeed, miRNAs and Ago proteins have been found in the nucleus of human cells (see below) [8, 168, 169].

Mature miRNAs serve as guides, directing RISC to mRNAs containing their complementary sequences (Fig. 8.1). Ago proteins serve as platforms for interacting proteins making base-specific contacts with the first nucleotides of the miRNA guide strand and, consequently, contributing to target recognition [110, 102]. This recognition between miRNA and mRNA targets mainly involves a limited base-pairing between the 5′-end '*seed*' region (2–8 nucleotides from the 5′-end) of the miRNA and the complementary sequences present in the 3′ untranslated region (UTR) of their mRNA targets. In some cases, miRNAs can efficiently bind to the 5′ UTR of the target mRNA, or even coding sequences [4, 73, 94, 80]; however, these non-canonical miRNA binding sites are less frequently used and less effective in suppressing mRNA function, possibly because the silencing complexes can be displaced by the translation machinery [4].

Reduced protein expression is brought about by mRNA translational repression, mRNA destabilization, or a combination of the two [37, 81, 90]. Complete complementarity between a miRNA and a mRNA target site is rare in animals, and cleavage of a target mRNA can only occur if a catalytically active Ago is bound [37]. When targeted for silencing by miRNAs, mRNA can be sequestered from the translational machinery, degraded or stored in large cytoplasmic foci, named processing bodies (P-bodies). The P-bodies contain a wide range of enzymes involved in RNA turnover [36]. Recent evidence indicates mRNA degradation is the major determinant of miRNA activity, being responsible for >84 % of the effects on protein expression [51]. However, in some model systems a block of translation initiation may precede mRNA decay and mediate the miRNA repressor activity [7]. Evidence also indicates that miRNAs induce translational up-regulation of target mRNAs [37, 152].

The regulatory potential of each miRNA is rendered even more complex as indicated by recent findings showing that miRNAs can also display a decoy activity that interferes with the function of regulatory proteins [30]. Moreover, endogenous RNA, and long non-coding RNA (lncRNA) complementary to miRNAs, can act as competing endogenous RNAs (ceRNAs) in both normal and pathological conditions. By binding a common pool of miRNAs, ceRNAs prevent them from binding their mRNA targets, thus blocking their activities [18, 68, 140, 146]. Overall, these studies reveal a new level of regulation of gene expression, controlled by an extensive interacting network of both coding RNAs, lncRNAs and miRNAs, which can be predicted on the basis of the overlap of miRNA-binding sites [67].

Recently it has been shown that miRNAs are also secreted from cells through the exosomal pathway, suggesting a new potential action of miRNA in cell-to-cell communication during the complex events that regulate development and differentiation [108].

Current State of the Field

MicroRNA and Retinoic Acid Regulatory Networks in Embryo Development

Since their discovery in *Caenorhabditis elegans* mutants, miRNA activities have consistently turned up as players in cell fate determination in other animals, including mammals [81, 122]. Interestingly, RA-regulated miRNAs are now known to influence mammalian development and embryonic stem cell (ESC) self-renewal and differentiation.

RA and miRNAs may act through common transcriptional pathways to regulate the balance between ESC self-renewal or differentiation capacity. miR-145 expression can be induced by RA treatment of human ESC [63], leading to suppression of mRNAs encoding the transcription factors, octamer-binding transcription factor 4 (Oct4), Sox2, and Kruppel-like factor 4 (Klf4), which are associated with the preservation of pluripotency in stem cells. Re-expression of these factors in human somatic cells reprograms them to a pluripotent stem cell state [143, 164]. The molecular mechanism for terminating the pluripotent state of ESC is an example of the complex interactions between RA, miRNAs and TFs. In response to RA, CBP/p300 acetylates p53 at lysine 373, which leads to its dissociation from E3-ubiquitin ligases HDM2 and TRIM24 and causes p53 protein stabilization. This key step activates the production of miR-145 and miR-34a, which, leads to repression of Oct4, Klf4, Lin28a, and Sox2, which, in turn, prevents the hESC from backsliding to pluripotency [63]. Interestingly, Oct4 transcriptionally represses miR-145 [164], an indication that this circuit is under stringent reciprocal control.

However, Oct4 gene transcription is also downregulated by complex epigenetic mechanisms induced by RA as shown by studies where proliferating stem cells were exposed to RA [22, 128].

RA treatment also induces the expression of miR-134, which enhances mouse ESC differentiation along ectodermal lineages [147]. This is due, in part, to miR-134 direct translational attenuation of genes, including Nanog and LRH1, both of which act as positive regulators of Oct4/POU5F1 and ESC growth [147]. Overall, these data establish that both the RA and miRNA signalling pathways affect ESC differentiation through their potential to regulate, or target, multiple genes that play a central role in ESC maintenance and differentiation.

Homeobox Genes

Homeobox (Hox) genes are classic targets of RA signaling [74, 75, 118]. Precise temporal and spatial activation of their transcription is required for proper specification of regional identities along the body's main axes. The miR-10 and miR-196 families of miRNA genes are embedded within the vertebrate Hox clusters and their expression patterns are markedly similar to those of the Hox genes [76, 89, 97]. For example, the miR-196 gene is expressed in spatially non-overlapping domains with its conserved targets, Hoxa7, Hoxb8, Hoxc8 and Hoxd8 [97, 165]. Tabin's group made a conditional knockout of Dicer (a key enzyme required for producing functional mature miRNAs) that specifically removed Dicer from mouse limb buds. Using this approach, they found that miR-196 acts upstream of Hoxb8 and Sonic Hedgehog in mouse and chicken limb development [57]. They observed that expression of miR-196 is lower in the forelimb than in the hindlimb, where the miRNA acts as an inhibitor of Hoxb8, preventing it from being induced by RA [57]. Thus, in normal limb development, miR-196 appears to pre-empt inappropriate Hoxb8 induction by RA.

Over-expression of miR-196 alters Hox genes expression patterns that are required for the proper development of pectoral fin buds in zebrafish embryos [56]. Interestingly, this effect is a consequence of the direct activity of miR-196 on the RA signalling pathway. miR-196 targets the 3′UTR of the retinoic acid receptor ab (rarab). In fact, knocking down rarab mimicked the pectoral fin phenotype induced by miR-196 over-expression [56]. This is one of the few known examples of direct targeting of RAR transcripts by miRNAs.

miR-10 represses Hoxb1a and Hoxb3a within the spinal cord, and this repression works cooperatively with Hoxb4 [161]. Transcription of miR-10 is activated by RA. Overexpression of miR-10 induces phenotypes similar to those caused by the loss of Hoxb1a and Hoxb3a [161]. Interestingly, cluster genes of miR-10 Hox targets are located in close proximity to this miRNA gene, suggesting a coordinate wave of transcriptional and post-transcriptional regulation of Hox genes expression.

Heart and Muscle Development

The circulatory system is the first functional unit in the developing embryo, and the heart is the first functional organ. The essentiality of RA in cardiogenesis has been established by experiments demonstrating that heart-looping, development of posterior chambers, and differentiation of ventricular cardiomyocytes are all severely impaired in mice lacking RA [106] as well as by experiments showing that RA limits the cardiac progenitor pool [69].

RA and miRNAs display multiple interactions during cardiogenesis [162].

For example, disruption of miR-138 function leads to expansion of gene expression in the ventricular region that is normally restricted to the atrio-ventricular valve region. This aberrant expression of genes ultimately disrupts ventricular cardiomyocyte morphology and cardiac function [105]. In this study, miR-138 was found in specific domains of the zebrafish heart, where it was observed to repress the expression of the RA synthesis enzyme RALDH2 in the ventricle. This activity was complemented by miR-138-mediated ventricular repression of the gene encoding versican (cspg2), the core protein of the chondroitin sulfate proteoglycan, which is positively regulated by RA overall suggesting an antagonism between miR-138 and RA signalling pathway [105].

During the development of zebrafish ventricles, miR-143 expression is dependent on heartbeat. Like miR-138, miR-143 negatively controls mRNA expression of RALDH2 and retinoid x receptor alpha b (rxrab), which affect the correct development of the heart tubes. Hence, miR-143 and RA signalling pathways are targets of heartbeat-dependent physical control, highlighting heartbeat as an essential epigenetic factor during cardiogenesis [104]. Interestingly, transcriptional networks that establish heart chamber-specific gene expression are highly conserved across species, from zebrafish to humans, implying that this mechanism may be active in mammals [135].

Analysis of miRNA profiling showed that miR-10a and miR-1 expression is gradually increased during in vitro RA-induced differentiation of ESC into smooth muscle cells. Functional studies showed that, subsequently, miR-10a and miR-1 repress histone deacetylase 4 (HDAC4) and Klf4, respectively, and play critical roles in the determination of smooth muscle cell fate [59, 163].

Polycomb Group Proteins

miR-214 expression impacts transcription controlled by polycomb group proteins (PcGs) [66]. PcGs are a group of proteins that contribute to cell commitment and differentiation by repressing the transcription of genes that regulate development [82]. In undifferentiated skeletal muscle cells, PcG proteins such as Suz12-polycomb repressive complex 2 subunit, embryonic ectoderm development (Eed), and Bmi1-polycomb ring finger oncogene, occupy and repress transcription of miR-214 which is a direct target of MyoD and myogenin. PcGs are released in differentiating myoblasts and during RA-induced ESC differentiation leading to the recruitment of MyoD and myogenin to the miR-214 genomic region and subsequent transcription of miR-214. Transcribed miR-214, in turn, negatively feeds back on the enhancer of zeste homolog 2 (Ezh2) by directly inhibiting translation of its mRNA. Ezh2 is the catalytic subunit of the PcG complex that mediates histone 3 lysine 27 trimethylation (H3K27me3). Reduced levels of Ezh2 cause a derepression of developmental regulators that are PcG targets, which then leads to an accelerated differentiation of skeletal muscle cells [66]. This network may exist to increase the effectiveness of the system to rapidly reduce Ezh2 availability at critical stages, such as those regulating muscle cell differentiation.

Nervous System Development

The RA signaling pathway and miRNAs are implicated in many aspects of central nervous system development and function. Numerous miRNAs were revealed in a comprehensive analysis of RA-induced differentiation of human NT2 cells, an in

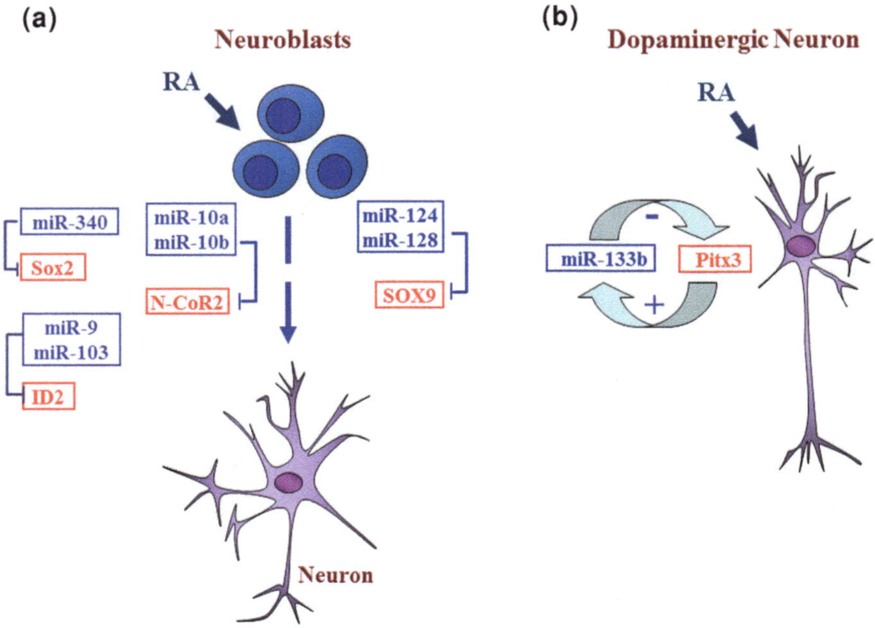

Fig. 8.2 miRNAs and neurogenesis. **a** Examples of RA-regulated miRNAs during neural differentiation (*blue*) and their targets (*red*). On the *right* miR-124 and miR-128, preferentially expressed in neurons, have been indicated as regulators of neural development and neuronal cell specification. miR-124 represses the expression of the transcription factor SOX9 that is linked to the stem cell phenotype. On the *left* miR-340, 10a, 10b, 9 and 103 act on genes involved in the maintenance of the undifferentiated state, thus favouring neural differentiation. **b** A miRNA based feed-back circuit regulates dopaminergic neurogenesis. miR-133b inhibits the expression of the transcription factor Pitx3, which, in turn, is a transcriptional inducer of miR-133b expression

vitro model of neurogenesis [131]. For example, miR-124 regulates adult mouse neurogenesis via suppression of Sox9 in the subventricular zone stem cell niche [21], miR-430 is associated with brain morphogenesis in zebrafish [47], and miR-124 and miR-128 play roles in neuronal cell specification [24, 130] (Fig. 8.2a).

miR-124 promotes neuronal differentiation by triggering brain-specific alternative pre-mRNA splicing [96]. *Splicing regulator polypyrimidine tract binding Protein 1* (PTB1) blocks the expression of *neural-specific splicing regulator* (PTB2) by introducing an alternative exon carrying a stop codon into the PTB2 transcript. miR-124 suppresses PTB1 which indirectly activates PTB2. In the presence of RA, miR-124 levels are increased, resulting in widespread splicing regulation of essential neural mRNAs that trigger neural differentiation (Fig. 8.3).

miR-133b is expressed in adult mammalian midbrain dopaminergic neurons where it regulates their maturation and function [71]. miR-133b takes part in a negative feedback circuit that includes the paired-like homeodomain transcription factor 3 (Pitx3), a known regulator of neuronal gene expression [71]. Moreover,

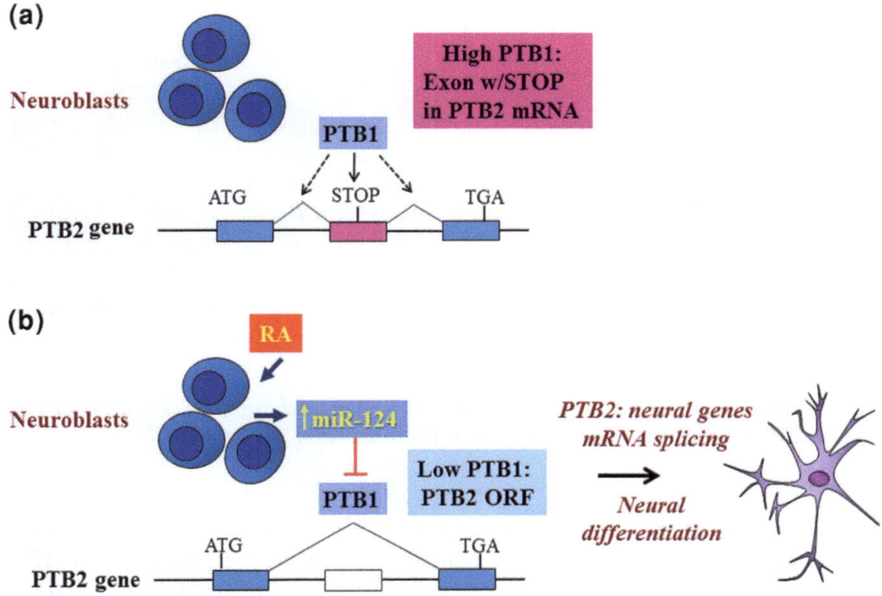

Fig. 8.3 miRNAs and neuroblasts. **a** In neuroblasts, the splicing regulator PTB1 introduces an alternative exon carrying a stop codon in the mRNA of the neural-specific splicing regulator PTB2, resulting in a block of PTB2 protein expression. **b** RA increases miR-124 levels in neuroblasts. This miRNA represses PTB1 protein expression, resulting in a change in mRNA splicing of the PTB2 gene, generating an open reading frame (*ORF*). PTB2, induces neuronal differentiation by triggering brain-specific alternative pre-mRNA splicing

miR-133b is a direct transcriptional target of Pitx3, and its induction post-transcriptionally suppresses Pitx3, resulting in a functional fine-tuning circuit for dopaminergic behaviours such as locomotion (Fig. 8.2b).

Relevance

MicroRNA and Retinoic Acid Regulatory Networks in Adult Life and Disease

Many of the differentiation promoting and growth suppressing activities of RA appear to utilize miRNAs as fundamental components of their molecular mechanisms. Most of the data on RA-regulated miRNAs during adulthood derive from two biological processes regulated by RA: hematopoiesis and neuronal differentiation. Research in these two fields is relevant to two neoplastic diseases, acute promyelocytic leukemia (APL) and neuroblastoma, where patients

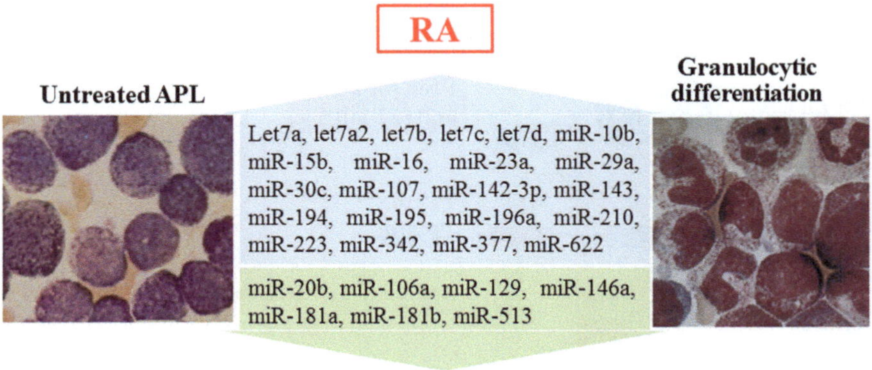

Fig. 8.4 miRNAs that are either up-regulated (*azure box*) or down-regulated (*green box*) in APL blasts undergoing granulocytic differentiation by RA treatment

being treated with RA have shown a notable clinical response. The introduction of all-*trans*-RA treatment has radically improved the prognosis of APL, whereas *13-cis*-RA significantly improves survival of neuroblastoma patients [98, 126]. The existence of two relevant disease model systems has encouraged research on effects of RA on target genes, including miRNA genes. The mechanism of miRNA regulation, be it direct or indirect, could have great functional relevance to the pathogenesis and, possibly, cure, of APL and other neoplastic diseases at the level of transcriptional repression of miRNA expression by PML/RARα or transcriptional activation by RA (Fig. 8.4). Moreover, the therapeutic effects of RA on neuroblastoma could be mediated in part by miRNA regulation.

Direct or Indirect Mechanisms: Insights from the RA-Responsive Leukemia Models

Many studies of RA signaling pathways regulated by miRNAs were performed using NB4 cells, an APL-derived cell line [78]. The results obtained in this cell line have been replicated in APL clinical samples and appear to be highly reliable [6, 16, 41, 45, 127]. For example, the repression by PML/RARα and up-regulation by RA of let-7c, miR-23a, miR-107, miR-210 and miR-342 have been confirmed in fresh APL blasts in two separate studies [16, 127], further suggesting that these miRNA activities are relevant in the biology of APL.

A number of miRNAs change their expression levels when hematopoietic cells are exposed to RA. The mechanism underlying the regulation of these miRNAs can be both direct and indirect, since RA-induced differentiation causes dramatic phenotypic changes in the cells.

Relatively few miRNAs have been shown to be directly regulated by RA through the activity of RARs or the APL-associated PML/RARα fusion oncoprotein on their target gene promoters [16, 127]. The let-7 family of miRNAs, commonly reported as RA-induced, does appear to be directly regulated by PML/RARα in APL cells [16]. Similarly, binding of the fusion protein, PML/RARα, to the miR-342 promoter in APL cells is induced by RA [16]. Thus, these miRNAs are examples of direct transcriptional regulation by RARs.

The regulation of miRNAs is also dependent upon indirect mechanisms, which are nonetheless affected by RA. The activity of the transcription factor *nuclear factor kappa B* (NFkB) is increased by RA treatment of APL cells [99]. One NFkB proximal binding site appears essential for the transactivation of the let-7a-3/let-7b cluster gene [45]. miR-342 is also under the control of the transcription factor PU.1, which is induced by RA treatment and binds to its responsive element on miR-345 gene promoter [27]. Indirect mechanisms are also involved in miRNA repression. For example, miR-145 is repressed during RA induced granulocytic differentiation. This miRNA gene is a transcriptional target of p73, which in turn is activated by RA treatment [6].

Other PML/RARα-regulated miRNAs were discovered through an *in-silico* search for potential PML/RARα binding sites in their gene promoters [127]. Sixty five intergenic miRNA genes were found to carry a predicted PML/RARα binding site in their promoter. However, experimental proof of regulation was obtained for relatively few of these miRNAs (miR-10b, miR-23a, miR-194, miR-195, miR-196a, miR-210, miR-377, and miR-622). The presence of PML/RARα on their promoter, indicating a direct regulation by the oncoprotein, was only shown for miR-23a and miR-210 [127].

The myeloblastic leukemia cell line HL60 is a classic model for RA-induced granulocytic differentiation [14]. miRNAs regulated by RA treatment in HL60 cells are largely the same as those regulated in APL blasts, although some miRNAs that are up-regulated (miR-22, miR-29a, miR-142-3p, miR-363, miR-494, miR-663) or down-regulated (miR-10a, miR-125b, miR-612) by RA in HL60 cells are not reported as modified by RA in the APL system. This possibly indicates the specificity of PML/RARα-mediated regulation in the APL context [91, 64].

Normal Hematopoiesis

Leukemia models of RA-controlled miRNAs have provided insight into their relevance in normal hematopoiesis. Several miRNAs that are induced by RA in leukemia cell lines increase their expression in differentiating normal $CD34^+$ hematopoietic progenitor cells (HPCs). Up-regulation of miR-29a and miR-142-3p was reported as a consequence of RA-induced differentiation of the myeloid cell lines, HL60, THP1 and NB4. The same miRNAs are also up-regulated during growth factor-induced myeloid differentiation of $CD34^+$ HPCs, suggesting that their regulation is differentiation-dependent rather than RA-specific [156]. Along the same lines, miR-15/16, which are up-regulated by RA in leukemia cell lines, display a higher expression level in mature peripheral blood leukocytes isolated from APL and Acute Myeloid Leukemia (AML) patients who are in clinical disease remission [44].

One significant example of an RA-induced miRNA is miR-223. This miRNA has been widely studied in normal hematopoiesis and leukemia models. miR-223 was originally identified as up-regulated in mouse mature bone marrow hematopoietic cells [19]. It is also induced by RA during granulocytic differentiation of human APL cells [41]. Moreover, miR-223 null mice develop a myeloproliferative hematopoietic disorder [65].

miR-223 up-regulation in myeloid precursors is indirectly caused by the RA-induced activation of RARs. In fact, RA treatment triggers a regulatory circuit involving miR-223 and two transcription factors, the CCAAT/enhancer-binding protein-alpha (C/EBPα) and the Nuclear Factor I-A (NFI-A). These transcription factors compete for binding to the miR-223 promoter. NFI-A maintains miR-223 at low levels, whereas its replacement by RA-induced C/EBPα up-regulates miR-223 expression [41]. NFI-A is also a post-transcriptional miR-223 target. Its RA-induced down-regulation is required to fulfill miR-223 key role in granulopoieisis [40, 41]. In fact, miR-223 is sufficient to reprogram granulocytic differentiation in distinct myeloid leukemia subtypes, independently from the presence of a specific genetic lesion [40, 41], whereas NFI-A acts as a novel developmental gene directing HSC/HPC lineage choice. NFI-A, per se, is indeed able to induce an erythroid transcriptional program in both primary HSC/HPCs and human myeloid cell lines where it acts directly at the proximal promoter regions of fundamental myeloid genes [138, 139]. NFI-A is also a predicted target of miR-107 whose expression is induced by RA in APL cells, perhaps contributing to transcription factor down-regulation during myeloid differentiation [45].

The fine tuning of miR-223 expression levels can dictate lineage fate decision and differentiation/maturation of CD34 + HPCs into erythroid, granulocytic and monocytic/macrophagic lineages. A high expression of miR-223 increases granulopoiesis and impairs erythroid—and monocytic/macrophagic—differentiation. Monocytic/macrophagic differentiation occurs when miR-223 levels are moderately increased, whereas erythroid commitment and differentiation require stably low levels of miR-223 [155].

miRNA-Mediated Tumor Suppressive Activities of RA in the Hematopoietic System

RA may act through miRNAs to control the proliferation and self-renewal of hematopoietic cells. The failure of this regulation may lead to leukemogenesis or an increase in leukemia malignant behaviour. The expression level of the urokinase-type plasminogen activator receptor (uPAR) protein, which is involved in the regulation of chemotaxis, adhesion and proteolysis, correlates with a significant lower remission rate after chemotherapy and a higher risk for relapse in

AML patients [48]. uPAR was experimentally shown to be a target of miR-195, miR-377 and miR-622, which are among the miRNAs up-regulated by RA in APL [127] (Fig. 8.4). Thus, by regulating these miRNAs, RA activity could contribute to the blockage of factors that are implicated in crucial pathways linked to leukemogenesis.

RA-induced miRNAs also suppress other pathways involved in neoplastic transformation or progression. The let-7c miRNA targets the pre-B-cell leukemia homeobox 2 (PBX2) mRNA, which encodes a homeodomain transcription factor, forming complexes with Hox and Meis. This complex contributes to leukemia cell proliferation and stemness [119], but is suppressed by the RA-induced let-7c. Moreover, the let-7 family members target Ras and Bcl-2 mRNAs, thus potentially decreasing proliferation and increasing apoptosis of hematopoietic cells [45]. Several other RA-regulated miRNAs (miR-27a, miR-196a, miR-377, miR-520d, miR-524) are predicted to target Hoxb8, impairing the stem properties of hematopoietic cells [127]. Conversely, RA down-regulates miR-146a, which targets Smad4 [171]. Thus RA could increase Smad4 expression, a target gene of miR-146a, increasing the growth suppressive effect of TGFβ signaling and APL cells proliferation [171].

Normal and Neoplastic Neural Differentiation

Similar to the hematopoietic system, RA activity in neural tissue has been studied in depth by taking advantage of neoplastic model systems. RA is able to induce in vitro differentiation of neuroblastoma cells. This finding has a clinical counterpart in the response of patients with neuroblastoma to *13-cis*-RA. Retinoid treatment produces a significant improvement in patient survival and is now a clinical standard. Thus, it is not surprising that an abundant body of literature has developed on the effects of retinoids on miRNAs in neural tissue differentiation. As recently summarized in an excellent review [136], these studies employed a fairly heterogeneous array of methodologies that yielded different results. Despite the discrepancies, the overall data show that a large amount of miRNA is regulated during RA-induced differentiation of neuroblastoma cells. Interestingly, many RA-regulated miRNAs are the same as those modified in the hematopoietic system.

Several RA-regulated miRNAs in the neural differentiation system deserve further discussion since their targets are well defined and relevant (Fig. 8.2a). miR-10a and miR-10b are powerfully induced by RA, although a specific RARE in their promoters has not been identified. miR-10a targets include the nuclear receptor co-repressor 2 (NCoR2), one of the co-repressors bound by RARα on DNA [42, 142]. In the absence of RA, this co-repressor drives HDACs on the promoter of RARα target genes, maintaining an inactive chromatin structure. In addition, in smooth muscle cells [59], it has been shown that miR-10a targets HDAC4, and in T-cells, the repressor Bcl-6 [142]. Thus, the induction of miR10a and miR10b by RA may trigger an activation loop by suppressing a co-repressor (NCoR2) and an effector (HDAC), favouring the recruitment and activity of co-activator complexes upon RA binding to RARα. In pancreatic cancer cells, miR-10a also targets Hoxb1 and Hoxb3 [160], possibly contributing to blockade of their anti-differentiation activity.

During neuroblastoma cell differentiation induced by RA, miR-340 is upregulated by demethylation of an upstream genomic region and directly represses the SOX2, a transcription factor linked to the stem cell phenotype [25]. The induction by RA of miR-9 and miR-103, both targeting the mRNA encoding the Helix-Loop-Helix transcription factor ID2, a differentiation inhibitor, contributes to the maintenance of the stem cell pool in the nervous system [3].

The miR 17-5p-92 cluster is suppressed by RA in Glioma and Neuroblastoma cell differentiation [10, 34]. This miRNA cluster is well known for its proliferative, anti-apoptotic and differentiation suppressive activity, an effect obtained by the coordinated suppression of a complex target network [112]. This underlines the potential role of miR 17-5p-92 in tumour development and maintenance and the effect of RA treatment in the regulation of this circuitry.

An interesting network elicited by RA in neuroblastoma cell differentiation involves the induction of miR-152. Overexpression of this miRNA decreases neuroblastoma cell invasiveness and growth. Interestingly, miR-152 targets DNA methyltransferase 1 (DNMT1). This may help to explain the changes in DNA methylation that occur during RA treatment of neuroblastoma cells [26]. Overall the activity of RA-induced miRNAs in neural cells seems to play a role in the suppression of the stem state and proliferation, while supporting differentiation. These biological functions are remarkably similar to those exerted by RA-regulated miRNAs in hematopoiesis.

Other Biological Model Systems

RA affects miRNA expression levels and functions in many other biological model systems. Retinoids have been used for their differentiation and anti-proliferative activities in several tumours, although their clinical activity is relatively limited outside APL and neuroblastoma. The activity of retinoids on miRNAs in other model systems is complex and sometimes difficult to correlate to molecular and biological networks. Sometimes the results appear contradictory.

In estrogen receptor positive breast cancer cells, RA induces the expression of miR-21, which is believed to increase cell proliferation [148]. However, the cells display a reduced motility, and proliferation possibly due to the suppression of various miR-21 targets, including tissue plasminogen activator, (t-PA), interleukin 1β (IL1B), intracellular adhesion molecule 1 (ICAM1) and maspin. On the other hand, the expression levels of Maspin, a "*non-inhibitory*" member of serin protease inhibitors (SERPIN), may serve as a prognostic marker for clinical outcomes in some tumours such as breast cancer [11].

In pancreatic cancer cells, RA increases the expression of miR-10a, which targets Hoxb1 and Hoxb3. The expression levels of these two genes is increased in cells derived from pancreatic cancer metastasis. However, RA-antagonists also suppress their expression levels, implying that inhibition of RA activity could be desirable in pancreatic cancer [160].

Recently, the activity of RA in inducing miR-10a has been studied in T(reg) lymphocytic cells. The repression of the miR-10a targets, Bcl-6 and NCoR2, decreased conversion of inducible T(reg) cells into the follicular T(H)17 subset of helper T cells [142]. This effect limits the plasticity of helper T-cells, indicating that RA acts through miRNAs in the determination of cell fate in the immune system.

miRNAs could play an important role in the differentiation activity of RA on spermatogonial stem cells. The induction of let-7 miRNAs by RA [149] may decrease the expression of the proliferative miRNA downstream targets Mycn, Ccnd1, and Col1α2, while the repression by RA of the miRNA clusters 17-5p-92 and 106b-25 may contribute to increasing their protein expression levels [150]. The sum of these effects is possibly, the induction of spermatogonial differentiation.

Finally, modulation of miRNAs by RA has unexpected effects on viral infection. RA-induced miR-23b targets the very low density lipoprotein (VLDL) receptor, which is a receptor for a minor group of rhinoviruses, resulting in decreased sensitivity to the infection by these viruses [113].

Retinoic Acid-Induced Nuclear Roles of miRNAs

Hundreds of miRNAs are distributed both in the nucleus and in the cytoplasm of normal and cancer cells [88, 116, 121]. Some miRNAs can even display a preferential nuclear localization [88]. In fact, the miRNA-associated RISC proteins Ago and Dicer have been found in the nucleus [17, 92, 153, 157], and contribute to control of transcription through the epigenetic regulation of chromatin structure [8, 168, 169]. Interestingly, importin 8, a specificity factor in the miRNA pathway, required for binding of Ago proteins to a variety of mRNA targets, also affects the nuclear localization of Ago proteins [159].

The nuclear activity of miRNAs can result in both gene transcriptional activation or silencing [60, 70, 114, 120, 144, 167, 168, 169, 170]. A series of interesting observations indicates that the transcriptional functions of miRNA depend on miRNA-complementary sequences present in the DNA of target gene promoters [8, 114, 168, 169]. It remains to be determined whether miRNAs directly interact with DNA or, alternatively, with nascent RNA sequences originating from the promoter of transcribed genes [141].

RA can be the trigger for this novel mechanism of transcriptional regulation by miRNAs. We have shown [168, 169] that during RA-induced granulocytic differentiation of immature myeloid precursor cells, miR-223 translocates to the nucleus,

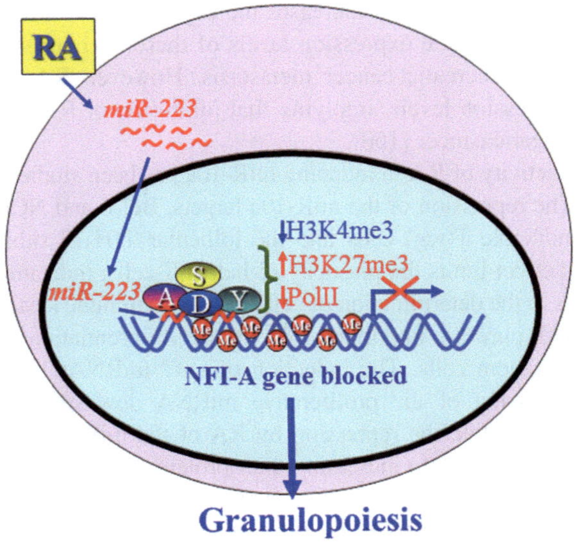

Fig. 8.5 Transcriptional activity of miR-223. Upon RA treatment of myeloid precursors, the expression level of miR-223 increases. A fraction of this miRNA translocates into the nucleus where it binds, either directly or indirectly, to the promoter region of the NFI-A gene. miR-223 aggregates with a transcriptional complex including Ago1 (*A*), Dicer1 (*D*) and the Polycomb Group proteins, Suz12 (*S*) and YY1 (*Y*). This increases H3K27 methylation, decreases H3K4 methylation, increases DNA methylation (*Me*) and blocks NFI-A PolII-dependent transcription, allowing granulocytic maturation of the cells

where it binds chromosomal chromatin. We have studied the transcriptional effect of miR-223 on the gene encoding NFI-A, a post-transcriptional target of its activity [168, 169]. miR-223 causes transcriptional repression of this gene. This repressor activity occurs through miR-223 binding to DNA sequences on the NFI-A promoter that are complementary to those of the miR-223 "*seed*" region. At these sites, miR-223 recruits a protein complex involving Ago1, Dicer1 and the PcG proteins YY1 and Suz12. The final effect is DNA hypermethylation and a change in chromatin epigenetic status, causing an increase in lysine 27 of histone 3 (H3K27) methylation and a decrease in H3K4 methylation (Fig. 8.5). "*Bivalent*" chromatin marks, depicted by the simultaneous presence of activator H3K4 trimethylation (H3K4me3) and repressive H3K27me3 histone modifications characterize developmentally regulated gene promoters and are also present in the NFI-A promoter. In the early phases of myeloid differentiation, they are repressive due to the RA-induced activity of miR-223, and the NFI-A gene transcription is inhibited. The biological consequence of these molecular events is progression towards myeloid differentiation and blockade of entry into the erythroid lineage [168, 169].

Among the other RA-regulated miRNAs, the let-7 family has shown the ability to mediate transcriptional gene silencing. In senescent human fibroblasts, endogenous let-7f contributes to transcriptional gene silencing of E2F target genes acting on those promoters where let-7f complementary sequences were found. This miRNA contributes to aggregate a repressor complex including the RISC protein Ago2 and pRb1 [8]. This complex increases H3K27me3 and DNA methylation, resulting in transcriptional repression of target genes. Notably, diverse let-7 family members are induced by RA in myeloid hematopoietic cells [16, 45, 119] (Fig. 8.4) and some of them are targets of the APL-associated PML/RARα fusion protein [16]. In principle, therefore, a transcriptional effect of the let-7 may mediate part of the physiological and pathologic biological effects produced respectively by RARα or its leukemic mutant.

Future Directions

Retinoids and miRNAs are morphogens and essential regulators of embryogenesis, normal cell differentiation, and proliferation. During adult life, both are required for proper organ functioning. Considering the importance of miRNA— and RA-induced regulatory networks, the deregulation of any factor or pathway is highly likely to contribute to disease development. Currently, many aspects of their interactions remain unexplored and a number of questions are still unanswered. For instance, the complex sequence of miRNA biogenesis can be affected by retinoids at multiple steps, and any component of the retinoid pathway could be a target of the activity of miRNAs, as suggested by the initial findings reported in this chapter and further addressed, below.

MicroRNA Targeting of Retinoid Receptors

miRNA targeting and regulation of RARs and RXRs is almost completely unexplored. This is therefore a major area of future research. Various transcription factors, including nuclear receptors, are involved in complex feedback circuits with miRNAs, and it is likely that retinoid receptors are no exception [23, 29, 115]. miRNAs regulating the expression of RARs and/or RXRs may affect any biological area of retinoid activities. Thus, this research theme deserves further attention and will probably provide new, interesting information that will help to complete the picture of retinoid functions.

Most certainly, miRNAs make fundamental contributions to the fine tuning of developmental and differentiation pathways that are regulated by retinoids. However, very few miRNAs have been shown to act directly on the major RARs and proteins that contribute to RA signaling. Table 8.1 reports the results of a bioinformatic survey showing miRNA targeting of the major RARs and RXRs that is limited to conserved sites in miRNA families of mammals and vertebrates. Table 8.1 is based on two algorithms (PicTar and TargetScan Human 6.2). Shown in bold are the few experimentally verified miRNAs with activity on RARs and RXRs as indicated by the Tarbase 6.0 database [154]. Accepted verification

Table 8.1 Predicted and validated miRNAs targeting major RA receptor mRNAs

Gene name	Refseq ID	3′UTR length (nt)	miRNA families broadly conserved among vertebrates (conserved sites)	miRNA families conserved among mammals (conserved sites)
RARA	NM_000964	1,417	miR-27abc/27a-3p, miR-128/128ab, miR-135ab/135a-5p, miR-138/138ab, miR-194, miR-205/205ab, **miR-218/218-5p** [53], *miR-337, miR-199a, mir-199b, miR-220*	
RARB	NM_000965	1,315	miR-1ab/206/613, miR-15abc, miR-16-2-3p [58], miR-16abc/195/322/424/497/1907, miR-29abcd, miR-30abcdef/30abe-5p/384-5p, miR-34ac/34bc-5p/449abc/449c-5p, miR-133abc, miR-141/200a, miR-216a, miR-135ab/135a-5p, miR-144, miR-17/17-5p/20ab/20b-5p/93/106ab/427/518a-3p/519d, *miR-16, miR-101, miR-103, miR-107, miR-133b, miR-136, miR-146, miR-193, miR-195, miR-199a, miR-200a, miR-368*	miR-376c/741-5p
RARG	NM_000966	1,107	miR-24/24ab/24-3p, miR-30abcdef/30abe-5p/384-5p, miR-96/507/1271, **miR-182** [87], miR-142-3p, **miR-124/124ab/506** [90], miR-143/1721/4770, **miR-335-5p** [145], *miR-22, miR-34abc,* miR320, miR-331, miR-9*	miR-285p/708/1407/1653/3139, miR-326/330/330-5p
RXRA	NM_002957	3,971	miR-9/9ab, miR-27abc/27a-3p miR-128/128ab, miR-124/124ab/506, miR-216b/216b-5p, miR-219-5p/508/508-3p/4782-3p	miR-320abcd/4429, miR-197, miR-876-5p/3167
RXRB	NM_021976	1,094	miR-101/101ab, miR-144, miR-490-3p, *miR-17-3p, miR-199a**, **miR-26b-5p** [46], **miR-197-3p** [158], **miR-346** [158]	miR-488*
RXRG	NM_006917	341	miR-23abc/23b-3p	

The indicated genes (Refseq ID sequences) were analysed with TargetScan human 6.2 (http://www.targetscan.org/) and PicTar (http://pictar.mdc-berlin.de/). miRNAs identified by both algorithms are underlined. miRNAs identified only by PicTar are in italics. miRNAs, derived from the Tarbase 6.0 database [154] are in boldface; studies reporting some of these miRNAs are indicated in parenthesis

methods include microarray screenings of mRNA expression upon overexpression of miRNA or knockdown of their function [46, 56, 87, 90, 145, 158] and high throughput screenings of mRNA co-immunoprecipitated with miRNAs by several RNA-binding proteins (RBPs) including Ago protein [53]. Thus, a direct confirmation of miRNA interactions with the 3′UTR of these receptor mRNAs is not required by this database. As previously described, miR-196 does indeed target the 3′UTR of rarab in zebrafish [56].

In summary, the evidence supporting a functional interplay between miRNAs and RA signaling appears to be robust. However, our understanding of the molecular basis for and physiological significance of such interactions is far from complete. Numerous conceptual and experimental questions remain. Future research should focus on understanding whether and how retinoid and miRNA actions can trigger a cascade of dynamic events that result in fine-tuned, specific, control of gene expression. We will also be on the watch for important new developments that involve the recently discovered RA-induced nuclear functions of miRNAs.

References

1. Altuvia Y, Landgraf P, Lithwick G, Elefant N, Pfeffer S, Aravin A, Brownstein MJ, Tuschl T, Margalit H (2005) Clustering and conservation patterns of human microRNAs. Nucleic Acids Res 33:2697–2706
2. Ambros V, Lee RC (2004) Identification of microRNAs and other tiny noncoding RNAs by cDNA cloning. Methods Mol Biol 265:131–158
3. Annibali D, Gioia U, Savino M, Laneve P, Caffarelli E, Nasi S (2012) A new module in neural differentiation control: two microRNAs upregulated by retinoic acid, miR-9 and -103, target the differentiation inhibitor ID2. PLoS One 7:e40269
4. Bartel DP (2009) MicroRNAs: target recognition and regulatory functions. Cell 136:215–233
5. Baskerville S, Bartel DP (2005) Microarray profiling of microRNAs reveals frequent coexpression with neighboring miRNAs and host genes. RNA 11:241–247
6. Batliner J, Buehrer E, Fey MF, Tschan MP (2012) Inhibition of the miR-143/145 cluster attenuated neutrophil differentiation of APL cells. Leuk Res 36:237–240
7. Bazzini AA, Lee MT, Giraldez AJ (2012) Ribosome profiling shows that miR-430 reduces translation before causing mRNA decay in zebrafish. Science 336:233–237
8. Benhamed M, Herbig U, Ye T, Dejean A, Bischof O (2012) Senescence is an endogenous trigger for microRNA-directed transcriptional gene silencing in human cells. Nat Cell Biol 14:266–275
9. Berezikov E, Cuppen E, Plasterk RH (2006) Approaches to microRNA discovery. Nat Genet 38(Suppl):S2–S7
10. Beveridge NJ, Tooney PA, Carroll AP, Tran N, Cairns MJ (2009) Down-regulation of miR-17 family expression in response to retinoic acid induced neuronal differentiation. Cell Signal 21:1837–1845
11. Bodenstine TM, Seftor RE, Khalkhali-Ellis Z, Seftor EA, Pemberton PA, Hendrix MJ (2012) Maspin: molecular mechanisms and therapeutic implications. Cancer Metastasis Rev 31:529–551
12. Bohnsack MT, Czaplinski K, Gorlich D (2004) Exportin 5 is a RanGTP-dependent dsRNA-binding protein that mediates nuclear export of pre-miRNAs. RNA 10:185–191
13. Borchert GM, Lanier W, Davidson BL (2006) RNA polymerase III transcribes human microRNAs. Nat Struct Mol Biol 13:1097–1101

14. Breitman TR, Selonick SE, Collins SJ (1980) Induction of differentiation of the human promyelocytic leukemia cell line (HL-60) by retinoic acid. Proc Natl Acad Sci USA 77:2936–2940
15. Cai X, Hagedorn CH, Cullen BR (2004) Human microRNAs are processed from capped, polyadenylated transcripts that can also function as mRNAs. RNA 10:1957–1966
16. Careccia S, Mainardi S, Pelosi A, Gurtner A, Diverio D, Riccioni R, Testa U, Pelosi E, Piaggio G, Sacchi A, Lavorgna S, Lo-Coco F, Blandino G, Levrero M, Rizzo MG (2009) A restricted signature of miRNAs distinguishes APL blasts from normal promyelocytes. Oncogene 28:4034–4040
17. Cernilogar FM, Onorati MC, Kothe GO, Burroughs AM, Parsi KM, Breiling A, Lo SF, Saxena A, Miyoshi K, Siomi H, Siomi MC, Carninci P, Gilmour DS, Corona DF, Orlando V (2011) Chromatin-associated RNA interference components contribute to transcriptional regulation in Drosophila. Nature 480:391–395
18. Cesana M, Cacchiarelli D, Legnini I, Santini T, Sthandier O, Chinappi M, Tramontano A, Bozzoni I (2011) A long noncoding RNA controls muscle differentiation by functioning as a competing endogenous RNA. Cell 147:358–369
19. Chen C-Z, Li L, Lodish HF, Bartel DP (2004) MicroRNAs modulate hematopoietic lineage differentiation. Science 303:83–86
20. Chendrimada TP, Gregory RI, Kumaraswamy E, Norman J, Cooch N, Nishikura K, Shiekhattar R (2005) TRBP recruits the Dicer complex to Ago2 for microRNA processing and gene silencing. Nature 436:740–744
21. Cheng LC, Pastrana E, Tavazoie M, Doetsch F (2009) miR-124 regulates adult neurogenesis in the subventricular zone stem cell niche. Nat Neurosci 12:399–408
22. Chuang YS, Huang WH, Park SW, Persaud SD, Hung CH, Ho PC, Wei LN (2011) Promyelocytic leukemia protein in retinoic acid-induced chromatin remodelling of Oct 4 gene promoter. Stem Cells 29:660–669
23. Cochrane DR, Cittelly DM, Richer JK (2011) Steroid receptors and microRNAs: relationships revealed. Steroids 76:1–10
24. Conaco C, Otto S, Han JJ, Mandel G (2006) Reciprocal actions of REST and a microRNA promote neuronal identity. Proc Natl Acad Sci USA 103:2422–2427
25. Das S, Bryan K, Buckley PG, Piskareva O, Bray IM, Foley N, Ryan J, Lynch J, Creevey L, Fay J, Prenter S, Koster J, van Sluis P, Versteeg R, Eggert A, Schulte JH, Schramm A, Mestdagh P, Vandesompele J, Speleman F, Stallings RL (2013) Modulation of neuroblastoma disease pathogenesis by an extensive network of epigenetically regulated microRNAs. Oncogene 32:2927–2936
26. Das S, Foley N, Bryan K, Watters KM, Bray I, Murphy DM, Buckley PG, Stallings RL (2010) MicroRNA mediates DNA demethylation events triggered by retinoic acid during neuroblastoma cell differentiation. Cancer Res 70:7874–7881
27. De Marchis ML, Ballarino M, Salvatori B, Puzzolo MC, Bozzoni I, Fatica A (2009) A new molecular network comprising PU.1, interferon regulatory factor proteins and miR-342 stimulates ATRA-mediated granulocytic differentiation of acute promyelocytic leukemia cells. Leukemia 23:856–862
28. Denli AM, Tops BB, Plasterk RH, Ketting RF, Hannon GJ (2004) Processing of primary microRNAs by the Microprocessor complex. Nature 432:231–235
29. Eendebak RJ, Lucassen PJ, Fitzsimons CP (2011) Nuclear receptors and microRNAs: Who regulates the regulators in neural stem cells? FEBS Lett 585:717–722
30. Eiring AM, Harb JG, Neviani P, Garton C, Oaks JJ, Spizzo R, Liu S, Schwind S, Santhanam R, Hickey CJ, Becker H, Chandler JC, Andino R, Cortes J, Hokland P, Huettner CS, Bhatia R, Roy DC, Liebhaber SA, Caligiuri MA, Marcucci G, Garzon R, Croce CM, Calin GA, Perrotti D (2010) miR-328 functions as an RNA decoy to modulate hnRNP E2 regulation of mRNA translation in leukemic blasts. Cell 140:652–665
31. Elkayam E, Kuhn CD, Tocilj A, Haase AD, Greene EM, Hannon GJ, Joshua-Tor L (2012) The structure of human argonaute-2 in complex with miR-20a. Cell 150:100–110
32. ENCODE Project Consortium (2011) A user's guide to the encyclopedia of DNA elements (ENCODE). PLoS Biol 9:e1001046

33. ENCODE explorer—nature (2012). http://www.nature.com/encode/threads
34. Ernst A, Campos B, Meier J, Devens F, Liesenberg F, Wolter M, Reifenberger G, Herold-Mende C, Lichter P, Radlwimmer B (2010) De-repression of CTGF via the miR-17-92 cluster upon differentiation of human glioblastoma spheroid cultures. Oncogene 29:3411–3422
35. Esquela-Kerscher A, Slack FJ (2006) Oncomirs—microRNAs with a role in cancer. Nat Rev Cancer 6:259–269
36. Eulalio A, Behm-Ansmant I, Schweizer D, Izaurralde E (2007) P-body formation is a consequence, not the cause, of RNA-mediated gene silencing. Mol Cell Biol 27:3970–3981
37. Eulalio A, Huntzinger E, Izaurralde E (2008) Getting to the root of miRNA-mediated gene silencing. Cell 132:9–14
38. Farh KK, Grimson A, Jan C, Lewis BP, Johnston WK, Lim LP, Burge CB, Bartel DP (2005) The widespread impact of mammalian MicroRNAs on mRNA repression and evolution. Science 310:1817–1821
39. Fazi F, Nervi C (2008) MicroRNA: basic mechanisms and transcriptional regulatory networks for cell fate determination. Cardiovasc Res 79:553–561
40. Fazi F, Racanicchi S, Zardo G, Starnes LM, Mancini M, Travaglini L, Diverio D, Ammatuna E, Cimino G, Lo-Coco F, Grignani F, Nervi C (2007) Epigenetic silencing of the myelopoiesis regulator microRNA-223 by the AML1/ETO oncoprotein. Cancer Cell 12:457–466
41. Fazi F, Rosa A, Fatica A, Gelmetti V, De Marchis ML, Nervi C, Bozzoni I (2005) A minicircuitry comprised of microRNA-223 and transcription factors NFI-A and C/EBPalpha regulates human granulopoiesis. Cell 123:819–831
42. Foley NH, Bray I, Watters KM, Das S, Bryan K, Bernas T, Prehn JH, Stallings RL (2011) MicroRNAs 10a and 10b are potent inducers of neuroblastoma cell differentiation through targeting of nuclear receptor corepressor 2. Cell Death Differ 18:1089–1098
43. Friedman RC, Farh KK, Burge CB, Bartel DP (2009) Most mammalian mRNAs are conserved targets of microRNAs. Genome Res 19:92–105
44. Gao SM, Yang J, Chen C, Zhang S, Xing CY, Li H, Wu J, Jiang L (2011) miR-15a/16-1 enhances retinoic acid-mediated differentiation of leukemic cells and is up-regulated by retinoic acid. Leuk Lymphoma 52:2365–2371
45. Garzon R, Pichiorri F, Palumbo T, Visentini M, Aqeilan R, Cimmino A, Wang H, Sun H, Volinia S, Alder H, Calin GA, Liu CG, Andreeff M, Croce CM (2007) MicroRNA gene expression during retinoic acid-induced differentiation of human acute promyelocytic leukemia. Oncogene 26:4148–4157
46. Gennarino VA, Sardiello M, Avellino R, Meola N, Maselli V, Anand S, Cutillo L, Ballabio A, Banfi S (2009) MicroRNA target prediction by expression analysis of host genes. Genome Res 19:481–490
47. Giraldez AJ, Cinalli RM, Glasner ME, Enright AJ, Thomson JM, Baskerville S, Hammond SM, Bartel DP, Schier AF (2005) MicroRNAs regulate brain morphogenesis in zebrafish. Science 308:833–838
48. Graf M, Reif S, Hecht K, Pelka-Fleischer R, Pfister K, Schmetzer H (2005) High expression of urokinase plasminogen activator receptor (UPA-R) in acute myeloid leukemia (AML) is associated with worse prognosis. Am J Hematol 79:26–35
49. Gregory RI, Yan KP, Amuthan G, Chendrimada T, Doratotaj B, Cooch N, Shiekhattar R (2004) The Microprocessor complex mediates the genesis of microRNAs. Nature 432:235–240
50. Guil S, Caceres JF (2007) The multifunctional RNA-binding protein hnRNP A1 is required for processing of miR-18a. Nat Struct Mol Biol 14:591–596
51. Guo H, Ingolia NT, Weissman JS, Bartel DP (2010) Mammalian microRNAs predominantly act to decrease target mRNA levels. Nature 466:835–840
52. Haase AD, Jaskiewicz L, Zhang H, Laine S, Sack R, Gatignol A, Filipowicz W (2005) TRBP, a regulator of cellular PKR and HIV-1 virus expression, interacts with Dicer and functions in RNA silencing. EMBO Rep 6:961–967
53. Hafner M, Landthaler M, Burger L, Khorshid M, Hausser J, Berninger P, Rothballer A, Ascano M Jr, Jungkamp AC, Munschauer M, Ulrich A, Wardle GS, Dewell S, Zavolan M,

Tuschl T (2010) Transcriptome-wide identification of RNA-binding protein and microRNA target sites by PAR-CLIP. Cell 141:129–141
54. Hammond SM, Bernstein E, Beach D, Hannon GJ (2000) An RNA-directed nuclease mediates post-transcriptional gene silencing in Drosophila cells. Nature 404:293–296
55. Han J, Lee Y, Yeom KH, Nam JW, Heo I, Rhee JK, Sohn SY, Cho Y, Zhang BT, Kim VN (2006) Molecular basis for the recognition of primary microRNAs by the Drosha-DGCR8 complex. Cell 125:887–901
56. He X, Yan YL, Eberhart JK, Herpin A, Wagner TU, Schartl M, Postlethwait JH (2011) miR-196 regulates axial patterning and pectoral appendage initiation. Dev Biol 357:463–477
57. Hornstein E, Mansfield JH, Yekta S, Hu JK, Harfe BD, McManus MT, Baskerville S, Bartel DP, Tabin CJ (2005) The microRNA miR-196 acts upstream of Hoxb8 and Shh in limb development. Nature 438:671–674
58. Hu Y, Correa AM, Hoque A, Guan B, Ye F, Huang J, Swisher SG, Wu TT, Ajani JA, Xu XC (2011) Prognostic significance of differentially expressed miRNAs in esophageal cancer. Int J Cancer 128:132–143
59. Huang H, Xie C, Sun X, Ritchie RP, Zhang J, Chen YE (2010) miR-10a contributes to retinoid acid-induced smooth muscle cell differentiation. J Biol Chem 285:9383–9389
60. Huang V, Place RF, Portnoy V, Wang J, Qi Z, Jia Z, Yu A, Shuman M, Yu J, Li LC (2012) Upregulation of Cyclin B1 by miRNA and its implications in cancer. Nucleic Acids Res 40:1695–1707
61. Hutvagner G, McLachlan J, Pasquinelli AE, Balint E, Tuschl T, Zamore PD (2001) A cellular function for the RNA-interference enzyme Dicer in the maturation of the let-7 small temporal RNA. Science 293:834–838
62. Jacob F, Monod J (1961) Genetic regulatory mechanisms in the synthesis of proteins. J Mol Biol 3:318–356
63. Jain AK, Allton K, Iacovino M, Mahen E, Milczarek RJ, Zwaka TP, Kyba M, Barton MC (2012) p53 regulates cell cycle and microRNAs to promote differentiation of human embryonic stem cells. PLoS Biol 10:e1001268
64. Jian P, Li ZW, Fang TY, Jian W, Zhuan Z, Mei LX, Yan WS, Jian N (2011) Retinoic acid induces HL-60 cell differentiation via the upregulation of miR-663. J Hematol Oncol 4:20
65. Johnnidis JB, Harris MH, Wheeler RT, Stehling-Sun S, Lam MH, Kirak O, Brummelkamp TR, Fleming MD, Camargo FD (2008) Regulation of progenitor cell proliferation and granulocyte function by microRNA-223. Nature 451:1125–1129
66. Juan AH, Kumar RM, Marx JG, Young RA, Sartorelli V (2009) Mir-214-dependent regulation of the polycomb protein Ezh2 in skeletal muscle and embryonic stem cells. Mol Cell 36:61–74
67. Karreth FA, Pandolfi PP (2013) ceRNA cross-talk in cancer: when ce-bling rivalries go awry. Cancer Discov 3:1113–1121
68. Karreth FA, Tay Y, Perna D, Ala U, Tan SM, Rust AG, DeNicola G, Webster KA, Weiss D, Perez-Mancera PA, Krauthammer M, Halaban R, Provero P, Adams DJ, Tuveson DA, Pandolfi PP (2011) In vivo identification of tumor—suppressive PTEN ceRNAs in an oncogenic BRAF-induced mouse model of melanoma. Cell 147:382–395
69. Keegan BR, Feldman JL, Begemann G, Ingham PW, Yelon D (2005) Retinoic acid signaling restricts the cardiac progenitor pool. Science 307:247–249
70. Kim DH, Saetrom P, Snove O Jr, Rossi JJ (2008) MicroRNA-directed transcriptional gene silencing in mammalian cells. Proc Natl Acad Sci USA 105:16230–16235
71. Kim J, Inoue K, Ishii J, Vanti WB, Voronov SV, Murchison E, Hannon G, Abeliovich A (2007) A MicroRNA feedback circuit in midbrain dopamine neurons. Science 317:1220–1224
72. Kim VN, Nam JW (2006) Genomics of microRNA. Trends Genet 22:165–173
73. Kloosterman WP, Wienholds E, Ketting RF, Plasterk RH (2004) Substrate requirements for let-7 function in the developing zebrafish embryo. Nucleic Acids Res 32:6284–6291
74. Kmita M, Duboule D (2003) Organizing axes in time and space; 25 years of colinear tinkering. Science 301:331–333

75. Krumlauf R (1994) Hox genes in vertebrate development. Cell 78:191–201
76. Lagos-Quintana M, Rauhut R, Meyer J, Borkhardt A, Tuschl T (2003) New microRNAs from mouse and human. RNA 9:175–179
77. Lagos-Quintana M, Rauhut R, Yalcin A, Meyer J, Lendeckel W, Tuschl T (2002) Identification of tissue-specific microRNAs from mouse. Curr Biol 12:735–739
78. Lanotte M, Martin-Thouvenin V, Najman S, Ballerini P, Valensi F, Berger R (1991) NB4, a maturation inducible cell line with t(15;17) marker isolated from a human acute promyelocytic leukemia (M3). Blood 77:1080–1086
79. Lawrie CH, Gal S, Dunlop HM, Pushkaran B, Liggins AP, Pulford K, Banham AH, Pezzella F, Boultwood J, Wainscoat JS, Hatton CS, Harris AL (2008) Detection of elevated levels of tumour-associated microRNAs in serum of patients with diffuse large B-cell lymphoma. Br J Haematol 141:672–675
80. Lee I, Ajay SS, Yook JI, Kim HS, Hong SH, Kim NH, Dhanasekaran SM, Chinnaiyan AM, Athey BD (2009) New class of microRNA targets containing simultaneous 5′-UTR and 3′-UTR interaction sites. Genome Res 19:1175–1183
81. Lee RC, Feinbaum RL, Ambros V, The C (1993) Elegans heterochronic gene lin-4 encodes small RNAs with antisense complementarity to lin-14. Cell 75:843–854
82. Lee TI, Jenner RG, Boyer LA, Guenther MG, Levine SS, Kumar RM, Chevalier B, Johnstone SE, Cole MF, Isono K, Koseki H, Fuchikami T, Abe K, Murray HL, Zucker JP, Yuan B, Bell GW, Herbolsheimer E, Hannett NM, Sun K, Odom DT, Otte AP, Volkert TL, Bartel DP, Melton DA, Gifford DK, Jaenisch R, Young RA (2006) Control of developmental regulators by Polycomb in human embryonic stem cells. Cell 125:301–313
83. Lee Y, Ahn C, Han J, Choi H, Kim J, Yim J, Lee J, Provost P, Radmark O, Kim S, Kim VN (2003) The nuclear RNase III Drosha initiates microRNA processing. Nature 425:415–419
84. Lee Y, Hur I, Park SY, Kim YK, Suh MR, Kim VN (2006) The role of PACT in the RNA silencing pathway. EMBO J 25:522–532
85. Lee Y, Jeon K, Lee JT, Kim S, Kim VN (2002) MicroRNA maturation: stepwise processing and subcellular localization. EMBO J 21:4663–4670
86. Lee Y, Kim M, Han J, Yeom KH, Lee S, Baek SH, Kim VN (2004) MicroRNA genes are transcribed by RNA polymerase II. EMBO J 23:4051–4060
87. Li G, Luna C, Qiu J, Epstein DL, Gonzalez P (2009) Alterations in microRNA expression in stress-induced cellular senescence. Mech Ageing Dev 130:731–741
88. Liao JY, Ma LM, Guo YH, Zhang YC, Zhou H, Shao P, Chen YQ, Qu LH (2010) Deep sequencing of human nuclear and cytoplasmic small RNAs reveals an unexpectedly complex subcellular distribution of miRNAs and tRNA 3′ trailers. PLoS. One. 5:e10563
89. Lim LP, Glasner ME, Yekta S, Burge CB, Bartel DP (2003) Vertebrate microRNA genes. Science 299:1540
90. Lim LP, Lau NC, Garrett-Engele P, Grimson A, Schelter JM, Castle J, Bartel DP, Linsley PS, Johnson JM (2005) Microarray analysis shows that some microRNAs downregulate large numbers of target mRNAs. Nature 433:769–773
91. Lin KY, Zhang XJ, Feng DD, Zhang H, Zeng CW, Han BW, Zhou AD, Qu LH, Xu L, Chen YQ (2011) miR-125b, a target of CDX2, regulates cell differentiation through repression of the core binding factor in hematopoietic malignancies. J Biol Chem 286:38253–38263
92. Lippman Z, Gendrel AV, Black M, Vaughn MW, Dedhia N, McCombie WR, Lavine K, Mittal V, May B, Kasschau KD, Carrington JC, Doerge RW, Colot V, Martienssen R (2004) Role of transposable elements in heterochromatin and epigenetic control. Nature 430:471–476
93. Lund E, Guttinger S, Calado A, Dahlberg JE, Kutay U (2004) Nuclear export of microRNA precursors. Science 303:95–98
94. Lytle JR, Yario TA, Steitz JA (2007) Target mRNAs are repressed as efficiently by microRNA-binding sites in the 5′ UTR as in the 3′ UTR. Proc Natl Acad Sci USA 104:9667–9672
95. MacRae IJ, Ma E, Zhou M, Robinson CV, Doudna JA (2008) In vitro reconstitution of the human RISC-loading complex. Proc Natl Acad Sci USA 105:512–517

96. Makeyev EV, Zhang J, Carrasco MA, Maniatis T (2007) The MicroRNA miR-124 promotes neuronal differentiation by triggering brain-specific alternative pre-mRNA splicing. Mol Cell 27:435–448
97. Mansfield JH, Harfe BD, Nissen R, Obenauer J, Srineel J, Chaudhuri A, Farzan-Kashani R, Zuker M, Pasquinelli AE, Ruvkun G, Sharp PA, Tabin CJ, McManus MT (2004) MicroRNA-responsive 'sensor' transgenes uncover Hox-like and other developmentally regulated patterns of vertebrate microRNA expression. Nat Genet 36:1079–1083
98. Masetti R, Biagi C, Zama D, Vendemini F, Martoni A, Morello W, Gasperini P, Pession A (2012) Retinoids in pediatric onco-hematology: the model of acute promyelocytic leukemia and neuroblastoma. Adv Ther 29:747–762
99. Mathieu J, Giraudier S, Lanotte M, Besancon F (2005) Retinoid-induced activation of NF-kappaB in APL cells is not essential for granulocytic differentiation, but prolongs the life span of mature cells. Oncogene 24:7145–7155
100. Matranga C, Tomari Y, Shin C, Bartel DP, Zamore PD (2005) Passenger-strand cleavage facilitates assembly of siRNA into Ago2-containing RNAi enzyme complexes. Cell 123:607–620
101. Mayr C, Hemann MT, Bartel DP (2007) Disrupting the pairing between let-7 and Hmga2 enhances oncogenic transformation. Science 315:1576–1579
102. Meister G (2013) Argonaute proteins: functional insights and emerging roles. Nat Rev Genet 14:447–459
103. Mitchell PS, Parkin RK, Kroh EM, Fritz BR, Wyman SK, Pogosova-Agadjanyan EL, Peterson A, Noteboom J, O'Briant KC, Allen A, Lin DW, Urban N, Drescher CW, Knudsen BS, Stirewalt DL, Gentleman R, Vessella RL, Nelson PS, Martin DB, Tewari M (2008) Circulating microRNAs as stable blood-based markers for cancer detection. Proc Natl Acad Sci USA 105:10513–10518
104. Miyasaka KY, Kida YS, Banjo T, Ueki Y, Nagayama K, Matsumoto T, Sato M, Ogura T (2011) Heartbeat regulates cardiogenesis by suppressing retinoic acid signaling via expression of miR-143. Mech Dev 128:18–28
105. Morton SU, Scherz PJ, Cordes KR, Ivey KN, Stainier DY, Srivastava D (2008) MicroRNA-138 modulates cardiac patterning during embryonic development. Proc Natl Acad Sci USA 105:17830–17835
106. Niederreither K, Vermot J, Messaddeq N, Schuhbaur B, Chambon P, Dolle P (2001) Embryonic retinoic acid synthesis is essential for heart morphogenesis in the mouse. Development 128:1019–1031
107. Nishi K, Nishi A, Nagasawa T, Ui-Tei K (2013) Human TNRC6A is an Argonaute-navigator protein for microRNA-mediated gene silencing in the nucleus. RNA 19:17–35
108. Ohshima K, Inoue K, Fujiwara A, Hatakeyama K, Kanto K, Watanabe Y, Muramatsu K, Fukuda Y, Ogura S, Yamaguchi K, Mochizuki T (2010) Let-7 microRNA family is selectively secreted into the extracellular environment via exosomes in a metastatic gastric cancer cell line. PLoS One 5:e13247
109. Okada C, Yamashita E, Lee SJ, Shibata S, Katahira J, Nakagawa A, Yoneda Y, Tsukihara T (2009) A high-resolution structure of the pre-microRNA nuclear export machinery. Science 326:1275–1279
110. Okamura K, Ishizuka A, Siomi H, Siomi MC (2004) Distinct roles for Argonaute proteins in small RNA-directed RNA cleavage pathways. Genes Dev. 18:1655–1666
111. Okamura K, Phillips MD, Tyler DM, Duan H, Chou YT, Lai EC (2008) The regulatory activity of microRNA* species has substantial influence on microRNA and 3′ UTR evolution. Nat Struct Mol Biol 15:354–363
112. Olive V, Jiang I, He L (2010) Mir-17-92, a cluster of miRNAs in the midst of the cancer network. Int J Biochem Cell Biol 42:1348–1354
113. Ouda R, Onomoto K, Takahasi K, Edwards MR, Kato H, Yoneyama M, Fujita T (2011) Retinoic acid-inducible gene I-inducible miR-23b inhibits infections by minor group rhinoviruses through down-regulation of the very low density lipoprotein receptor. J Biol Chem 286:26210–26219

114. Pagano F, De Marinis E, Grignani F, Nervi C (2013) Epigenetic role of miRNAs in normal and leukemic hematopoiesis. Epigenomics 5:539–552
115. Pandey DP, Picard D (2010) Multidirectional interplay between nuclear receptors and microRNAs. Curr Opin Pharmacol 10:637–642
116. Park CW, Zeng Y, Zhang X, Subramanian S, Steer CJ (2010) Mature microRNAs identified in highly purified nuclei from HCT116 colon cancer cells. RNA Biol 7:606–614
117. Pasquinelli AE, Reinhart BJ, Slack F, Martindale MQ, Kuroda MI, Maller B, Hayward DC, Ball EE, Degnan B, Muller P, Spring J, Srinivasan A, Fishman M, Finnerty J, Corbo J, Levine M, Leahy P, Davidson E, Ruvkun G (2000) Conservation of the sequence and temporal expression of let-7 heterochronic regulatory RNA. Nature 408:86–89
118. Pearson JC, Lemons D, McGinnis W (2005) Modulating Hox gene functions during animal body patterning. Nat Rev Genet 6:893–904
119. Pelosi A, Careccia S, Lulli V, Romania P, Marziali G, Testa U, Lavorgna S, Lo-Coco F, Petti MC, Calabretta B, Levrero M, Piaggio G, Rizzo MG (2013) miRNA let-7c promotes granulocytic differentiation in acute myeloid leukemia. Oncogene 32:3648–3654
120. Place RF, Li LC, Pookot D, Noonan EJ, Dahiya R (2008) MicroRNA-373 induces expression of genes with complementary promoter sequences. Proc Natl Acad Sci USA 105:1608–1613
121. Politz JC, Hogan EM, Pederson T (2009) MicroRNAs with a nucleolar location. RNA 15:1705–1715
122. Reinhart BJ, Slack FJ, Basson M, Pasquinelli AE, Bettinger JC, Rougvie AE, Horvitz HR, Ruvkun G (2000) The 21-nucleotide let-7 RNA regulates developmental timing in Caenorhabditis elegans. Nature 403:901–906
123. Rodriguez A, Griffiths-Jones S, Ashurst JL, Bradley A (2004) Identification of mammalian microRNA host genes and transcription units. Genome Res 14:1902–1910
124. Ruvkun G, Wightman B, Ha I (2004) The 20 years it took to recognize the importance of tiny RNAs. Cell 116(S93–6):2
125. Saito K, Ishizuka A, Siomi H, Siomi MC (2005) Processing of pre-microRNAs by the Dicer-1-Loquacious complex in Drosophila cells. PLoS Biol 3:e235
126. Sanz MA, Lo-Coco F (2011) Modern approaches to treating acute promyelocytic leukemia. J Clin Oncol 29:495–503
127. Saumet A, Vetter G, Bouttier M, Portales-Casamar E, Wasserman WW, Maurin T, Mari B, Barbry P, Vallar L, Friederich E, Arar K, Cassinat B, Chomienne C, Lecellier CH (2009) Transcriptional repression of microRNA genes by PML-RARA increases expression of key cancer proteins in acute promyelocytic leukemia. Blood 113:412–421
128. Schoorlemmer J, Van PA, Van DEN EM, Jonk L, Pals C, Kruijer W (1994) Characterization of a negative retinoic acid response element in the murine Oct4 promoter. Mol Cell Biol 14:1122–1136
129. Schwarz DS, Hutvagner G, Du T, Xu Z, Aronin N, Zamore PD (2003) Asymmetry in the assembly of the RNAi enzyme complex. Cell 115:199–208
130. Smirnova L, Grafe A, Seiler A, Schumacher S, Nitsch R, Wulczyn FG (2005) Regulation of miRNA expression during neural cell specification. Eur J Neurosci 21:1469–1477
131. Smith B, Treadwell J, Zhang D, Ly D, McKinnell I, Walker PR, Sikorska M (2010) Large-scale expression analysis reveals distinct microRNA profiles at different stages of human neurodevelopment. PLoS One 5:e11109
132. Song JJ, Smith SK, Hannon GJ, Joshua-Tor L (2004) Crystal structure of Argonaute and its implications for RISC slicer activity. Science 305:1434–1437
133. Sontheimer EJ, Carthew RW (2005) Silence from within: endogenous siRNAs and miRNAs. Cell 122:9–12
134. Sood P, Krek A, Zavolan M, Macino G, Rajewsky N (2006) Cell-type-specific signatures of microRNAs on target mRNA expression. Proc Natl Acad Sci USA 103:2746–2751
135. Srivastava D (2006) Making or breaking the heart: from lineage determination to morphogenesis. Cell 126:1037–1048

136. Stallings RL, Foley NH, Bray IM, Das S, Buckley PG (2011) MicroRNA and DNA methylation alterations mediating retinoic acid induced neuroblastoma cell differentiation. Semin Cancer Biol 21:283–290
137. Stark A, Brennecke J, Bushati N, Russell RB, Cohen SM (2005) Animal MicroRNAs confer robustness to gene expression and have a significant impact on 3′UTR evolution. Cell 123:1133–1146
138. Starnes LM, Sorrentino A, Ferracin M, Negrini M, Pelosi E, Nervi C, Peschle C (2010) A transcriptome-wide approach reveals the key contribution of NFI-A in promoting erythroid differentiation of human CD34(+) progenitors and CML cells. Leukemia 24:1220–1223
139. Starnes LM, Sorrentino A, Pelosi E, Ballarino M, Morsilli O, Biffoni M, Santoro S, Felli N, Castelli G, De Marchis ML, Mastroberardino G, Gabbianelli M, Fatica A, Bozzoni I, Nervi C, Peschle C (2009) NFI-A directs the fate of hematopoietic progenitors to the erythroid or granulocytic lineage and controls beta-globin and G-CSF receptor expression. Blood 114:1753–1763
140. Sumazin P, Yang X, Chiu HS, Chung WJ, Iyer A, Llobet-Navas D, Rajbhandari P, Bansal M, Guarnieri P, Silva J, Califano A (2011) An extensive microRNA-mediated network of RNA-RNA interactions regulates established oncogenic pathways in glioblastoma. Cell 147:370–381
141. Taft RJ, Glazov EA, Cloonan N, Simons C, Stephen S, Faulkner GJ, Lassmann T, Forrest AR, Grimmond SM, Schroder K, Irvine K, Arakawa T, Nakamura M, Kubosaki A, Hayashida K, Kawazu C, Murata M, Nishiyori H, Fukuda S, Kawai J, Daub CO, Hume DA, Suzuki H, Orlando V, Carninci P, Hayashizaki Y, Mattick JS (2009) Tiny RNAs associated with transcription start sites in animals. Nat Genet 41:572–578
142. Takahashi H, Kanno T, Nakayamada S, Hirahara K, Sciume G, Muljo SA, Kuchen S, Casellas R, Wei L, Kanno Y, O'Shea JJ (2012) TGF-beta and retinoic acid induce the microRNA miR-10a, which targets Bcl-6 and constrains the plasticity of helper T cells. Nat Immunol 13:587–595
143. Takahashi K, Yamanaka S (2006) Induction of pluripotent stem cells from mouse embryonic and adult fibroblast cultures by defined factors. Cell 126:663–676
144. Tan Y, Zhang B, Wu T, Skogerbo G, Zhu X, Guo X, He S, Chen R (2009) Transcriptional inhibiton of Hoxd4 expression by miRNA-10a in human breast cancer cells. BMC Mol Biol 10:12
145. Tavazoie SF, Alarcon C, Oskarsson T, Padua D, Wang Q, Bos PD, Gerald WL, Massague J (2008) Endogenous human microRNAs that suppress breast cancer metastasis. Nature 451:147–152
146. Tay Y, Kats L, Salmena L, Weiss D, Tan SM, Ala U, Karreth F, Poliseno L, Provero P, Di CF, Lieberman J, Rigoutsos I, Pandolfi PP (2011) Coding-independent regulation of the tumor suppressor PTEN by competing endogenous mRNAs. Cell 147:344–357
147. Tay YM, Tam WL, Ang YS, Gaughwin PM, Yang H, Wang W, Liu R, George J, Ng HH, Perera RJ, Lufkin T, Rigoutsos I, Thomson AM, Lim B (2008) MicroRNA-134 modulates the differentiation of mouse embryonic stem cells, where it causes post-transcriptional attenuation of Nanog and LRH1. Stem Cells 26:17–29
148. Terao M, Fratelli M, Kurosaki M, Zanetti A, Guarnaccia V, Paroni G, Tsykin A, Lupi M, Gianni M, Goodall GJ, Garattini E (2011) Induction of miR-21 by retinoic acid in estrogen receptor-positive breast carcinoma cells: biological correlates and molecular targets. J Biol Chem 286:4027–4042
149. Tong MH, Mitchell D, Evanoff R, Griswold MD (2011) Expression of Mirlet7 family microRNAs in response to retinoic acid-induced spermatogonial differentiation in mice. Biol Reprod 85:189–197
150. Tong MH, Mitchell DA, McGowan SD, Evanoff R, Griswold MD (2012) Two miRNA clusters, Mir-17-92 (Mirc1) and Mir-106b-25 (Mirc3), are involved in the regulation of spermatogonial differentiation in mice. Biol Reprod 86:72
151. Tsang J, Zhu J, van Oudenaarden A (2007) MicroRNA-mediated feedback and feedforward loops are recurrent network motifs in mammals. Mol Cell 26:753–767

152. Vasudevan S, Tong Y, Steitz JA (2007) Switching from repression to activation: microRNAs can up-regulate translation. Science 318:1931–1934
153. Verdel A, Jia S, Gerber S, Sugiyama T, Gygi S, Grewal SI, Moazed D (2004) RNAi-mediated targeting of heterochromatin by the RITS complex. Science 303:672–676
154. Vergoulis T, Vlachos IS, Alexiou P, Georgakilas G, Maragkakis M, Reczko M, Gerangelos S, Koziris N, Dalamagas T, Hatzigeorgiou AG (2012) TarBase 6.0: capturing the exponential growth of miRNA targets with experimental support. Nucleic Acids Res 40:D222–D229
155. Vian L, Di Carlo M, Pelosi E, Fazi F, Santoro S, Cerio AM, Boe A, Rotilio V, Billi M, Racanicchi S, Testa U, Grignani F, Nervi C (2014) Transcriptional fine-tuning of microRNA-223 levels directs lineage choice of human hematopoietic progenitors. Cell Death Differ 21:290–301
156. Wang XS, Gong JN, Yu J, Wang F, Zhang XH, Yin XL, Tan ZQ, Luo ZM, Yang GH, Shen C, Zhang JW (2012) MicroRNA-29a and microRNA-142-3p are regulators of myeloid differentiation and acute myeloid leukemia. Blood 119:4992–5004
157. Wassenegger M (2005) The role of the RNAi machinery in heterochromatin formation. Cell 122:13–16
158. Weber F, Teresi RE, Broelsch CE, Frilling A, Eng C (2006) A limited set of human MicroRNA is deregulated in follicular thyroid carcinoma. J Clin Endocrinol Metab 91:3584–3591
159. Weinmann L, Hock J, Ivacevic T, Ohrt T, Mutze J, Schwille P, Kremmer E, Benes V, Urlaub H, Meister G (2009) Importin 8 is a gene silencing factor that targets argonaute proteins to distinct mRNAs. Cell 136:496–507
160. Weiss FU, Marques IJ, Woltering JM, Vlecken DH, Aghdassi A, Partecke LI, Heidecke CD, Lerch MM, Bagowski CP (2009) Retinoic acid receptor antagonists inhibit miR-10a expression and block metastatic behavior of pancreatic cancer. Gastroenterology 137:2136–2145
161. Woltering JM Durston AJ (2008) MiR-10 represses HoxB1a and HoxB3a in zebrafish. PLoS One 3:e1396
162. Xavier-Neto J, Rosenthal N, Silva FA, Matos TG, Hochgreb T, Linhares VL (2001) Retinoid signaling and cardiac anteroposterior segmentation. Genesis 31:97–104
163. Xie C, Huang H, Sun X, Guo Y, Hamblin M, Ritchie RP, Garcia-Barrio MT, Zhang J, Chen YE (2011) MicroRNA-1 regulates smooth muscle cell differentiation by repressing Kruppel-like factor 4. Stem Cells Dev 20:205–210
164. Xu N, Papagiannakopoulos T, Pan G, Thomson JA, Kosik KS (2009) MicroRNA-145 regulates OCT4, SOX2, and KLF4 and represses pluripotency in human embryonic stem cells. Cell 137:647–658
165. Yekta S, Shih IH, Bartel DP (2004) MicroRNA-directed cleavage of HOXB8 mRNA. Science 304:594–596
166. Yi R, Qin Y, Macara IG, Cullen BR (2003) Exportin-5 mediates the nuclear export of pre-microRNAs and short hairpin RNAs. Genes Dev 17:3011–3016
167. Younger ST, Corey DR (2011) Transcriptional gene silencing in mammalian cells by miRNA mimics that target gene promoters. Nucleic Acids Res 39:5682–5691
168. Zardo G, Ciolfi A, Vian L, Billi M, Racanicchi S, Grignani F, Nervi C (2012) Transcriptional targeting by microRNA-Polycomb complexes: A novel route in cell fate determination. Cell Cycle 11:3543–3549
169. Zardo G, Ciolfi A, Vian L, Starnes LM, Billi M, Racanicchi S, Maresca C, Fazi F, Travaglini L, Noguera N, Mancini M, Nanni M, Cimino G, Lo-Coco F, Grignani F, Nervi C (2012) Polycombs and microRNA-223 regulate human granulopoiesis by transcriptional control of target gene expression. Blood 119:4034–4046
170. Zhang C (2009) Novel functions for small RNA molecules. Curr Opin Mol Ther 11:641–651
171. Zhong H, Wang HR, Yang S, Zhong JH, Wang T, Wang C, Chen FY (2010) Targeting Smad4 links microRNA-146a to the TGF-beta pathway during retinoid acid induction in acute promyelocytic leukemia cell line. Int J Hematol 92:129–135

Chapter 9
Integrative Genomics to Dissect Retinoid Functions

Marco-Antonio Mendoza-Parra and Hinrich Gronemeyer

Abstract Retinoids and rexinoids, as all other ligands of the nuclear receptor (NR) family, act as ligand-regulated *trans*-acting transcription factors that bind to *cis*-acting DNA regulatory elements in the promoter regions of target genes (for reviews see [12, 22, 23, 26, 36]). Ligand binding modulates the communication functions of the receptor with the intracellular environment, which essentially entails receptor-protein and receptor-DNA or receptor-chromatin interactions. In this communication network, the receptor simultaneously serves as both intracellular sensor and regulator of cell/organ functions. Receptors are "*intelligent*" mediators of the information encoded in the chemical structure of a nuclear receptor ligand, as they interpret this information in the context of cellular identity and cell-physiological status and convert it into a dynamic chain of receptor-protein and receptor-DNA interactions. To process input and output information, they are composed of a modular structure with several domains that have evolved to exert particular molecular recognition functions. As detailed in other chapters in this volume, the main functional domains are the DNA-binding (DBD) and ligand-binding (LBD) [5–7, 38, 56, 71]. The LBD serves as a dual input-output information processor. Inputs, such as ligand binding or receptor phosphorylations, induce allosteric changes in receptor surfaces that serve as docking sites for outputs, such as subunits of transcription and epigenetic machineries or enzyme complexes. The complexity of input and output signals and their interdependencies is far from being understood.

M.-A. Mendoza-Parra · H. Gronemeyer (✉)
Department of Functional Genomics and Cancer, Institut de Génétique et de Biologie Moléculaire et Cellulaire (IGBMC), CNRS, INSERM, Université de Strasbourg, BP 10142, 67404 Illkirch Cedex, France
e-mail: hg@igbmc.fr

M. A. Asson-Batres and C. Rochette-Egly (eds.), *The Biochemistry of Retinoic Acid Receptors I: Structure, Activation, and Function at the Molecular Level*, Subcellular Biochemistry 70,
DOI: 10.1007/978-94-017-9050-5_9, © Springer Science+Business Media Dordrecht 2014

Abbreviations

9-*cis* RA	9-*cis* retinoid acid
AM580	RARα-selective synthetic ligand
apoNR	Non-liganded NR
at-RA	All-*trans* retinoic acid
BMS753	RARα-selective synthetic ligand
BMS961	RARγ-selective synthetic ligand
CD437	RARγ-selective synthetic ligand
Cistrome	The total set of genes in a given cell that contains *cis*-acting DNA binding/response/target sites for a given TF; generally defined by ChIP-seq and related technologies
ChIP	Chromatin immunoprecipitation
ChIP-seq	ChIP coupled to massive parallel sequencing
CoA	Co-activator
CoR	Co-repressor
DBD	DNA-binding domain
ECC	Embryo carcinoma cell (e.g., F9 or P19)
ESC	Embryonic stem cells
Epigenome	General term to describe the patterns of post-translational modification of chromatin histones alone the genome and the modification of DNA, such as methylation or hydroxymethylation of cytosines
HDAC	Histone deacetylase
IP	Immunoprecipitation
IPed	Immunoprecipitated
Isotype	Three RAR and RXR receptors expressed from distinct genes (RARα, RARβ RARγ; RXRα, RXRβ and RXRγ)
LBD	Ligand-binding domain
MEF	Mouse embryonic fibroblast
NR	Nuclear receptor
RAR α, β, γ	Retinoic acid receptor α, β, γ
RXR α, β, γ	Retinoid X receptor α, β, γ
TF	Transcription factor
Transcriptome	All transcribed RNAs produced in one or a population of cells.

Retinoid receptors, RARs and RXRs, are each expressed from the three isotypic genes (α, β and γ), which express isoforms by differential promoter usage and splicing [36]. RAR and RXR isotypes form heterodimers, and RAR isotype-selective and RXR-selective ligands have been developed [12]. While RAR-RXR heterodimers respond to cognate RAR ligands, RXR ligand action requires prior RAR ligand binding (termed RXR '*subordination*', a phenomenon that is molecularly well understood [24] but does not apply to all RXR heterodimers.

The early steps of nuclear receptor function and the physiological impact of retinoic acid receptor (RAR) heterodimers are rather well understood. Numerous molecular, structural, and structure-activity relationship studies have informed us

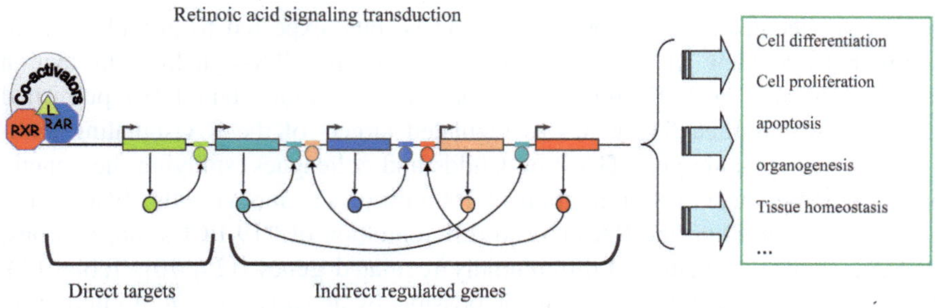

Fig. 9.1 Schematic representation of the retinoic acid signaling transduction process

about the sequence of events that follows ligand binding, and we understand how these events can be modulated by ligand design [9, 26, 53]. However, how a single ligand, such as retinoic acid (RA), that binds to multiple receptors regulates a plethora of cell-specific dynamic networks of genes and how the epigenome contributes to transcriptional regulation that ultimately reads out as a cell-level, physiological phenomenon, is still a black box (Fig. 9.1). It is our view that it is time to develop a *systems biology of nuclear receptor action*. Due to advances in massive parallel sequencing and bioinformatics analyses of genome-wide data sets, such a quest is possible. Indeed, it is now possible to integrate data on global transcription factor binding, epigenetic chromatin histone and DNA modification patterns with transcriptome and 3-dimensional chromatin structure data. *Decision* points that govern temporal control points in gene networks that are the ultimate genetic readouts of the RA- (or, more generally, NR-) induced physiological phenomena can be extracted and deciphered from these integrations. Here, we will discuss the chronology of the development of increasingly larger data sets for RA action and provide an overview of present attempts to integrate a multitude of genome wide data sets in the context of a 4-dimensional appreciation of chromatin structure and activity.

History: Retinoic Acid Signaling in the Post-Genomic Era

With the publication of the first draft of the human genome sequence in 2001 and subsequently, various other model organisms, the molecular genetics behind organismal homeostasis has entered into a new era. In fact, studies on the regulation of biological phenomena are now generally performed in the context of available genome sequences and high throughput technologies for diverse applications, such as the expression of all nascent or specific classes of RNAs, the global binding of a transcription factor ('cistrome'), the genome-wide post-translational modification of chromatin ('epigenome') or the 4-dimensional organization of chromatin in space and time.

Importantly, this new way of interrogating the genome generates a greater number of significant targets than those identified in previous years by standard genetics/molecular biology approaches, and is thus, expected to provide a more comprehensive view of the regulatory events during RA-signaling. In fact, a review published in 2002 summarized the efforts of more than 1,191 published articles that classified 532 genes as regulated targets of the RA-signaling pathway [4]. In that same year, Geoffrey Childs and colleagues, studying the genetic basis for RA-induced differentiation of F9 embryo carcinoma cells (ECCs) into parietal endoderm and the RA-induced differentiation of P19 ECCs into neurons, characterized more than 500 differentially regulated genes ([29, 70]; Table 9.1). This increased discovery rate was made possible by an early version of microarray technology (cDNA PCR-spotted microarrays; reviewed in [40]. Interestingly, the two studies assessed differential gene expression at different time-points during RA-induced cell fate transitions. The corresponding functional genome annotation and temporal gene expression patterns offered a first insight into the signal transduction pathways involved in endodermal and neuronal differentiation.

To shed light on whether the expression of the genes was directly dependent on or regulated by the presence of a liganded and DNA-bound, RAR/RXR heterodimer, Harris and Childs [29] selected immediate response genes by inducing F9 differentiation with all-*trans*-RA (*at*-RA) for 6 h in the presence of the protein biosynthesis inhibitor, cycoheximide. Under these conditions, they identified 109 genes that displayed significant differential induction. Of these, only 22 were validated in a 9 day *at*-RA-exposure, time-course assay, suggesting that the other genes were false-positives due to the cycloheximide treatment.

Subsequent global transcriptomics studies were carried out using different model systems to try to discriminate putative primary/direct and secondary/indirect RA-responsive genes by systematically querying early and late treatment time points [16]. Others took advantage of the RAR-specificity of the *at*-RA synthetic analog, TTNPB, as a way to further increase the specificity of the assay [2, 42]. Although the underlying rationale was that the pan-RAR agonist, TTNPB, would more specifically identify RAR-responsive genes by decreasing the potential to inadvertently identify genes that were responding to permissive *9-cis*-RA-bound RAR/RXR heterodimers rather than *at*-RA-bound receptor complexes, the use of TTNPB does obviously not discriminate between the contributions of the different RAR/RXR isotype heterodimers.

Using RAR isotype-selective knock-out F9 ECC cells, Lorraine Gudas and her colleagues studied the role of RARγ by performing global gene expression profiling with wild-type and RARγ-/- cells in the presence or absence of *at*-RA [63]. Earlier studies [10, 11, 65] had demonstrated that the use of a RARγ specific ligand, and not those targeting the RARα or RARβ isotypes, drives *at*-RA induced F9 cell differentiation. The global gene expression study by Su and Gudas demonstrated that wild type and mutated RARγ$^{-/-}$ cells presented with similar proliferation and morphological characteristics in the absence of RA treatment, but displayed important differences in their gene expression profiles. A similar observation was made with mutated RARα$^{-/-}$ F9 cells [37], suggesting that RARα and RARγ possess *ligand-independent* gene regulatory functions. Notably, these studies, in concordance with earlier studies [10, 11], reveal there are serious limitations in using RAR knockout cell lines to decipher the specific roles of RAR

Table 9.1 Published studies focused on dissecting retinoid function by applying "omics" approaches

Publication source	Relevant feature	Omics methodology
Harris and Childs [29]	Global gene expression of RA/dibutyryl cAMP-induced F9 parietal endoderm differentiation (kinetics over 9 days of treatment)	PCR-spotted microarrays (8,900 mouse cDNAs)
Wei et al. [70]	Global gene expression of RA-induced P19 neural differentiation (kinetics over 8 days of treatment)	PCR-spotted microarrays (9,000 mouse cDNAs)
Arima et al. [2]	Global analysis of RAR-responsive genes in the Xenopus embryo treated with the RAR-specific ligand TTNPB	Xenopus EST microarray (EST clones from NIBB Mochii normalized Xenopus neurula library (19,200 clones) and tail bud library (23,040 clones)
Eifert et al. [16]	Global gene expression of RA-induced F9 primitive endoderm differentiation. Two time-points (8 and 24 h under ATRA treatment) were evaluated relative to the absence of treatment condition	Atlas mouse 1.2 cDNA expression array (BD biosciences clontech; 1,176 mouse cDNAs plus 9 housekeeping cDNAs in a nylon membrane format); Affymetrix murine genome U74Av2 gene ChIP oligonucleotide microarrays (12,488 unique genes per array)
Mamoon et al. [42]	RA-responsive genes assessed in murine hepatocyte cell line (AML12) treated with ATRA as well as with the RAR-specific ligand TTNPB	Affymetrix Genechip mouse genome 430 2.0 microarrays (39,000 transcripts represented)
Su and Gudas [63]	Global gene expression in wild type and RARγ-/- F9 cells with and without RA treatment	Affymetrix Genechip® arrays
Hua et al. [30]	Global gene expression and e-GFP tagged RARα/RARγ binding sites assessed in MCF7 breast cancer cells under RA treatment	Agilent human genome oligo microarrays (gene expression); Affymetrix GeneChIP® Human tiling 2.0R arrays (ChIP-chip)
Delacroix et al. [13]	Overexpressed RARα and RARγ chromatin localization in MEF and ES cells under ATRA treatment	Agilent promoter arrays (ChIP-chip)
Mahony et al. [41]	Global gene expression and pan-RAR localization in mES cells before and after RA-treatment	Affymetrix mouse genome 430 2.0 microarrays (gene expression); Solexa sequencing (ChIP-seq)
Delacroix et al. [48]	Global gene expression in F9 cells under ATRA and RAR specific agonists. RXRα and RARγ binding sites assessed under ATRA treatment (kinetics over 48 h)	Affymetrix mouse GeneChIP® microarrays (gene expression); Solexa sequencing (ChIP-seq)

isotypes. This is because there can be artifactual ligand responses by RAR/RXR heterodimers in RAR isotype knockouts. This suggests that the normal role(s) of each RAR isotype, as deduced from studies of RAR-RXR knockout cell models, requires cautious interpretation.

Genome-Wide Mapping of Retinoic Acid Receptors Binding Sites

Dissection of the effects of RA on physiological processes requires a comprehensive mapping of the chromatin interaction of the various RAR/RXR heterodimers before and after ligand exposure (Fig. 9.1). Previous in vitro binding and transactivation studies demonstrated that RAR/RXR heterodimers bind efficiently to inverted (IR) or direct repeat (DR) sequences of the hexameric motif (A/G)G(G/T)TCA, often spaced by 5, 2 or 1 nucleotides (DR5, DR2, DR1) due to the dimerization characteristics of the DNA binding domain [54, 67, 74, 75]. While this characteristic RA Response Element (RARE) could in principle allow the identification of all potential RAR/RXR binding sites in a given genome, the sequence motifs of some RAREs associated with well-known RA-induced genes demonstrated major divergence from the consensus motif [35]. Available evidence indicates that although consensus RAREs may be efficient and correspond to high affinity binding sites, they rarely occur in natural RA target genes. In addition, relying on consensus sequences to identify RA-regulated genes does not take into consideration additional epigenetic mechanisms that regulate access of RAR/RXR heterodimers and transcription factors to chromatin [33, 34], the action of "*pioneer transcription factors*" [73], or synergistic interactions with or tethering to other NR/TFs [57].

At present, the methods of choice for comprehensive and unbiased mapping of protein-chromatin interactions are a combination of chromatin immunoprecipitation (ChIP) with high throughput profiling approaches like ChIP-chip and more recently, ChIP-seq (Fig. 9.2). DNA biochips, also called DNA arrays, were first described in 1995. This technology provides a method for interrogating *protein-DNA-chromatin* associations by first, co-immunoprecipitating protein-DNA fragments, and then hybridizing the captured DNA on a solid support coated with an arrangement of single-stranded DNA molecules (commonly referred to as probes) covering, for instance, the complete genome sequence of a selected species. The assay resolution of this approach, commonly called *ChIP-chip* (Fig. 9.2) is directly related to the number of genomic probes that can be spotted onto the solid support and its sensitivity depends on the minimal amount of co-immunoprecipitated DNA fragments required per assay. ChIP-chip was widely used until the arrival of second generation genome sequencers that were able to provide faster, less expensive, and more direct ways to evaluate the diversity of co-immunoprecipitated DNA fragments. Both assay resolution and sensitivity were highly increased using *ChIP-seq* technology because it involves direct sequencing of the immunoprecipitated fragments (Fig. 9.2), thus accounting for the current overwhelming use of this approach.

Among the first studies to apply global approaches for mapping chromatin localization of RARs were those carried out by Delacroix and colleagues [13] using mouse embryo fibroblasts (MEFs) and by Hua and colleagues [30] using human MCF-7 breast cancer cells. The MEF study attempted to discriminate between direct and indirect RA-regulated targets using Taf4$^{lox/-}$ MEFs, which undergo morphological changes upon RA treatment accompanied by changes in the expression of more than 1,000 genes [19]. After integration of 3xFlag-HA tagged RARα or RARγ isotypes, which allowed immunoprecipitation with anti-Flag and anti-HA antibodies, they performed ChIP assays and hybridized the immunoprecipitated DNA to Agilent arrays coated with DNA encoding the promoter regions of around 17,000 genes (ChIP-chip) [13]. They identified ~300 RAR-occupied sites of which less than 25 % corresponded to differentially expressed RA target genes. In part, the low correlation between RAR occupancy and actual functional relevance can be explained by the design of the assay; that is, ChIP-chip was performed with MEFs treated for 2 h with RA, while the transcription profiling was done with MEFs treated for 24 h. This highlights the potential risks in comparing different sources of global information and the need to rigorously design assays with appropriate attention to matched conditions of targets and probes and normalization of datasets.

Prior to the MEF study, using a conceptually similar approach, eGFP-tagged RARα or RARγ were integrated in human MCF-7 breast cancer cells to allow characterization of the role of specific RAR isotypes in mediating the anti-proliferative and apoptotic effects of RA [30]. In this case, immunoprecipitated chromatin was hybridized to tiling arrays containing more than 40 million oligonucleotide probes that represented the entire human genome. More than 3,000 RARγ and more than 7,000 RARα binding sites were found under these conditions. Importantly, more than 85 % of the identified sites were located in intronic or promoter-distal intergenic regions. In addition, the transcriptional response in the MCF-7 model system was evaluated with *at*-RA, as well as with the RARα-specific agonist, AM580 and the RARγ specific-agonist, CD437. Moreover, the authors used RARα and RARγ isotype-specific RNA interference to link the differential gene expression seen with isotype-selective ligands with presence of the corresponding RAR isotype, and found a high degree of correlation. This study also demonstrated an unexpected competition between RAR/RXR heterodimers and the estrogen receptor ERα for binding sites, suggesting there may be antagonistic transcription regulation for up to 71 % of the evaluated target genes.

Development of the Field: Identification of Decision Points and Key Factors that Diversify and Dynamically Regulate RA-Induced Gene Expression

More recently, a similar study was carried out with mouse embryonic stem cells (ESC) to identify the RA-dependent gene programs involved in neuronal differentiation [41]. A pan-RAR antibody was used to map endogenous RAR chromatin

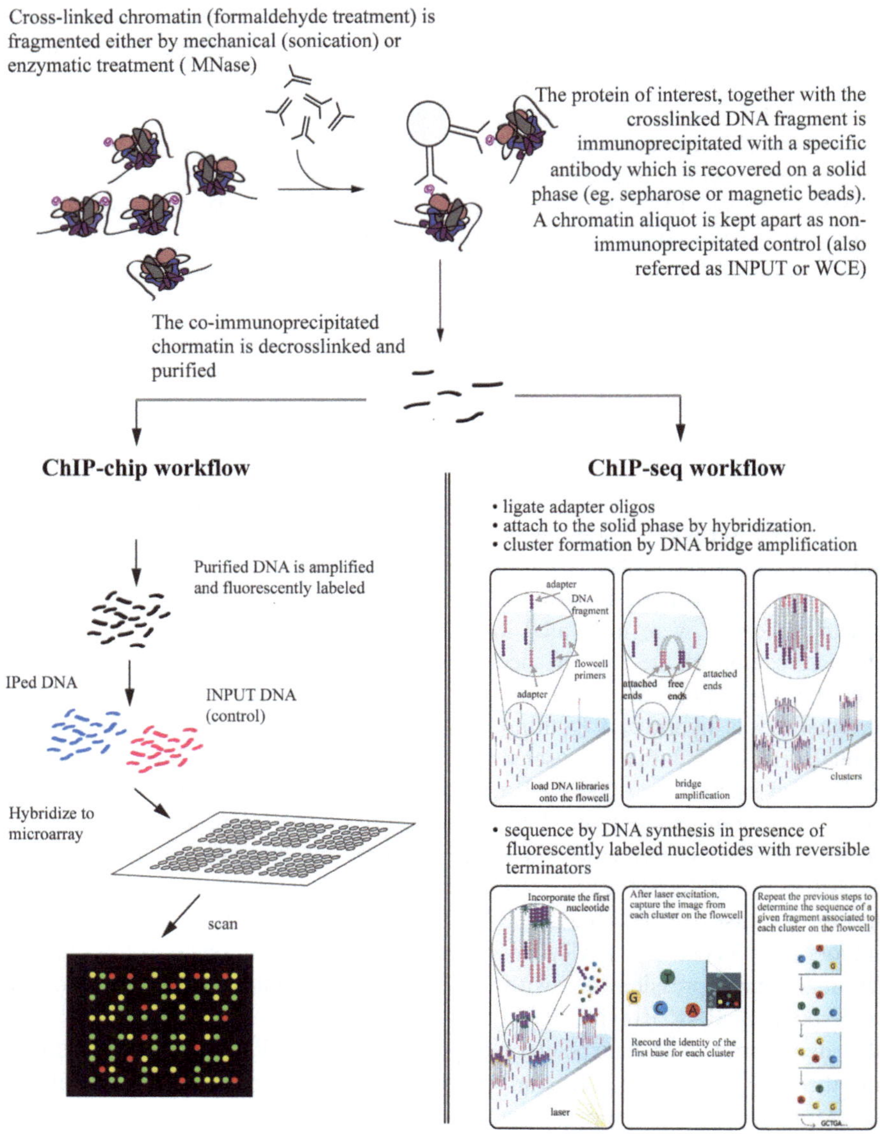
Cross-linked chromatin (formaldehyde treatment) is fragmented either by mechanical (sonication) or enzymatic treatment (MNase)
The protein of interest, together with the crosslinked DNA fragment is immunoprecipitated with a specific antibody which is recovered on a solid phase (eg. sepharose or magnetic beads). A chromatin aliquot is kept apart as non-immunoprecipitated control (also referred as INPUT or WCE)
The co-immunoprecipitated chormatin is decrosslinked and purified
ChIP-chip workflow
Purified DNA is amplified and fluorescently labeled
IPed DNA
INPUT DNA (control)
Hybridize to microarray
scan
ChIP-seq workflow
• ligate adapter oligos
• attach to the solid phase by hybridization.
• cluster formation by DNA bridge amplification
adapter
DNA fragment
flowcell primers
adapter
load DNA libraries onto the flowcell
attached ends
free ends
attached ends
bridge amplification
clusters
• sequence by DNA synthesis in presence of fluorescently labeled nucleotides with reversible terminators
Incorporate the first nucleotide
laser
After laser excitation, capture the image from each cluster on the flowcell
G
T
C
A
Record the identity of the first base for each cluster
Repeat the previous steps to determine the sequence of a given fragment associated to each cluster on the flowcell
GCTGA...

Fig. 9.2 Schematic comparison between chromatin immunoprecipitation assays evaluated by hybridization onto DNA biochips (ChIP-chip) and massive parallel sequencing (ChIP-seq). In ChIP-chip assays, the immunoprecipitated DNA is hybridized with a solid surface (biochip) previously coated with single-stranded DNA molecules (referred as probes) representing for instance, a complete genome. In the illustrated example (*left panel*), the immunoprecipitated DNA and the non-immunoprecipitated control (also referred to as INPUT or WCE for whole cell extract) are labeled with two different fluorophores and then hybridized together on a single DNA biochip. An alternative approach is based on hybridization on two different DNA biochips, followed by a computational comparison of the imaged fluorescent levels. ChIP-seq assays are based on the direct sequencing of the immunoprecipitated DNA by using a massive parallel sequencing approach. Briefly it consists of the incorporation of adapter sequences at both ends of the immunoprecipitated DNA; then such adapters are used for attaching the DNA molecules onto a solid surface coated with single-strand DNA molecules representing the complementary sequence to the adapters in use. The attached molecules are amplified by following several rounds of "bridge DNA amplification" based on the alternate attachment of the adapters in use to the solid surface (*right panel*). Bridge amplification produces DNA clusters formation, which are then sequenced by DNA synthesis in the presence of fluorescently labeled nucleotides with reversible terminators. This illustrated procedure corresponds to that developed by the company Solexa. Other massive parallel sequencing approaches have also been developed for this purpose

binding sites, thus avoiding the potential for over-expressed or tagged constructs to identify artifactual binding sites. The assay, performed on cells exposed to *at*-RA for 8 h, revealed both constitutive and RA-induced, de novo binding sites. ChIP-seq studies were performed to further assess the role of the RAR binding sites on transcription regulation, global microarray-based gene expression, and RNA polymerase II initiation and elongation. The number of genes differentially regulated by RA was estimated using a 5 kb proximity criterion that predicted a link between RAR binding sites and *proximal* transcriptionally active, coding regions. Clearly, RA-responsive genes can also, in principle, be regulated by *distal* enhancers, which cannot be identified by such a simplified binding site proximity criterion.

Our own recent study used the well-established F9 embryonal carcinoma cell line (ECC) model to dissect the gene regulatory pathways responsible for RA-induced endodermal differentiation. This was done by integrating global RAR binding and gene regulation information from samples collected at five different time-points over the course of 48 h exposure of the cells to *at*-RA or RARα, β, or γ-specific agonists [48].

In contrast with results obtained using MCF7 cells [30], F9 cells treated with an RARγ-specific agonist, BMS961 (but not those exposed to an RARα-specific agonist, BMS753), induced a pattern of differential gene expression that was similar to that induced by *at*RA [48]. This is consistent with previous studies demonstrating that the RARγ agonist (but not those of RARα or RARβ) induces an F9 cell differentiation phenotype that is indistinguishable from the one induced by the natural ligand, *at*-RA, suggesting that RARγ is driving differentiation in F9 cells [65].

Given the decisive role of RARγ in F9 cell differentiation, we identified the chromatin binding sites of RARγ/RXRα heterodimers by mapping each receptor separately at all 5 time-points. Overall, RXRα displayed more binding sites than RARγ, as was expected from the promiscuous heterodimerization of RXRα with

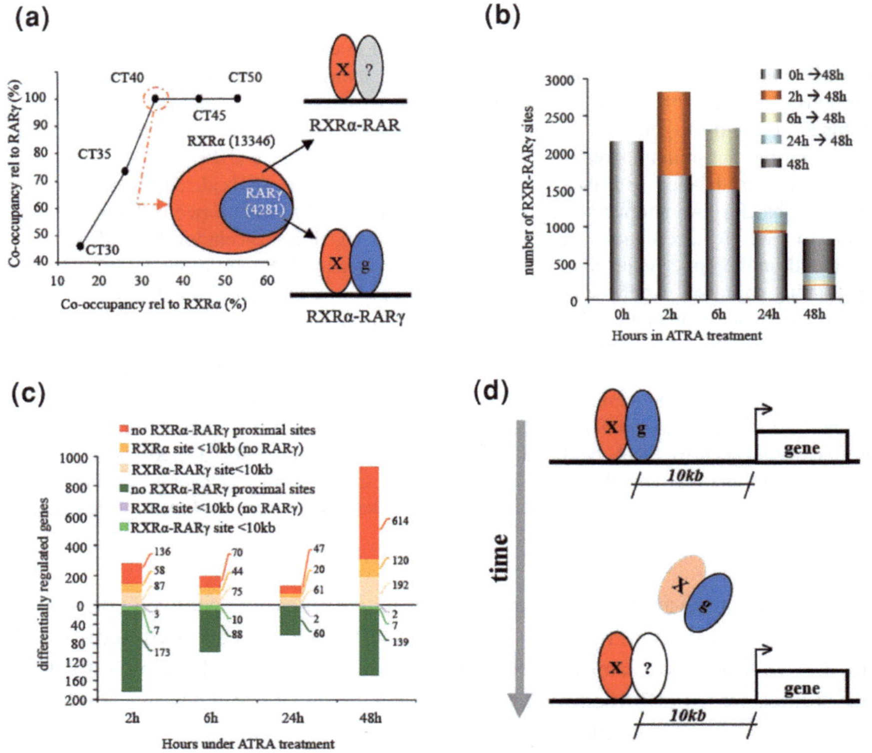

Fig. 9.3 RXRα and RARγ nuclear receptors present a highly dynamic binding to chromatin during ATRA-induced F9 differentiation. **a** The percent of RXRα and RARγ co-occupancy relative to the total number of RXRα or RARγ binding sites retrieved over all time-series evaluated profiles is illustrated for different P-value confidence thresholds (CT = −10*log (P-value)). The inset (Venn diagram) shows that at CT = 40 all identified RARγ sites are found co-occupied with RXRα. This subset of binding sites is considered bona fide RXRα/RARγ heterodimer binding sites and has been used for all further analysis. **b** The RXRα/RARγ binding sites identified in (**a**) are illustrated in the context of their temporal recruitment, duration of occupancy and dissociation. RXRα/RARγ co-occupied sites per time point are subclassified based on their recruitment intervals and depicted by colour coding. **c** Genes exhibiting ATRA-induced or repressed mRNA levels at the indicated time points during F9 cell differentiation (induced genes ≥1.8-fold; repressed genes ≤0.5-fold relative to vehicle) were classified as putative target genes if at least one RXRα or RXRα/RARγ binding site was located in proximity (≤10 kb distance). **d** Schematic model illustrating the progressive loss of RARγ but not of RXRα from chromatin binding sites observed during ATRA-induced F9 differentiation

multiple partners (Fig. 9.3a). Merging the datasets and extracting binding sites that were common to both RXRα and RARγ revealed a population of constitutive RARγ/RXRα binding sites plus a population that was highly dynamic during *at*-RA treatment. The overall number of RARγ/RXRα binding sites decreased during F9 differentiation (~2,000 sites in the absence of treatment and less than 1,000 sites after 48 h exposure of cells to *at*-RA), and we detected significant amounts of de novo recruited heterodimers even after 24 h of *at*-RA treatment (Fig. 9.3b, c). These results suggested a sustained and highly dynamic interaction of the

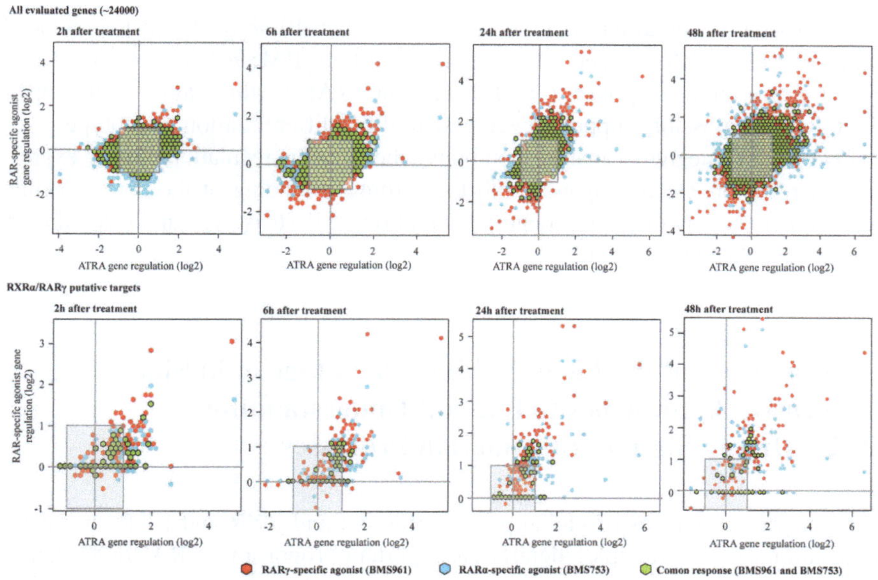

Fig. 9.4 Differential gene expression response induced by ATRA treatment in comparison to that induced by the RAR-specific agonists. The *upper panels* illustrates the gene expression response induced in all evaluated coding regions (24,000 genes; Affymetrix mouse GeneChIP® microarrays), while the *lower panels* displays the response in the characterized RXRα/RARγ putative target genes. Gene regulation response induced either by the RARγ-specific agonist BMS961 (*red*), by the RARα-specific agonist BMS753 (*light blue*) or by both ligands (*green*) are displayed in the context of the ATRA-induced response. The *central box* in each panel delineates a gene expression response area lower than 2 folds, thus the significant gene regulation responses are found outside of this delimited surface

RARγ/RXRα heterodimer with chromatin targets during this cell physiological process. The observed decrease in binding sites of RARγ/RXRα heterodimers relative to the total number of RXRα-occupied sites may result from an exchange of RXRα with other NR partners during the differentiation (Fig. 9.3d).

We found that more than 50 % of the genes induced during the first 24 h of *at*-RA treatment showed a RXRα or an RARγ/RXRα binding site within 10 kb proximity (Fig. 9.3c). In contrast, most of the down-regulated genes lacked such sites. Importantly, more than 70 % of the mapped RXRα sites could not be associated to an annotated coding region, suggesting that they might regulate transcription through 3-dimensional chromatin structures or may regulate as yet non-annotated transcripts. To further confirm direct transcriptional regulation by RARγ/RXRα binding sites, we compared transcriptional responses in cells exposed to *at*-RA or RAR-specific agonists (Fig. 9.4) [48]. Approximately 60 % of the *at*-RA-induced putative RARγ/RXRα targets did respond similarly to the differentiation competent RARγ agonist BMS961. Surprisingly, however, ~40 % responded also to the RARα agonist BMS753. This suggests that (1) ~40 % of the *at*-RA-induced putative RARγ/RXRα targets that did not respond to the BMS961 treatment require, or can operate in a redundant manner with other RAR isotypes, are not essential for differentiation, and (2) another 40 % of the *at*-RA-induced putative RARγ/RXRα targets that responded to both BMS961 and BMS753 display a promiscuous response to both RARα and RARγ, which alone is not sufficient, but may possibly support development of the differentiation phenotype.

Overall, the integrative analysis combining the dynamic regulation of gene expression by receptor-selective ligands with the chromatin binding of the corresponding heterodimers largely facilitates identification of direct RAR isotype and heterodimer-selective regulated target genes.

Current State of the Field: Taking Advantage of in Silico Integrative Approaches to Expand Understanding of RA-Driven Signal Transduction Processes

The integrative analysis of global gene expression and RXR and RAR chromatin association can, in principle, identify an important proportion of RAR/RXR heterodimer-mediated gene regulatory events. Including a temporal dimension revealed a highly dynamic target gene expression profile and dynamic occupancy of chromatin by pre-existing and de novo recruited RAR/RXR heterodimers, as well as heterodimer replacement or even partner swaps.

The gain of information remains restricted to directly regulated RAR/RXR heterodimer-targets, which represent only a small fraction of all differentially RA-regulated genes. The remaining majority of regulated genes are generally referred to as indirectly or secondarily regulated genes. It is reasonable to assume a hierarchical order of transcription regulation in which the direct targets are in the front line of the signaling process (‘*initiator program*’) and the downstream layers comprise temporally specified (‘*executor*’) gene programs that result in amplification, diversification and specification of secondary gene programs that determine cell fate and ultimately, cell differentiation. The initiation phase is mediated mainly, albeit not exclusively, by TFs. Our studies have confirmed that many TFs are among the early genes. Therefore, the reconstruction of the executor programs may profit from the characterization of the cascade of TFs that propagate the signal transduction and diversification process. In past years, the chromatin location of various TFs has been mapped in several model systems by ChIP-chip or ChIP-seq approaches and released to public repositories, thus generating an important resource for in silico omics dataset integration. Importantly, the integration of TF target gene information into time course gene expression data is a powerful method for identifying downstream regulatory events during a signal transduction process [18]. To deconvolute RA signaling pathways during F9 induced differentiation, we have integrated TF target gene annotation, including identified direct putative RARγ/RXRα targets, into the *at*-RA–induced gene programming [48]. For this we used the recently developed Dynamic Regulatory Events Miner (DREM; [18], which uses input data for temporal alteration of gene expression at given time points and transcription factor-target gene interactions. The underlying

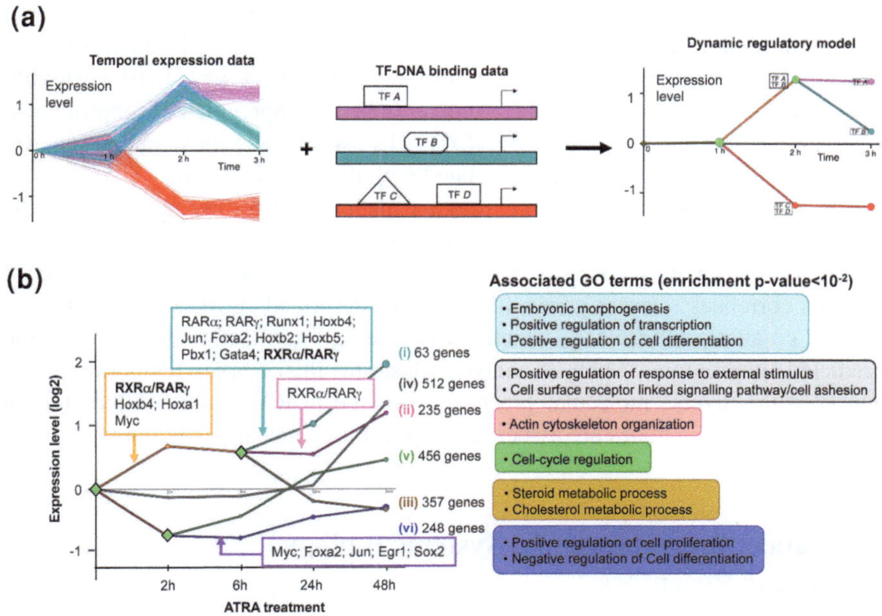

Fig. 9.5 Reconstructing a dynamic regulatory map for the RA-driven transcriptome. **a** Shematic representation of the integrative approach used by DREM. Temporal gene expression information (*left panel*) is combined with transcription factors-DNA binding annotations (*middle panel*) to infer a dynamic regulatory model. In the illustrated example, co-expressed genes are classified in three major paths (coloured in *pink*, *green* and *red* respectively) which in addition can be associated to a defined TFs based on the TF-DNA binding annotations (i.e. *pink* genes are regulated by TF A, *green* genes by TF B and *red* genes by TF C and TF D). In this manner, DREM aims at assessing how likely is that a given group of co-expressed genes may be transcriptionally regulated by a given TF. **b** DREM co-expression analysis is represented by *colour*-coded paths that summarize common characteristics. *Diamonds* indicate the predicted bifurcation points giving rise to the different co-expression paths and transcription factors whose target genes are overenriched in a given path are also illustrated. The number of genes per co-expression path, as well as their relevant gene ontology terms is displayed in aside. Panel **a** has been adapted from Ernst [18]

hypothesis is that various co-expression events (described here as co-expression paths) derive from defined transcriptional regulatory decisions. Integrating the information retrieved from TF target gene annotations from time course analyses of gene expression patterns will allow predictions for the involvement of a given TF in the formation of a defined co-expression path (i.e., at bifurcation points, as illustrated in Fig. 9.5). In the case of RA-induced differentiation of F9 ECC, DREM analysis predicted six distinct gene co-expression paths that recapitulate the different subprograms generated during RA-induced signal transduction. DREM allows evaluation of whether a given co-expression path is enriched for genes that are annotated as targets of a specific TF and whose actions contribute to the predicted bifurcation. In our analysis, three bifurcation points leading to signal diversification were identified along with candidate TFs predicted to cause diversification. As proof-of-principle, DREM correctly associated RARγ/RXRα with upregulated subprograms validated by differential gene expression and the chromatin-binding pattern of RARγ/RXRα (Fig. 9.5). Notably, DREM predicted that homeobox family transcription factors (e.g., Hoxa1, Hoxb2, Hoxb4, Hoxb5) and others like RARα or Foxa2, were enriched in upregulated subprograms, whereas TFs like Egr1 [50] and Sox2 [52] (TFs associated with stem cell renewal rather than differentiation), were associated with the repressed path.

The predicted RA-induced co-expression paths were further evaluated in the context of bibliographic gene co-citation interactions in order to construct predicted RA-driven RARγ/RXRα-mediated signaling networks [48]. This type of analysis correlates relevant genes, like the ones described above with their bibliographic co-citation '*partners*', which helps in assigning functional features to the predicted subprograms. Such an analysis illustrates the complex temporal coordination of the diverse molecular processes involved in RA-induced differentiation and predicts critical nodes are associated with cell fate transitions initiated by RA.

Relevance: Importance of a Systems Biology of Nuclear Receptors

Early studies in *Drososphila* paved the way toward a systems biology view of NR action. These studies characterized the temporal programming of gene activation induced by the steroid hormone, ecdysterone, to initiate molting and metamorphosis [3]. This was possible because gene activation manifests itself as local, reversible, alterations (*puffs*) of the polytene chromosomes that comprise approximately 1,000 chromatids in the interphase nuclei in the salivary glands of 3rd instar larvae. Temporal alteration of the puffing pattern provided a readout of sequential activation of gene programs. Early puffs corresponded to direct activation of TF-encoding, target genes by the ecdysone receptor [59, 66]. The ecdysone receptor turned out to be a heterodimer [72]. These features of the fly gene program share similarity with RAR/RXR-mediated activation of genetic sub-programs [48]. Thus, it is likely that the principles of the temporal gene programming seen for the ecdysone and retinoid receptors correspond to a general mechanism for signal diversification and temporal programming of hierarchical, downstream gene programs.

However, there are several additional factors and regulatory paradigms that impact program execution. One involves the surprising dynamics of RAR/RXR heterodimer binding [48]. These dynamics are characterized by heterodimer binding and dissociation at all time-points during the observation period of 48 h. As well, extensive RXRα partner swapping is observed, with either one RXRα heterodimer dissociating and being replaced by another one, or with a '*partner swap*' occurring while the heterodimer is bound to chromatin by an as yet, unknown, mechanism [48]. The effects of such heterodimer swapping on the dynamics of co-regulator-receptor complexes at target chromatin during cell differentiation has not been addressed with global approaches. This missing piece deserves attention, particularly in view of potentially distinct preferences of RXR heterodimers for co-regulators that are recruited to resulting multi-protein (epigenetic) complexes. Receptor-co-regulator interaction may correspond to a mechanism of target gene specification, and, vice versa, upon binding, the target DNA may allosterically alter receptor structure and function [43, 76].

Some, if not most, TFs that are regulated during the sequence of events governing cell fate determination have the ability to act as pioneer factors [73]. Such factors *open* chromatin structure thereby generating DNAse I hypersensitive sites that are otherwise not accessible to TFs that lack pioneering activity [69]. Obviously, one salient feature of the pioneering concept is that it provides a basis for the hierarchical and temporal order for the execution of gene programs.

Future Directions: An Integrative Genomics Era

How can the structural information present in a simple chemical molecule like *at*-RA set-up the sequence of temporally controlled events that finally lead to a differentiated cell? The F9 ECC global gene profiling studies have provided for the first time a systems biology view that RA-induced signaling comprises a diverse series of events that set in motion different regulatory decisions which occur in a time-defined manner throughout cell differentiation [44, 48]. Yet this view is far from comprehensive. In part, this is due to technical constraints related to the complexity of a system that operates with up to six receptors and multiple heterodimers. But it is also a consequence of the reduced number of molecular events that can currently be imported into spatio-temporal omics dataset analyses.

Multiple RXR: RAR Heterodimers Mediate RA-Signaling

The first level of signal diversification results from the multiplicity of RAR/RXR complexes that can be formed, which is a function of the expression levels of the six different RAR/RXR isotypes in a given cell (RARα, RARβ, RARγ, RXRα, RXRβ, RXRγ). Our recent study showed that it is possible to dissect the gene program regulated by the RARγ/RXRα heterodimer in a specific manner; however, the contribution of other RAR/RXR heterodimers RXRα partner swapping remain to be elucidated. Exploring the role of other RAR/RXR heterodimers depends on the availability of high quality "ChIP-seq grade" antibodies and ChIP-seq profiles. Note that in this respect there may be significant differences in the quality of ChIP-seq profiles calling for rigorous quality control assessment of data sets to allow integrative data analysis [47]. While this limitation can be overcome by using stably expressing epitope-tagged receptors, the risk of an altered functionality imposed by tagged receptors and potential interference of these modified receptors with regulation by endogenous receptors cannot be rigorously excluded. Notably, the heterodimer swap we observed between RARγ/RXRα and RARα/RXRα at the RARβ2 promoter is a note of caution for the use of exogenous modified factors.

Although RA signaling is mediated by heterodimers, this does not imply that both partners are located in the same chromatin region at the same time and in the same cell. Therefore, subsequent *reChIP-seq* assays, i.e. sequential ChIPs with antibodies directed against the two partners, followed by massive parallel sequencing, are necessary to provide reliable information about co-occupancy of the evaluated heterodimer partners at a given chromatin site. While reChIP assays have been shown to be a powerful method for evaluating simultaneous co-occupancy events in a locus-centric manner [8, 49], the low yields are not compatible with the requirements for global ChIP-seq assays. To overcome this problem we have recently combined reChIP assays with linear DNA amplification and sequencing (LinDA-reChIP-seq) in order to define the global binding pattern of co-occupied RXRα and RARγ chromatin sites to predict heterodimer binding patterns [46, 60, 61]. Using such strategies, the complexity of RAR/RXR heterodimers can be deconvoluted to reveal the contributions of different receptor pairs. As these studies go forward, it will be important to remember that RXR heterodimers with partners other than RARs may be involved in regulating networks that are initially set up by *at*-RA.

Pioneers, Epigenetic Modifications, and Co-Regulators Establish Regulatory Principles Affecting RA-Regulated Gene Programs Upstream and Downstream of RAR/RXR Heterodimer Action

It is well established that at given times and in particular cells, TFs bind to only a small fraction of their possible target sites in the genome. Pioneer TF remodelling and epigenetic chromatin modification can regulate TF access to certain chromatin sites (for a recent review see [73]). RAR/RXRs themselves may pioneer, for example in the context of RA-induced differentiation, the sub-programs regulated by other TFs. The interplay between the epigenetic status of target gene chromatin and RA regulation has been demonstrated in gene-centric studies with Polycomb proteins and the H3K27me3 mark [1, 25, 34]. Other epigenetic modifications may also regulate receptor recruitment and/or access, and the epigenetic action of co-activator/co-integrators recruited by liganded RAR/RXR heterodimers may exert pioneering activities for downstream programs. Comprehensive analysis of multi-dimensional omics-derived information together with bioinformatics tools that retrieve and integrate data describing RAR/RXR chromatin binding patterns, epigenomes, and transcriptomes will elucidate dynamic gene regulatory networks and provide a framework for experimental confirmation of the molecular mechanisms, key factors, and decision points that define cell fate decisions brought about by RA signaling.

The Dynamic Role of Three-Dimensional Chromatin Organization

The designation of RA target genes from ChIP-seq studies is generally based on proximity criteria which define genes 5 kb or 10 kb away from receptor-binding sites as candidate target genes. Using this definition, a large majority of binding sites are located in intergenic regions, and thus, only a small fraction of all identified binding events are considered in subsequent analyses. The function of these intergenic binding sites has become much clearer from recent studies interrogating the 3-dimensional organization of chromatin in the nucleus. It is now well accepted that the chromatin architecture, i.e. the organization of chromatin in *loops*, *domains* and possibly, *factories* with dedicated functionalities [64], corresponds to a structural organization that regulates the physical interaction between promoters and distant regulatory elements, sometimes with the involvement of non-coding RNAs. This view suggests the entire nucleus to be considered as a regulatory network of its own [20]. Technologies have been developed to analyze this architecture globally [Circular chromosome conformation capture [15, 62, 77]; Hi-C [31, 51, 68]; TCC [32], or with emphasis on a particular signaling or regulatory/processing component [ERα [21]; CTCF [28]; RNA polymerase II [39]. Yet, the dynamic aspect of nuclear architecture in processes like RA-induced differentiation or the changes in nuclear architecture associated with pathologic effects on signaling in diseased cells or organs has not been addressed. It is interesting to note that links between chromatin architecture and features of cancer cells are emerging [55, 58].

Computational Challenges for Omics Data Processing and Integration

The rapid development of next-generation sequencing (NGS) technologies poses multiple challenges for the bioinformatics analyses of enormous amounts of data that are being gathered in massive parallel sequencing projects. At the level of data gathering one of the critical issues, which is still largely ignored in the field, is the need for a generally applicable numerical quality control system; such a system is prerequisite for multi-dimensional data analyses (for a discussion and a recently developed quality control system see [47]. While several computational efforts have aimed to assess the local enrichment confidence in the single NGS-generated profiles that have been reported (for a recent comparison of peak finding algorithms see [45], methodologies for multi-profile comparisons are still in their infancy [17, 27]. The use of integrative genomics approaches may become the methodology of choice for decorticating RA-driven signal transduction events and thus, there is a real need to develop and standardize computational methods with a focus on enhancing the confidence factor in omics datasets. Importantly,

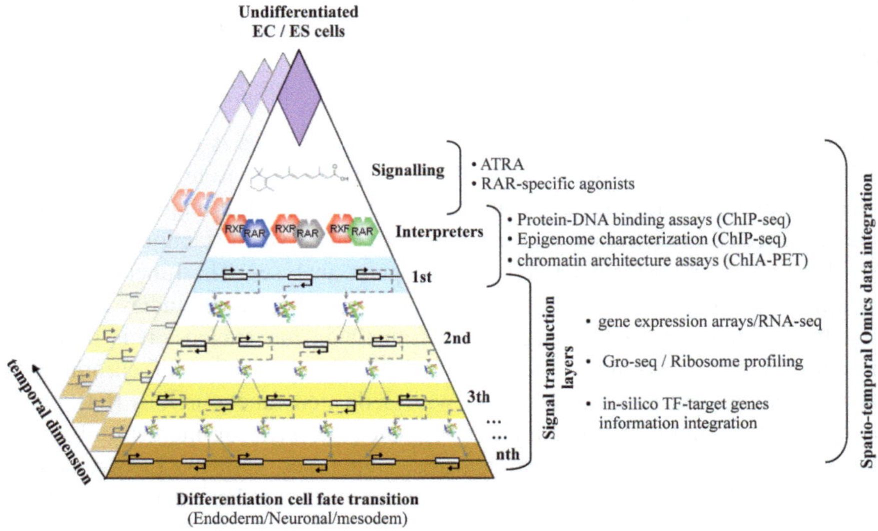

Fig. 9.6 Schematic overview of the spatio-temporal omics data integration designed to study the RA-induced signaling pathway diversification. From *top* to *bottom* The signal induction applied to undifferentiated Embryonic carcinoma (EC)/embryonic stem (ES) cells (ATRA or RAR-specific agonists) is diversified through the interpreters (RXR/RAR nuclear receptors) which may activate several signal transduction layers giving rise to the corresponding differentiation stages. The methodology in use for assessing the presence of the different components involved in this process is displayed in a side. Importantly, two major axes are taken for this analysis: (i) the three-dimensional chromatin structure assessed by proximity-ligation based methodologies (like ChIA-PET) and (ii) the assessment of these events at different time-points during differentiation

future dataset integration in RA-driven differentiation studies will be performed by integrating two major additional elements: (i) the three-dimensional chromatin structure revealed by methodologies like Hi-C or ChIA-PET (see above) and (ii) the temporal nature of the evaluated events throughout the induction process. Importantly, such spatio-temporal analyses will integrate information coming from RAR/RXR binding to chromatin, the chromatin modification status and nucleosome occupancy, and the observed differential transcriptional/translational activity. In addition, computational methods for reconstructing the dynamic regulatory gene networks will be applied with the hope of inferring the temporally defined regulatory decisions that underlie diversification of *at*-RA-induced signaling during developmental processes [14, 18, 48]. Studies done to date have provided initial insight into the enormous complexity that we are facing in stem cell model systems. These data are summarized in a schematic illustration depicting a current view of the molecular and mechanistic 4-dimensional hierarchies (Fig. 9.6) that govern cell fate transitions initiated by a single inducer.

References

1. Amat R, Gudas LJ (2010) RARgamma is required for correct deposition and removal of Suz12 and H2A.Z in embryonic stem cells. J Cell Physiol 226:293–298
2. Arima K, Shiotsugu J, Niu R, Khandpur R, Martinez M, Shin Y, Koide T, Cho KW, Kitayama A, Ueno N et al (2005) Global analysis of RAR-responsive genes in the Xenopus neurula using cDNA microarrays. Dev Dyn 232:414–431
3. Ashburner M, Chihara C, Meltzer P, Richards G (1974) Temporal control of puffing activity in polytene chromosomes. Cold Spring Harb Symp Quant Biol 38:655–662
4. Balmer JE, Blomhoff R (2002) Gene expression regulation by retinoic acid. J Lipid Res 43:1773–1808
5. Bourguet W, Germain P, Gronemeyer H (2000) Nuclear receptor ligand-binding domains: three-dimensional structures, molecular interactions and pharmacological implications. Trends Pharmacol Sci 21:381–388
6. Bourguet W, Ruff M, Chambon P, Gronemeyer H, Moras D (1995) Crystal structure of the ligand-binding domain of the human nuclear receptor RXR-α. Nature 375:377–382
7. Bourguet W, Vivat V, Wurtz JM, Chambon P, Gronemeyer H, Moras D (2000) Crystal structure of a heterodimeric complex of RAR and RXR ligand-binding domains. Mol Cell 5:289–298
8. Chaya D, Hayamizu T, Bustin M, Zaret KS (2001) Transcription factor FoxA (HNF3) on a nucleosome at an enhancer complex in liver chromatin. J Biol Chem 276:44385–44389
9. Chen W, Roeder RG (2011) Mediator-dependent nuclear receptor function. Semin Cell Dev Biol 22:749–758
10. Chiba H, Clifford J, Metzger D, Chambon P (1997) Distinct retinoid X receptor-retinoic acid receptor heterodimers are differentially involved in the control of expression of retinoid target genes in F9 embryonal carcinoma cells. Mol Cell Biol 17:3013–3020
11. Chiba H, Clifford J, Metzger D, Chambon P (1997) Specific and redundant functions of retinoid X Receptor/Retinoic acid receptor heterodimers in differentiation, proliferation, and apoptosis of F9 embryonal carcinoma cells. J Cell Biol 139:735–747
12. de Lera AR, Bourguet W, Altucci L, Gronemeyer H (2007) Design of selective nuclear receptor modulators: RAR and RXR as a case study. Nat Rev Drug Discov 6:811–820
13. Delacroix L, Moutier E, Altobelli G, Legras S, Poch O, Choukrallah MA, Bertin I, Jost B, Davidson I (2010) Cell-specific interaction of retinoic acid receptors with target genes in mouse embryonic fibroblasts and embryonic stem cells. Mol Cell Biol 30:231–244
14. Dimitrakopoulou K, Tsimpouris C, Papadopoulos G, Pommerenke C, Wilk E, Sgarbas KN, Schughart K, Bezerianos A (2011) Dynamic gene network reconstruction from gene expression data in mice after influenza A (H1N1) infection. J Clin Bioinform 1:27
15. Dostie J, Richmond TA, Arnaout RA, Selzer RR, Lee WL, Honan TA, Rubio ED, Krumm A, Lamb J, Nusbaum C et al (2006) Chromosome conformation capture carbon copy (5C): a massively parallel solution for mapping interactions between genomic elements. Genome Res 16:1299–1309
16. Eifert C, Sangster-Guity N, Yu LM, Chittur SV, Perez AV, Tine JA, McCormick PJ (2006) Global gene expression profiles associated with retinoic acid-induced differentiation of embryonal carcinoma cells. Mol Reprod Dev 73:796–824
17. Ernst J, Kellis M (2010) Discovery and characterization of chromatin states for systematic annotation of the human genome. Nat Biotech 28:817–825
18. Ernst J, Vainas O, Harbison CT, Simon I, Bar-Joseph Z (2007) Reconstructing dynamic regulatory maps. Mol Syst Biol 3:74
19. Fadloun A, Kobi D, Delacroix L, Dembele D, Michel I, Lardenois A, Tisserand J, Losson R, Mengus G, Davidson I (2008) Retinoic acid induces TGFbeta-dependent autocrine fibroblast growth. Oncogene 27:477–489

20. Felsenfeld G, Dekker J (2012) Genome architecture and expression. Curr Opin Genet Dev 22:59–61
21. Fullwood MJ, Liu MH, Pan YF, Liu J, Xu H, Mohamed YB, Orlov YL, Velkov S, Ho A, Mei PH et al (2009) An oestrogen-receptor-alpha-bound human chromatin interactome. Nature 462:58–64
22. Germain P, Chambon P, Eichele G, Evans RM, Lazar MA, Leid M, De Lera AR, Lotan R, Mangelsdorf DJ, Gronemeyer H (2006) International Union of Pharmacology. LX. Retinoic acid receptors. Pharmacol Rev 58:712–725
23. Germain P, Chambon P, Eichele G, Evans RM, Lazar MA, Leid M, De Lera AR, Lotan R, Mangelsdorf DJ, Gronemeyer H (2006) International Union of Pharmacology. LXIII. Retinoid X receptors. Pharmacol Rev 58:760–772
24. Germain P, Iyer J, Zechel C, Gronemeyer H (2002) Co-regulator recruitment and the mechanism of retinoic acid receptor synergy. Nature 415:187–192
25. Gillespie RF, Gudas LJ (2007) Retinoid regulated association of transcriptional co-regulators and the polycomb group protein SUZ12 with the retinoic acid response elements of Hoxa1, RARbeta(2), and Cyp26A1 in F9 embryonal carcinoma cells. J Mol Biol 372:298–316
26. Gronemeyer H, Gustafsson JA, Laudet V (2004) Principles for modulation of the nuclear receptor superfamily. Nat Rev Drug Discov 3:950–964
27. Halbritter F, Vaidya HJ, Tomlinson SR (2012) GeneProf: analysis of high-throughput sequencing experiments. Nat Methods 9:7–8
28. Handoko L, Xu H, Li G, Ngan CY, Chew E, Schnapp M, Lee CW, Ye C, Ping JL, Mulawadi F et al (2011) CTCF-mediated functional chromatin interactome in pluripotent cells. Nat Genet 43:630–638
29. Harris TM, Childs G (2002) Global gene expression patterns during differentiation of F9 embryonal carcinoma cells into parietal endoderm. Funct Integr Genomics 2:105–119
30. Hua S, Kittler R, White KP (2009) Genomic antagonism between retinoic acid and estrogen signaling in breast cancer. Cell 137:1259–1271
31. Jin F, Li Y, Dixon JR, Selvaraj S, Ye Z, Lee AY, Yen CA, Schmitt AD, Espinoza CA, Ren B (2013) A high-resolution map of the three-dimensional chromatin interactome in human cells. Nature 503:290–294
32. Kalhor R, Tjong H, Jayathilaka N, Alber F, Chen L (2012) Genome architectures revealed by tethered chromosome conformation capture and population-based modeling. Nat Biotechnol 30:90–98
33. Kashyap V, Gudas LJ (2010) Epigenetic regulatory mechanisms distinguish retinoic acid-mediated transcriptional responses in stem cells and fibroblasts. J Biol Chem 285:14534–14548
34. Kashyap V, Gudas LJ, Brenet F, Funk P, Viale A, Scandura JM (2011) Epigenomic reorganization of the clustered Hox genes in embryonic stem cells induced by retinoic acid. J Biol Chem 286:3250–3260
35. Lalevee S, Anno YN, Chatagnon A, Samarut E, Poch O, Laudet V, Benoit G, Lecompte O, Rochette-Egly C (2011) Genome-wide in silico identification of new conserved and functional retinoic acid receptor response elements (direct repeats separated by 5 bp). J Biol Chem 286:33322–33334
36. Laudet V, Gronemeyer H (2002) The nuclear receptor facts book. Academic Press, San Diego
37. Laursen KB, Wong PM, Gudas LJ (2012) Epigenetic regulation by RARalpha maintains ligand-independent transcriptional activity. Nucleic Acids Res 40:102–115
38. le Maire A, Teyssier C, Erb C, Grimaldi M, Alvarez S, de Lera AR, Balaguer P, Gronemeyer H, Royer CA, Germain P, Bourguet W (2010) A unique secondary-structure switch controls constitutive gene repression by retinoic acid receptor. Nat Struct Mol Biol 17:801–807
39. Li G, Fullwood MJ, Xu H, Mulawadi FH, Velkov S, Vega V, Ariyaratne PN, Mohamed YB, Ooi HS, Tennakoon C et al (2010) ChIA-PET tool for comprehensive chromatin interaction analysis with paired-end tag sequencing. Genome Biol 11:R22
40. Lyons P (2003) Advances in spotted microarray resources for expression profiling. Brief Funct Genomic Proteomic 2:21–30

41. Mahony S, Mazzoni EO, McCuine S, Young RA, Wichterle H, Gifford DK (2011) Ligand-dependent dynamics of retinoic acid receptor binding during early neurogenesis. Genome Biol 12:R2
42. Mamoon A, Ventura-Holman T, Maher JF, Subauste JS (2008) Retinoic acid responsive genes in the murine hepatocyte cell line AML 12. Gene 408:95–103
43. Meijsing SH, Pufall MA, So AY, Bates DL, Chen L, Yamamoto KR (2009) DNA binding site sequence directs glucocorticoid receptor structure and activity. Science 324:407–410
44. Mendoza-Parra MA, Gronemeyer H (2013) Genome-wide studies of nuclear receptors in cell fate decisions. Semin Cell Dev Biol 24:706–715
45. Mendoza-Parra MA, Nowicka M, Van Gool W, Gronemeyer H (2013) Characterising ChIP-seq binding patterns by model-based peak shape deconvolution. BMC Genom 14:834
46. Mendoza-Parra MA, Shankaranarayanan P, Hinrich G (2012) Sequential chromatin immuno-precipitation protocol for global analysis through massive parallel sequencing (reChIP-seq)
47. Mendoza-Parra MA, Van Gool W, Mohamed Saleem MA, Ceschin DG, Gronemeyer H (2013) A quality control system for profiles obtained by ChIP sequencing. Nucleic Acids Res 41:e196
48. Mendoza-Parra MA, Walia M, Sankar M, Gronemeyer H (2011) Dissecting the retinoid-induced differentiation of F9 embryonal stem cells by integrative genomics. Mol Syst Biol 7:538
49. Metivier R, Penot G, Hubner MR, Reid G, Brand H, Kos M, Gannon F (2003) Estrogen receptor-alpha directs ordered, cyclical, and combinatorial recruitment of cofactors on a natural target promoter. Cell 115:751–763
50. Min IM, Pietramaggiori G, Kim FS, Passegue E, Stevenson KE, Wagers AJ (2008) The transcription factor EGR1 controls both the proliferation and localization of hematopoietic stem cells. Cell Stem Cell 2:380–391
51. Nagano T, Lubling Y, Stevens TJ, Schoenfelder S, Yaffe E, Dean W, Laue ED, Tanay A, Fraser P (2013) Single-cell Hi-C reveals cell-to-cell variability in chromosome structure. Nature 502:59–64
52. Orkin SH, Wang J, Kim J, Chu J, Rao S, Theunissen TW, Shen X, Levasseur DN (2008) The transcriptional network controlling pluripotency in ES cells. Cold Spring Harb Symp Quant Biol
53. Perissi V, Jepsen K, Glass CK, Rosenfeld MG (2010) Deconstructing repression: evolving models of co-repressor action. Nat Rev Genet 11:109–123
54. Rastinejad F, Perlmann T, Evans RM, Sigler PB (1995) Structural determinants of nuclear receptor assembly on DNA direct repeats. Nature 375:203–211
55. Rickman DS, Soong TD, Moss B, Mosquera JM, Dlabal J, Terry S, MacDonald TY, Tripodi J, Bunting K, Najfeld V et al (2012) Oncogene-mediated alterations in chromatin conformation. Proc Nat Acad Sci USA 109:9083–9088
56. Rochel N, Ciesielski F, Godet J, Moman E, Roessle M, Peluso-Iltis C, Moulin M, Haertlein M, Callow P, Mely Y et al (2011) Common architecture of nuclear receptor heterodimers on DNA direct repeat elements with different spacings. Nat Struct Mol Biol 18:564–570
57. Ross-Innes CS, Stark R, Holmes KA, Schmidt D, Spyrou C, Russell R, Massie CE, Vowler SL, Eldridge M, Carroll JS (2010) Cooperative interaction between retinoic acid receptor-alpha and estrogen receptor in breast cancer. Genes Dev 24:171–182
58. Schuster-Bockler B, Lehner B (2012) Chromatin organization is a major influence on regional mutation rates in human cancer cells. Nature 488:504–507
59. Segraves WA, Hogness DS (1990) The E75 ecdysone-inducible gene responsible for the 75B early puff in Drosophila encodes two new members of the steroid receptor superfamily. Genes Dev 4:204–219
60. Shankaranarayanan P, Mendoza-Parra MA, van Gool W, Trindade LM, Gronemeyer H (2012) Single-tube linear DNA amplification for genome-wide studies using a few thousand cells. Nat Protoc 7:328–338
61. Shankaranarayanan P, Mendoza-Parra MA, Walia M, Wang L, Li N, Trindade LM, Gronemeyer H (2011) Single-tube linear DNA amplification (LinDA) for robust ChIP-seq. Nat Meth 8:565–567

62. Soler E, Andrieu-Soler C, de Boer E, Bryne JC, Thongjuea S, Stadhouders R, Palstra RJ, Stevens M, Kockx C, van Ijcken W et al (2010) The genome-wide dynamics of the binding of Ldb1 complexes during erythroid differentiation. Genes Dev 24:277–289
63. Su D, Gudas LJ (2008) Gene expression profiling elucidates a specific role for RARgamma in the retinoic acid-induced differentiation of F9 teratocarcinoma stem cells. Biochem Pharmacol 75:1129–1160
64. Sutherland H, Bickmore WA (2009) Transcription factories: gene expression in unions? Nat Rev Genet 10:457–466
65. Taneja R, Roy B, Plassat JL, Zusi CF, Ostrowski J, Reczek PR, Chambon P (1996) Cell-type and promoter-context dependent retinoic acid receptor (RAR) redundancies for RAR beta 2 and Hoxa-1 activation in F9 and P19 cells can be artefactually generated by gene knockouts. Proc Nat Acad Sci USA 93:6197–6202
66. Thummel CS (2001) Molecular mechanisms of developmental timing in C. elegans and Drosophila. Dev Cell 1:453–465
67. Umesono K, Murakami KK, Thompson CC, Evans RM (1991) Direct repeats as selective response elements for the thyroid hormone, retinoic acid, and vitamin D3 receptors. Cell 65:1255–1266
68. van Berkum NL, Lieberman-Aiden E, Williams L, Imakaev M, Gnirke A, Mirny LA, Dekker J, Lander ES (2010) Hi-C: a method to study the three-dimensional architecture of genomes. J Vis Exp
69. Voss TC, Schiltz RL, Sung MH, Yen PM, Stamatoyannopoulos JA, Biddie SC, Johnson TA, Miranda TB, John S, Hager GL (2011) Dynamic exchange at regulatory elements during chromatin remodeling underlies assisted loading mechanism. Cell 146:544–554
70. Wei Y, Harris T, Childs G (2002) Global gene expression patterns during neural differentiation of P19 embryonic carcinoma cells. Differentiation 70:204–219
71. Wurtz JM, Bourguet W, Renaud JP, Vivat V, Chambon P, Moras D, Gronemeyer H (1996) A canonical structure for the ligand-binding domain of nuclear receptors. Nat Struct Biol 3:206
72. Yao TP, Segraves WA, Oro AE, McKeown M, Evans RM (1992) Drosophila ultraspiracle modulates ecdysone receptor function via heterodimer formation. Cell 71:63–72
73. Zaret KS, Carroll JS (2011) Pioneer transcription factors: establishing competence for gene expression. Genes Dev 25:2227–2241
74. Zechel C, Shen XQ, Chambon P, Gronemeyer H (1994) Dimerization interfaces formed between the DNA binding domains determine the cooperative binding of RXR/RAR and RXR/TR heterodimers to DR5 and DR4 elements. EMBO J 13:1414–1424
75. Zechel C, Shen XQ, Chen JY, Chen ZP, Chambon P, Gronemeyer H (1994) The dimerization interfaces formed between the DNA binding domains of RXR, RAR and TR determine the binding specificity and polarity of the full-length receptors to direct repeats. EMBO J 13:1425–1433
76. Zhang J, Chalmers MJ, Stayrook KR, Burris LL, Wang Y, Busby SA, Pascal BD, Garcia-Ordonez RD, Bruning JB, Istrate MA et al (2011) DNA binding alters coactivator interaction surfaces of the intact VDR-RXR complex. Nat Struct Mol Biol 18:556–563
77. Zhao Z, Tavoosidana G, Sjolinder M, Gondor A, Mariano P, Wang S, Kanduri C, Lezcano M, Sandhu KS, Singh U et al (2006) Circular chromosome conformation capture (4C) uncovers extensive networks of epigenetically regulated intra- and interchromosomal interactions. Nat Genet 38:1341–1347

Chapter 10
Complexity of the RAR-Mediated Transcriptional Regulatory Programs

Zhijie Liu, Qidong Hu and Michael G. Rosenfeld

Abstract In the past several decades, intensive research in this field has uncovered a surprising number of regulatory factors and their associated enzymatic properties to reveal the network of complexes that function in activation and repression of the transcriptional programs mediated by nuclear receptors (NR). These factors and their associated complexes have been extensively characterized both biochemically and functionally [34, 87, 94]. Several principles have emerged: (1) It is widely recognized that ligand-dependent cofactor complexes mediating repression and activation exhibit ligand-dependent exchange. (2) These complexes mediate modifications of chromatin structure consequent to their binding at regulatory elements, particularly at promoter and enhancer sites. (3) The concept about the rapid exchange of coregulatory complexes at regulatory sites has been suggested [88]. Key questions in the NR field have included: (a) What are the cofactors and exchange complexes used to mediate the ligand and signaling network-dependent switches in gene regulation programs; (b) Do long non-coding RNAs (lncRNAs) serve as regulatory "factors" for ligand-dependent gene programs, and do enhancers actually regulate transcription units encoding enhancer non-coding RNAs (eRNAs) that might have functional significance; (c)

Z. Liu and Q. Hu made equal contributions.

Z. Liu (✉) · Q. Hu · M. G. Rosenfeld (✉)
Howard Hughes Medical Institute, Department of Medicine,
University of California, La Jolla, San Diego, CA, USA
e-mail: z9liu@ucsd.edu

M. G. Rosenfeld
e-mail: mrosenfeld@ucsd.edu

Q. Hu
e-mail: qihu@ucsd.edu

Q. Hu
Department of Anatomy, Yong Loo Lin School of Medicine,
National University of Singapore, Singapore, Singapore

M. A. Asson-Batres and C. Rochette-Egly (eds.), *The Biochemistry of Retinoic Acid Receptors I: Structure, Activation, and Function at the Molecular Level*, Subcellular Biochemistry 70,
DOI: 10.1007/978-94-017-9050-5_10,

What is the relationship between DNA damage repair machinery and transcriptional machinery? (d) Do Retinoic Acid Receptors (RAR) also regulate Pol III-dependent, non-coding repeat transcriptional units in stem cells? and (e) How have new technologies such as deep sequencing altered our ability to investigate transcriptional regulatory mechanisms utilized by NRs?

Abbreviations

3′UTR	3′ untranslated region
AF1	Active function 1 domain
AF2	Active function 2 domain
AP1	Activating protein 1
AR	Androgen receptor
*at*RA	All-*trans* retinoic acid
ChIP	Chromatin immunoprecipitation
CSB	Cockayne syndrome B protein
CTCF	CCCTC-binding factor
DBD	DNA binding domain
DCP1A	mRNA-decapping enzyme 1A
DCP2	mRNA-decapping enzyme 2
DR	Direct repeat
ER	Estrogen receptor
ER	Everted repeat
ERCC1	Excision repair cross-complementing protein 1
eRNA	Enhancer RNA
ESCs	Embryonic stem cells
GR	Glucocorticoid receptor
HMGCS2	3-Hydroxy-3-Methylglutaryl-Coenzyme A Synthase 2
HRE	Hormone response element
IR	Inverted repeat
LBD	Ligand-binding domain
LC-ESI-MS	Liquid chromatography-electrospray ionization–mass spectrometry
lncRNA	Long non-coding RNA
MAPK	Mitogen-activated protein kinase
NCoR	Nuclear receptor corepressor
ncRNA	Non-coding RNA
NER	Nucleotide excision repair
NR	Nuclear receptor
PKA	Protein kinase A
Pol II	RNA polymerase II
Pol III	RNA polymerase III
PPAR	Peroxisome proliferator-activated receptor
RA	Retinoic acid

RAR	Retinoic acid receptor
RARE	Retinoic acid response element
RPA	Replication protein A
RXR	Retinoid X receptor
SRA	Steroid receptor RNA activator
STAT3	Signal transducer and activator of transcription 3
TBL1	Transducin-beta-like protein 1
TBLR1	Transducin beta-like 1-related protein 1
TF3C	General transcription factor 3C
TF IIF	Transcription factor IIF
TR	Thyroid hormone receptor
XPA	Xeroderma pigmentosum, complementation group A
XPC	Xeroderma pigmentosum, complementation group C
XPF	Xeroderma pigmentosum, complementation group F
XPG	Xeroderma pigmentosum, complementation group G

History: Transcriptional Cofactors That Regulate RAR Transcriptional Activities

Coactivators and Corepressors

The definition of coactivators and corepressors has rapidly expanded as chromatin-associated factors that modify transcriptional programs based on their interactions with NRs. These cofactors have become recognized to include proteins, RNAs, and most recently, lipids. They often form large complexes with exchangeable components and their exchanges are regulated by covalent modifications including phosphorylation, ubiquitylation, SUMOylation, acetylation and methylation [42, 53, 86]. A partial list of this growing number of functional cofactors is provided in Table 10.1. For RAR, these cofactors can be divided into three different groups based on their function. (1) Coactivators that function with liganded RAR to activate transcription, often dependent on the presence of LXXLL motifs. RAR coactivators have a wide variety of functions, including serving as platforms for the assembly of coactivator complexes, or as enzymes that modify histone, RNA polymerase, other cofactors, or RAR itself. (2) Corepressors that function with unliganded RAR at target sites to repress gene transcription, often dismissed after RA binds to the receptor. However, in some instances, components present in corepressor complexes are retained and even required for activation functions, as exemplified by transducin-beta-like protein 1 (TBL1) and transducin beta-like 1-related protein 1 (TBLR1), with phosphorylation activating their ubiquitin ligase functions [86, 89]. (3) RA-dependent corepressors characterized by the presence of LXXLL protein motifs that have usually been identified from coactivators. Intriguingly, this third group of repressors unconventionally

Table 10.1 RAR cofactors that endow RAR with different transcriptional activities

Cofactor	Group	Protein function	References
SRC-1	I	P160 family, adaptor	[82]
SRC-2	I	P160 family, adaptor	[108]
SRC-3	I	P160 family, adaptor	[105]
CBP/P300	I	Histone acetylation	[6]
GCN5	I	Histone acetylation	[5]
ADA3	I	Histone acetylation	[5]
PCAF	I	Histone acetylation	[4]
Asxl1	I	Histone acetylation	[13]
NCOA7	I	Coiled-coil containing protein	[98]
SWI/SNF	I	Chromatin remodeling	[22]
BAF60c1	I	Chromatin remodeling	[22]
BAF60c2	I	Chromatin remodeling	[22]
CARM1	I	Histone methyltransferase	[8]
Mll5	I	H3K4 methylation	[31]
OGT	I	GlcNAcylation of Mll5	[31]
ASC-2	I	H3K4 methylation	[64]
Mll3	I	H3K4 methylation	[64]
Mll4	I	H3K4 methylation	[64]
TRAP220	I	Mediator complex	[99]
Med25	I	Mediator complex	[63]
TAFII 135	I	TBP-associated factors	[77]
PHF8	I	H3K9me2/1 demethylation	[91]
PARG	I	Cleaves ADP-ribose polymers	[60]
NCoR	II	Contains 2 SANT domains	[44]
SMRT	II	Contains 2 SANT domains	[9]
Sin3a	II	Contains 3 PAH domains	[79]
HDACs	II	Histone deacetylation	[30, 79]
CAC1	II	Contains Cullin domain	[78]
TBL1	II	Recruitment of proteasome complex	[86]
TBLR1	II	Recruitment of proteasome complex	[86]
Suz12	II	H3K27 methylation	[51]
CaMKIIγ	II	RAR kinase to enhance its repression	[101]
Ajuba	II	Lim domain protein	[45]
PLZF	II	Zinc finger protein	[43]
HP1α	II	Contains 2 chromo domains	[65]
LSD1	II	H3K4 demethylation	[65]
Asxl1	II	H3K4 demethylation	[65]
TNIP1	III	Coiled-coil containing protein	[35]
RIF1	III	Nuclear matrix protein	[67]
Trim24	III	Tripartite motif (TRIM) family	[52]
PRAME	III	Contains 4 LRR (leucine-rich) repeats	[26]
LCoR	III	Contains HTH DNA-binding domain	[29]
RIP140	III	Contains 9 LXXLL motifs	[47]

Partial list of RAR cofactors categorized by their different function. These cofactors are divided into three different groups: I: RA-dependent coactivators; II: Corepressors functioning with unliganded RAR; and III: RA-dependent corepressors

uses a ligand-dependent interaction strategy to allow them to repress, rather than activate, gene expression. The mechanisms for coactivator and corepressor functions are detailed in Chaps. 2 and 3 of this volume.

ncRNAs as RAR Cofactors

While prevailing research has long focused on the function of protein coregulators, more recent data argues for the involvement of non-coding RNAs (ncRNAs) as potential "*coregulators*" of transcriptional regulation by many classes of transcription factors (TF), including NRs. Several contemporary technologies beginning first with the use of quantitative PCR followed by the development and use of genome-wide deep sequencing technology and genome-wide transcription run-on assays (Global Run-On Sequencing or GRO-seq), have expanded our recognition of the presence of surprisingly massive transcription of genome-wide ncRNAs, which infers that the majority of the human genome is transcribed [3, 11, 38]. GRO-seq technology allows us to specifically detect only RNAs that are newly synthesized by RNA polymerase. These RNAs derive from both repeated and non-repeated genomic regions. Many thousands of relatively abundant lncRNAs have been identified [36, 37]. Many ncRNAs have also been found to be transcribed, but are present at much lower than coding genes levels of approximately 3–6 copies/cell. In some cases, the biological functions of the ncRNAs have already been established. Further studies have revealed that lncRNAs regulate the function of histone modification enzymes, and that they can often act *in trans*, at a distance far from the site of their transcription [104, 110, 113].

The first ncRNA gene shown to function in NR signaling, steroid receptor RNA activator (SRA), was identified from a screen for NR coactivators. The SRA ncRNA was reported to be capable of serving as a cofactor for RAR as well as other NRs [15, 16]. Several lines of evidence support the view that the functional gene product of the SRA gene is an RNA and not a protein. The biological function of this gene was not affected when the translation inhibitor, cycloheximide, was added to cell assays, and even when multiple translocation stop codons were introduced into SRA, its function in NR signaling was not impacted. The latter finding underscores the surprising revelation that SRA is an RNA without protein-coding potential. Co-fractionation experiments showed that SRA ncRNAs co-purify with SRC-1, a canonical coactivator of NRs. Studies have shown that SRA ncRNAs can enhance the activity of active function 1 domain (AF1) and active function 1 domain (AF2) in NRs. A pseudouridine synthase, mPus1p, can pseudouridylate SRA ncRNA, to enhance the transcriptional activation of RARγ [66, 116, 117], providing yet another layer regulation of NR-mediated transcriptional activity.

Based on these findings, we can now conclude that RAR binds at least one lncRNA, *SRA,* to exert its genomic regulatory effect, as shown in Fig. 1a. The

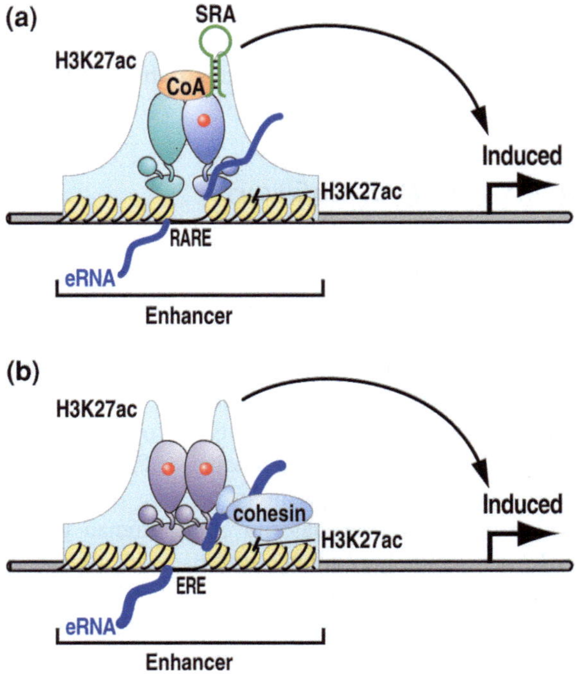

Fig. 1 Regulation of enhancers by liganded nuclear receptors and non-coding RNAs. **a** Proposed interaction of RAR with the lncRNA, SRA, following the binding of RAR/RXR heterodimers to RARE. **b** Effects of ERα in stimulating transcription of eRNAs at regulatory enhancers, which are also occupied by cohesin, resulting in target coding gene activation

sheer number of lncRNAs identified to date strongly implies that there are likely to be multiple ncRNA "*cofactors*" involved in the actions of NRs in addition to *SRA*. We therefore expect that additional ncRNA cofactors for RAR will be identified.

Enhancers as Transcription Units: Induction of Enhancer RNAs

Investigation of enhancers has revealed that ligand- or signal-induced activation is often accompanied by transcription of rather small (1–2 kb) transcripts, referred to as enhancer RNAs (eRNAs). One of the initial examples was provided in neurons, where signals such as potassium chloride (KCl)—a neuron stimulator—induced transcription of bidirectional eRNAs [54]. Similar eRNA production activities also have been observed on enhancers bound by NRs, including estrogen receptor (ER) and androgen receptor (AR), as shown in Fig. 1b [38, 68, 109]. The function of eRNAs is still controversial, but the data from several labs suggest that after hormone/ligand signaling induces their transcription, together with cohesins and components of the mediator complex, they help to mediate enhancer/promoter looping events [38, 54, 68, 109]. Knockdown of eRNAs and enhancer like lncRNAs affects the transcription of target coding genes [56, 57, 68].

DNA Damage Repair Components also Function as RAR Cofactors

Studies of NR-mediated gene activation have traditionally focused on the RNA Polymerase II (Pol II)-containing RNA synthesis machinery. In response to inductive signals, a cohort of factors, coactivators and Pol II, are recruited to the promoter or enhancer regulatory regions of the activated gene, leading to initiation of RNA synthesis. Recent studies suggest that components of the DNA damage repair machinery may also be required for efficient transcription of target genes [61, 62]. In response to exogenous or endogenous factors (such as irradiation and drugs) that induce DNA damage, a number of factors, including those having endonuclease activities, are recruited to repair DNA [70]. Among these factors are the nucleotide excision repair (NER) factors, XPC, CSB, RPA, XPA, XPG, XPF and ERCC1 [81]. These proteins function together to participate in the nucleotide excision repair (NER) process.

Le May et al. [62] showed that in the absence of genotoxic stress, XPF and XPG appear to play an unexpected role. Stimulation of HeLa cells with all-*trans* retinoic acid (*at*RA) leads to recruitment of XPF and XPG, along with Pol II and the transcription factor IIF (TF IIF), to the promoter region and gene coding region of the RARβ2 gene, a *at*RA target gene. In contrast, no *at*RA-induced recruitment of XPF or XPG is observed for the 3-Hydroxy-3-Methylglutaryl-Coenzyme A Synthase 2 (*HMGCS2*) gene, a peroxisome proliferator-activated receptor α (PPARα) target gene. However, in PPARα-overexpressing HeLa cells, treatment with PPARα ligand similarly induces NER factor enrichment at *HMGCS2*, suggesting that in the absence of genotoxic stress, NER recruitment at gene loci depends on the activation status of the gene. Consistent with this interpretation, when endogenous NER factors are knocked down in HeLa cells, *at*RA-induced activation of RARβ2 is dramatically impaired, indicating the presence of these NER factors is required for effective activation of *at*RA target genes. A requirement for NER factors is further suggested by close examination of the epigenetic changes that occur following their depletion. Loss of these factors leads to suboptimal DNA demethylation and histone post-transcriptional modifications, including H3K4/K9 methylation and H3K9/K14 acetylation at the promoter of RARβ2 resulting in poor expression of this gene.

When RARβ2 is activated, its promoter and terminator regions form a looping structure in a CCCTC-binding factor (CTCF)-dependent manner that promotes optimal expression of the gene. Le May et al. [61] found that, upon *at*RA stimulation, XPG and XPF were required for the proper assembly of the transcriptional machinery at both the promoter and terminator of RARβ2, and that CTCF recruitment preceded the docking of the transcriptional machinery. Using quantitative chromatin conformation capture (3C) assays, they observed an *at*RA-triggered chromatin looping between the promoter and terminator of RARβ2. However, when endogenous XPG or XPF was depleted, the looping was significantly suppressed, resulting in a dramatically decreased expression of RARβ2. They further proved that the participation of XPG and XPF in chromatin looping required the endonuclease activity of XPG and XPF. The catalytic activity induced DNA nicks or breaks and DNA demethylation, two events essential for efficient recruitment of CTCF and consequent chromatin looping. Thus, they revealed an essential role for XPG/XPF in *at*RA-triggered chromatin reorganization [61]. It will be of particular interest to find how general this strategy proves to be for the broader regulatory transcriptional program.

RAR Modifications also Affect its Interaction with Cofactors

RAR normally forms a heterodimeric structure with RXR, and ligand binding triggers canonical RAR/RXR signaling. A subsequent corepressor-coactivator switch is essential for RAR/RXR-regulated gene transcription [86]. Studies have shown that modification of RAR itself has a profound impact on its signaling activity in terms of heterodimerization, cofactor binding and transcriptional activity.

Rochette-Egly et al. [93] found that phosphorylation of RARα_1 by protein kinase A (PKA) was required for RA-induced parietal endodermal differentiation. In RARα_1 null F9 cells, RA-induced parietal endoderm differentiation was abolished but RARγ-regulated primitive endoderm differentiation was not impacted. Rescuing with wild-type RARα_1 restored parietal endoderm differentiation in the RARα_1 gene-knockout cells. However, RARα_1 with mutation at the PKA phosphorylation site could not efficiently rescue parietal endoderm differentiation in the mutant cells.

Phosphorylation of a different RAR isotype, RARγ_2, was also shown to be critical for its function. Upon ligand binding, RARγ_2 is normally degraded, a step that is required for its transactivation function. Gianni et al. [33] found that the AF1 and AF2 activation domains of RARγ_2 were involved in promoting the turnover of the receptor, and in particular, that the p38MAPK pathway phosphorylated the AF1 domain, thus facilitating the recognition and degradation of RARγ_2 by proteasomes [33]. When the phosphorylation of AF1 was blocked, the RARγ_2-mediated transactivation was dramatically impaired, supporting an important role of phosphorylation modification in NR signaling.

Other types of receptor modification with biological significance have also been investigated. With mass spectrometric analysis, Huq et al. [50] identified a trimethylation modification at Lys^{347} in the ligand binding domain (LBD) of murine RARα. This event is critical in promoting the dimerization of RARα and RXRα and the binding of cofactors to RARα, including CBP/p300 and RIP140. The ligand-dependent recruitment of these cofactors is essential for the transactivation activity of RARα. Interestingly, although trimethylation of Lys^{347} occurs within the LBD, the ligand binding kinetics is not affected. In another study, Huq et al. [49] identified two monomethylated residues, Lys^{109} and Lys^{171}, in RARα using an accurate and sensitive liquid chromatography-electrospray ionization/multistage mass spectrometry technique [LC-ESI-MS/MS] that can detect covalent modifications of proteins. These two new methylated residues were located within the DNA binding domain (DBD) and hinge regions of the receptor. Similar to the trimethylation of Lys^{347} in the RARα LBD, the monomethylation of Lys^{109} and Lys^{171} was found to facilitate the heterodimerization of RARα and RXRα. It also participates in the recruitment of cofactors to liganded RARα and promotes their transactivation activity. These studies have unveiled an important role of non-histone methylation events in NR-regulated transcription networks.

Development of the Field: Newly Developed Technologies Have Expanded Our Understanding of the RAR-Mediated Transcriptional Program

RAR Genome-Wide Binding Data Suggest More Complex RAR Transcriptional Programs at Both Promoter and Enhancer Sites

On a global level, a deeper understanding of NR transcriptional regulatory programs has been licensed by the rapid development of global genomic technologies based on next generation deep sequencing methodology. For example, Chromatin Immunoprecipitation (ChIP)-sequencing has allowed genome wide identification of potential interaction sites for DNA binding TFs and cofactors. Several such studies in different cell lines have identified RAR genome-wide binding sites [21, 48, 73, 95], and these data have altered our viewpoint of the most cogent regulatory elements for RAR action. GRO-seq analyses have also allowed us to visualize transcription events genome-wide and to delineate regions with transcription on both strands of DNA, differences in elongation, and promoter pausing events in transcriptional regulation [17, 71]. The new technology also permits determination of the location of lncRNAs in the genome. One of the most potent methods is Chromatin Isolation by RNA purification-sequencing (ChIRP-seq) [14], which permits investigation of the genomic regions interacting with lncRNAs and eRNAs. While the full impact of these new technologies has not yet been fully realized, one powerful aspect of these global technologies is that they have begun to reveal that different cohorts of regulated transcription units can use distinct molecular mechanisms in regulating different aspects of the full transcriptional program.

Before the availability of deep-sequencing technology, the identification of RAR-binding sites had focused only on the promoter and proximal promoter regions of RA targets [19, 24, 72, 76, 102]. Global genomic data analyses obtained with the newer ChIP-sequencing technology revealed a different picture. The data from the two groups conducting global genomic studies [48, 95] indicated that only a relatively small portion of RAR-binding sites were actually at proximal promoter regions; rather most RAR-binding sites were found in intronic or distal promoter intergenic regions. These results suggest that the regulation of RA targets likely involves the action of RAR receptors on regulatory elements that included enhancers marked with specific histone modifier binding marks, such as H3K4me1, H3K4me2, CBP/p300 and H4K16Ac [1, 12, 39, 41, 107, 111], as well as other potential distal regulatory sequences. To date, comprehensive functional studies of RAR-bound enhancers are still lacking, but we expect to see intensive investigation on this subject in the future. This is because enhancers participate in critical aspects of transcriptional regulation [40], alter chromatin interactions, and contribute to putative looping activities with promoters to deliver activating factors, such as components of the MLL complex [110].

RAR Binding is Dynamically Regulated During Differentiation

Although RAR can bind constitutively to target sites, several recent publications report a ligand-dependent shift in RAR binding sites during RA induced differentiation and the different RAR binding patterns in mouse embryonic fibroblasts and mouse embryonic stem cells (ESCs) [21, 73]. These data suggest that RAR binding is dynamically regulated by ligand treatment or cell differentiation status.

Using a pan-RAR antibody for ChIP-sequencing during RA induced differentiation, the David Gifford group found that RA treatment could cause widespread changes in RAR genome-wide binding during RA-induced neuronal differentiation [73]. Based on their RAR binding data, they concluded that only a small subset of RAR binding sites were constitutively bound, with two other sets of RAR binding sites present only in the absence or presence of RA. When they compared RAR binding sites occupied by unliganded and liganded receptors with the well-characterized TF regulatory network in mouse ES cells, they found the binding information of ES cell TFs and other TF regulatory proteins can accurately predict both constitutive and ligand-induced RAR binding. The binding of core ES cell regulators is highly correlated with unliganded RAR binding sites, and slightly less correlated with liganded RAR binding sites.

RAR ChIP-chip assays performed in both mouse embryonic fibroblasts and ES cell also revealed different RAR binding patterns in these two cell lines [21]. Because their ChIP-chip experiments were performed using extended promoter array (−5 to +2 kb of promoters) and we now realize that most RARs bind at intergenic enhancer regions [48, 95], they only found 354 binding peaks in MEFs and 462 peaks in ES cells [21]. They found only 58 common RAR binding peaks for both cell lines [21], suggesting that RARs have cell-type specific functions through binding to the different regulatory regions controlling different subsets of gene targets. It will be important in future studies to clarify whether the chromatin environment or other tissue-specific TFs, such as FoxA1, as reported by the Kevin White group in MCF7 cells [48], determines whether RAR binds to a specific locus.

New RAR Binding Motifs

Many studies have provided evidence that RAR/RXR heterodimers bind asymmetrically to retinoic acid response element (RARE) [74]. RAR genome-wide binding data give us a more comprehensive view of RAR binding patterns, and the data suggest that RAR binding is also enriched at other motifs [21, 48, 73]. The Kevin White group conducted an in-depth analysis for all possible hormone response element (HRE) motifs in RAR-binding sites and found that in addition to some well-known experimentally validated RAREs, such as direct repeat (DR)5 and DR2, there were some HREs not previously known as RAREs [48], such as everted repeat (ER)2. This leads to a very intriguing question—Do different RARE motifs confer different transcriptional regulation activities on RARs? Recently, the Pierre Chambon group found several special glucocorticoid receptor (GR) binding motifs (inverted repeat (IR)0, IR1 and IR2) function as negative response elements to mediate repression by agonist-liganded GR [100]. Further studies of these newly identified RAREs are expected to elucidate their relevance to RAR function under different conditions.

Trans-binding of RAR and Transcriptional Regulation

Some nuclear receptors, such as GR and PPARγ, have been reported to regulate gene expression through binding to other DNA-binding TFs, even if they do not bind directly to their DNA binding elements, which we termed "trans-binding" effect [58, 83, 84]. This was exemplified by the unexpected discovery that in response to ligand stimulation, PPARγ was recruited *in trans* to mediate transrepression in macrophages [83]. Ligand-dependent SUMOylation of PPARγ, or other nuclear receptors, permits their recruitment *in trans* to specific regulatory regions, repressing coding gene transcription (Shown in Fig. 2a) [32, 83, 106].

Recently, substantial data indicate that RAR also exhibits trans-binding activities by interacting with other signaling pathways, including estrogen/ERα signaling, Wnt signaling and activating protein 1 (AP1) transcription factor complex [25, 48, 55, 69, 80]. One well-established example is that RA-bound RAR represses the transcriptional activation of AP1 transcription factor complex, which consists of Fos and Jun [55, 69, 80]. Using various selective retinoids for RAR, researchers were able to dissociate its inhibition ability on AP1 from its classical RARE-binding transcriptional regulation activity [10, 28, 92], suggesting that RAR interferes with AP1 activity by a different functional mechanism from its regular DNA-binding function. Several models have been proposed to explain the transrepression of AP1 by RAR, including the fact that RAR directly interacts with Jun-Fos at their binding targets through trans-binding effect as shown in Fig. 2b [2, 20, 23, 96, 97, 103, 112, 118], even though we still do not have direct evidence to confirm that RAR can bind to AP1 *in trans* and repress its activation.

Fig. 2 Models of ligand-dependent transrepression by nuclear receptors. **a** Transrepression by liganded PPARγ. Shown is a model of liganded PPARγ that is SUMOylated (Su) with recruited nuclear corepressors. Here, the complex is shown inhibiting NK-κB gene activation. **b** Transrepression by liganded RAR. In this case, the SUMOylated RAR complex brings nuclear receptor corepressors (*NCoR*) to AP1 and represses AP1 target gene activation

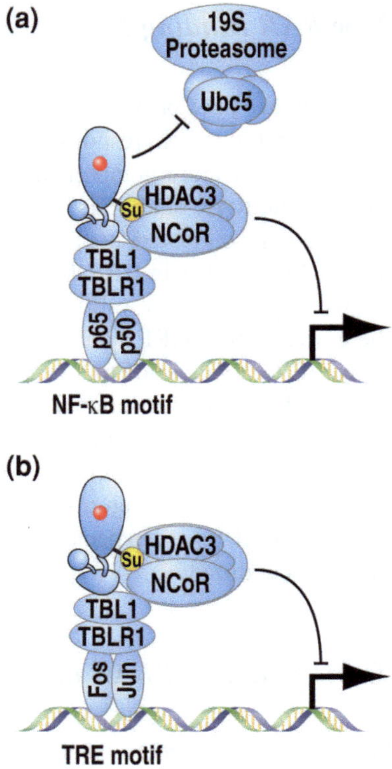

Current State of the Field: Deeper Understanding of RAR-Mediated Transcriptional Regulation

RAR Regulates Both Pol II and Pol III Transcriptional Programs

It is well documented that liganded RAR induces the expression of a cohort of the RNA Pol II-transcribed protein-coding genes [75], exemplified by the Hox genes, which are critically involved in development. Many of these genes harbor typical RAREs featuring a direct repeat cassette, DR2 or DR5, at their promoter or enhancer regions for efficient activation.

Bioinformatics analysis of over one million copies of human Alu repeats revealed that some of these repetitive elements contain canonical motifs for many TFs and NRs, such as NF-κB, RAR, ER and TR [90]. In particular, around one tenth of the human Alu repeats contain the DR2 cassette for RAR recognition and binding within the B box [59]. Given that the A/B boxes constitute an internal promoter for Pol III, it is possible that RA might trigger RAR to bind to the embedded DR2 cassette and thus drive Pol III-dependent transcription of this class of Alu repeats.

To further study how the RA/RAR signal regulates Pol III-mediated Alu repeat transcription, we have taken advantage of the RA-induced stem cell differentiation model. In human embryonic carcinoma "stem" cells, Ntera2/D1, and in human embryonic stem cells (H9), it has been found that *at*RA treatment dramatically enhanced the level of DR2 Alu transcripts [46]. By knocking down Pol III or blocking TF3C, it was confirmed that the *at*RA-induced DR2 Alu transcription was Pol III-dependent. It was also found that RAR, together with NCoR, binds to these DR2 Alu repeats in the absence of ligand, and that the corepressor complexes were dismissed upon RA treatment. The non-coding DR2 Alu transcripts were transported into the cytoplasm and became colocalized with the P bodies, the cytoplasmic machinery that contains Dicer and Argonaute (Ago) proteins and acts as RNA-processing hubs. It was also observed that the DR2 Alu transcripts were processed in the P bodies into a heterogeneously-sized population of RA-induced small (~30–65 nt) RNAs (riRNAs), initially requiring an unexpected, Dicer-dependent step.

To explore the biological function of riRNAs, bioinformatics analysis was performed to determine if riRNAs, like microRNAs, could potentially target complementary sequence in the 3′UTRs of a subunit of ES cell-expressed mRNAs, including those critical for stem cell maintenance, such as NANOG and TDGF-1. It was found that the treatment with *at*RA decreased the transcript levels of these genes in Ntera2 cells, and that the overexpression of DR2 Alu or riRNAs dramatically down-regulated these targets. And instead of initiating mRNA processing from the 3′ terminus as is the case for microRNA-mediated post-transcriptional regulation, riRNAs and associated Argonaute3 (AGO3) protein recruit decapping proteins, DCP1A and DCP2, to execute exonuclease cleavage from the 5′ terminus of targeted mRNAs. Thus, a new functional mechanism for RAR has been uncovered, in which the RAR and Pol III dependent DR2 Alu transcriptional events in stem cells functionally complement the Pol II-dependent neuronal transcriptional program (Fig. 3). This regulatory event provides a mechanism that helps to clear stem cell commitment transcripts, as RA induces the coding gene transcripts required for differentiation, and facilitates exit from the stem cell state. It is likely that other subsets of ALU repeats exert biological functions in many more differentiated cell types, and may have roles in cancer and aging.

It has also been found that the RA-inducible DR2 Alu repeats appear to be located close to (<10 kb) active Pol II transcription units, suggesting that there might be a critical architectural chromatin "domain" adjacent to active Pol II-transcribed coding gene loci required for the effective RA induction of Alu repeats by Pol III.

Cross-talk Between the RAR-Mediated Transcriptional Program and the Estrogen/Estrogen Receptor Pathway

The function of RAR-mediated transcriptional regulation in breast cancer, especially in estrogen receptor positive (ER+) breast cancer, has been the focus of several groups. The Jason Carroll and Kevin White laboratory groups set out to identify RAR genomic targets using ChIP and microarray gene expression

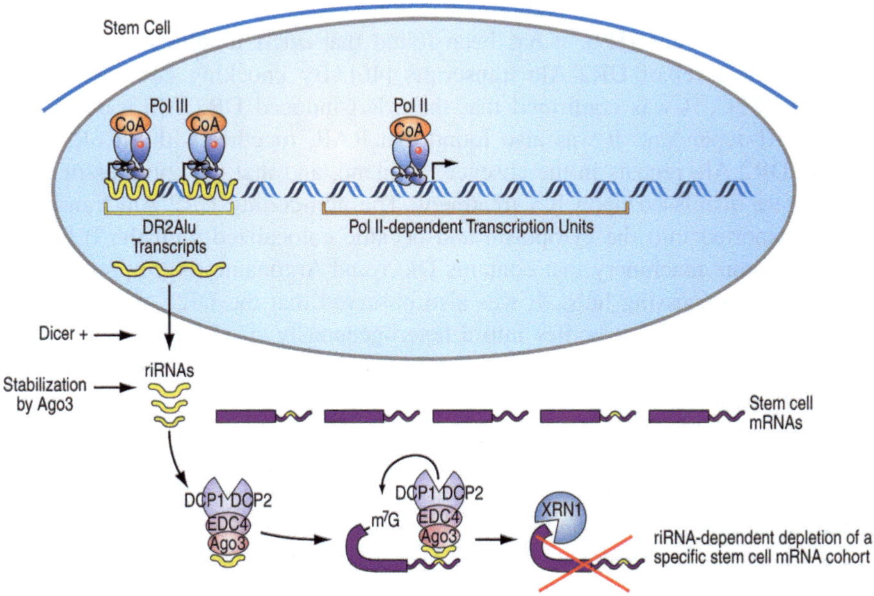

Fig. 3 Schematic diagram showing a proposed mechanism for Pol III transcriptional activation of a subclass of Alu repeats, referred to as DR2 Alu repeats, by the actions of liganded RAR in embryonic stem cells (*ESC*). Our recent studies have revealed that RAR has two transcription programs. One is the conventional RNA Pol II-driven program. The other is dependent on Pol III, in which liganded RAR, together with associated coactivators (*CoA*), drives the transcription of human DR2 Alu repeats. The resultant non-coding Alu RNAs are transported into the cytoplasm where they are processed into a new type of small RNAs, riRNAs, in a DICER-dependent manner. riRNAs require AGO3 to efficiently bind to complementary sequences in the 3′UTRs of many key stem cell mRNAs, which leads to recruitment of decapping complexes containing DCP1 and DCP2, resulting in the degradation of mRNAs by exonuclease XRN1

analysis in an MCF7 breast cancer cell line [48, 95]. Both studies shed light on how RAR regulates gene expression together with another dominant hormone signaling estrogen in breast cancer cells. However, although both groups found a large number of common binding sites for RAR and ERα, they came to totally different conclusions. One paper [95] suggests that RARα functions cooperatively with ERα to regulate the loading of coactivators at ERα enhancers [95], while the other paper [48] proposed that RARs and ERα actually compete to bind to common regulatory elements, thus mediating the genomic antagonism between RA and estrogen signaling in breast cancer [48].

The Carroll group found that estrogen induced RARα expression and that RARα was required for estrogen-induced growth of the MCF7 breast cancer cells [95]. They further performed RARα ChIP-sequencing in the MCF7 line. Their data indicate that RARα exhibits substantial co-occupancy with ERα at the genome-wide level, and that knockdown of ERα expression reduced RARα's binding at approximately half of these co-bound sites, suggesting a functional interaction between ERα and RARα. On the other hand, knockdown of RARα did not affect ERα binding, but did alter coactivator binding, such as p300, and affected histone H3 acetylation and RNA Pol II loading at the promoter regions of ERα targets. Hence, these authors hypothesized that besides its classic role as a heterodimeric partner of RXR proteins that respond to natural ligands such as RA, RARα can function cooperatively as an ERα-associated protein for maintaining ERα-cofactor interaction during estrogen-mediated gene transcription. Thus, the addition of RA ligand can competitively trigger the classic RARα role and inhibit estrogen target genes by affecting estrogen-ERα function. These data explain how RARα ligand can be used for an effective treatment, as well as provide a rationale for why both RARα agonists and antagonists inhibit breast cancer in animal models and preclinical trials.

The White group used GFP tag technology to map RARα and RARγ in MCF7 cell by ChIP-chip. They found that both RARα and RARγ binding is highly coincident with ERα [48]. Their gene expression data suggested: (1) Liganded RAR can both activate and repress different gene targets, while traditional view proposed that repressive function is mediated solely by unliganded RAR; and (2) The co-occupied RAR/ERα binding sites mediated the antagonistic actions between RA and estrogen on gene regulation. In contrast to the Carroll group's finding, their data suggested that instead of simultaneously binding to common sites, RARα and ERα compete to bind these sites. They also reported that FoxA1 and GATA3 TFs were recruited at RAR/ERα bound enhancers. Surprisingly, knockdown of FoxA1 affects the binding of RARs at these common binding sites, suggesting that RARs also function cooperatively with FoxA1 to gain access to their binding sites on enhancers or promoters.

The different conclusions of these studies in the breast cancer cell line raise many questions and we expect that additional studies will emerge to further characterize the functional interaction between RAR and ERα. Indeed, although the exact nature of the interaction between RA/RAR and estrogen/ERα in breast cancer cell line MCF7 is still obscure, some pilot studies, such as the genome-wide profiling of RAR and ERα bound sites, suggest a possible broad trans-binding between RAR or ERα. It will be instructive to further explore an appropriate working model to validate the RAR/ERα interaction in breast cancer cells, and to develop novel therapies for RA-resistant tumors.

Relevance: RAR-Mediated Transcriptional Regulation and Disease

Understanding the basic principles of gene regulation by NRs, exemplified by RAR, has particular importance for designing strategies that can ultimately alter transcriptional programs in development, homeostasis, disease, aging and DNA damage repair. For example, the realization of the critical roles of enhancers in NR transcriptional programs provides motivation for investigation of new strategies to block function of cell-type specific enhancers, perhaps by novel mutation or anti-eRNA approaches. Retinoids, through binding to its NRs (RAR), are physiological regulators of embryonic development, tissue homeostasis and cell differentiation, as well as mediating apoptosis and proliferation [7]. Because of their inhibitory effects on breast cancer cell lines and suppression of carcinogenesis in experimental animal models, retinoids occupy a prominent position among the chemopreventive agents that have been examined in preclinical studies and clinical trials [114, 115]. However, the clinical trials of retinoids in patients with advanced breast cancer were not as successful as initially expected. Thus, it is of prime importance to study the molecular mechanism of RAR-mediated transcriptional regulation in cancers, and the roles played by the three types of retinoic acid receptors in various cell types, to permit more effective strategies for harnessing the potential anti-cancer effects of retinoids. Even understanding at a molecular level why binding of retinoic acid receptors to some enhancers activate their target coding genes, while binding to other enhancers results in repression of their target coding genes, will provide new approaches to fine tuning these events in both health and disease.

This period of intensive investigation has undoubtedly pointed to a surprisingly large series of cofactors as critical components in the RAR signaling program under both the physiological and pathological conditions. In particular, the functional study of corepressors promises to enhance our understanding of the inefficiency of therapeutic application of RA in different cancer diseases. One example of such a cofactor is the human tumor antigen PRAME [18, 26, 27, 85]. Studies show that PRAME functions as a dominant repressor of RAR signaling by binding to RAR in the presence of RA and preventing ligand induced receptor activation through recruitment of Polycomb proteins [26]. Thus PRAME inhibits RA-induced differentiation, growth arrest, and apoptosis. Knockdown of PRAME expression by RNA interference in RA-resistant human melanoma restores RAR signaling and reinstates the sensitivity of tumor cells to the anti-proliferative effects of RA both in vitro and in vivo.

Future Directions: Future Questions on RAR-Mediated Transcriptional Regulation

With the current genome-wide profiling and interactome characterization, we can expect to see an ever-growing body of cofactors for the RAR program, including additional enzymes, ncRNAs, and other non-conventional RAR corepressors/coactivators. It will also be important to learn more about how DNA damage repair components, in concert with known coactivators at RAR enhancer and promoter sites, function in control of regulated transcription, looping and gene activation.

By harnessing the power of contemporary sequencing technologies, we are rapidly accumulating knowledge and gaining insight into how RAR mediates transcriptional regulation at a genome-wide level. We expect to see in-depth studies on

RAR function in different development, and disease models. These insights will, of course, answer many critical questions concerning normal development and pathological conditions in human, including:

1. How does liganded RAR function for both activation and repression as reported by recent genome-wide studies in breast cancer cell lines?
2. Does RAR globally use different types of RARE information to determine its function and to recruit different cofactors?
3. Does RAR act globally through trans-binding with other TFs by protein-protein interaction, and does the outcome require new functions of its DNA binding domain?

The era of molecular biology has brought us to a deep understanding of the biological roles and mechanisms of retinoic acid receptor function. In the near future, the era of global genomics will rapidly and significantly extend our knowledge for a clearer understanding of both the uniform and the distinct ways in which different cohorts of RA-regulated transcription units are transcribed.

Acknowledgments We thank members in the Rosenfeld laboratory for suggestions and comments on this manuscript, and the editing efforts of Rachel Pardee. M.G.R. is an investigator with HHMI.

References

1. Belikov S, Holmqvist PH, Astrand C, Wrange O (2012) FoxA1 and glucocorticoid receptor crosstalk via histone H4K16 acetylation at a hormone regulated enhancer. Exp Cell Res 318:61–74
2. Benkoussa M, Brand C, Delmotte MH, Formstecher P, Lefebvre P (2002) Retinoic acid receptors inhibit AP1 activation by regulating extracellular signal-regulated kinase and CBP recruitment to an AP1-responsive promoter. Mol Cell Biol 22:4522–4534
3. Birney E, Stamatoyannopoulos JA, Dutta A, Guigo R, Gingeras TR, Margulies EH, Weng Z, Snyder M, Dermitzakis ET, Thurman RE et al (2007) Identification and analysis of functional elements in 1 % of the human genome by the ENCODE pilot project. Nature 447:799–816
4. Blanco JC, Minucci S, Lu J, Yang XJ, Walker KK, Chen H, Evans RM, Nakatani Y, Ozato K (1998) The histone acetylase PCAF is a nuclear receptor coactivator. Genes Dev 12:1638–1651
5. Brown K, Chen Y, Underhill TM, Mymryk JS, Torchia J (2003) The coactivator p/CIP/SRC-3 facilitates retinoic acid receptor signaling via recruitment of GCN5. J Biol Chem 278:39402–39412
6. Chakravarti D, LaMorte VJ, Nelson MC, Nakajima T, Schulman IG, Juguilon H, Montminy M, Evans RM (1996) Role of CBP/P300 in nuclear receptor signalling. Nature 383:99–103
7. Chambon P (1996) A decade of molecular biology of retinoic acid receptors. Faseb J 10:940–954
8. Chen D, Ma H, Hong H, Koh SS, Huang SM, Schurter BT, Aswad DW, Stallcup MR (1999) Regulation of transcription by a protein methyltransferase. Science 284:2174–2177
9. Chen JD, Evans RM (1995) A transcriptional co-repressor that interacts with nuclear hormone receptors. Nature 377:454–457

10. Chen JY, Penco S, Ostrowski J, Balaguer P, Pons M, Starrett JE, Reczek P, Chambon P, Gronemeyer H (1995) RAR-specific agonist/antagonists which dissociate transactivation and AP1 transrepression inhibit anchorage-independent cell proliferation. EMBO J 14:1187–1197
11. Cheng J, Kapranov P, Drenkow J, Dike S, Brubaker S, Patel S, Long J, Stern D, Tammana H, Helt G et al (2005) Transcriptional maps of 10 human chromosomes at 5-nucleotide resolution. Science 308:1149–1154
12. Chepelev I, Wei G, Wangsa D, Tang Q, Zhao K (2012) Characterization of genome-wide enhancer-promoter interactions reveals co-expression of interacting genes and modes of higher order chromatin organization. Cell Res 22:490–503
13. Cho YS, Kim EJ, Park UH, Sin HS, Um SJ (2006) Additional sex comb-like 1 (ASXL1), in cooperation with SRC-1, acts as a ligand-dependent coactivator for retinoic acid receptor. J Biol Chem 281:17588–17598
14. Chu C, Qu K, Zhong FL, Artandi SE, Chang HY (2011) Genomic maps of long noncoding RNA occupancy reveal principles of RNA-chromatin interactions. Mol Cell 44:667–678
15. Colley SM, Leedman PJ (2011) Steroid receptor RNA activator—A nuclear receptor coregulator with multiple partners: insights and challenges. Biochimie 93:1966–1972
16. Cooper C, Vincett D, Yan Y, Hamedani MK, Myal Y, Leygue E (2011) Steroid receptor RNA activator bi-faceted genetic system: heads or tails? Biochimie 93:1973–1980
17. Danko CG, Hah N, Luo X, Martins AL, Core L, Lis JT, Siepel A, Kraus WL (2013) Signaling pathways differentially affect RNA polymerase II initiation, pausing, and elongation rate in cells. Mol Cell 50:212–222
18. De Carvalho DD, Binato R, Pereira WO, Leroy JM, Colassanti MD, Proto-Siqueira R, Bueno-Da-Silva AE, Zago MA, Zanichelli MA, Abdelhay E et al (2011) BCR-ABL-mediated upregulation of PRAME is responsible for knocking down TRAIL in CML patients. Oncogene 30:223–233
19. de The H, Vivanco-Ruiz MM, Tiollais P, Stunnenberg H, Dejean A (1990) Identification of a retinoic acid responsive element in the retinoic acid receptor beta gene. Nature 343:177–180
20. Dedieu S, Lefebvre P (2006) Retinoids interfere with the AP1 signalling pathway in human breast cancer cells. Cell Signal 18:889–898
21. Delacroix L, Moutier E, Altobelli G, Legras S, Poch O, Choukrallah MA, Bertin I, Jost B, Davidson I (2010) Cell-specific interaction of retinoic acid receptors with target genes in mouse embryonic fibroblasts and embryonic stem cells. Mol Cell Biol 30:231–244
22. Dilworth FJ, Fromental-Ramain C, Yamamoto K, Chambon P (2000) ATP-driven chromatin remodeling activity and histone acetyltransferases act sequentially during transactivation by RAR/RXR In vitro. Mol Cell 6:1049–1058
23. DiSepio D, Sutter M, Johnson AT, Chandraratna RA, Nagpal S (1999) Identification of the AP1-antagonism domain of retinoic acid receptors. Mol Cell Biol Res Commun: MCBRC 1:7–13
24. Durand B, Saunders M, Leroy P, Leid M, Chambon P (1992) All-trans and 9-cis retinoic acid induction of CRABPII transcription is mediated by RAR-RXR heterodimers bound to DR1 and DR2 repeated motifs. Cell 71:73–85
25. Easwaran V, Pishvaian M, Salimuddin, Byers S (1999) Cross-regulation of beta-catenin-LEF/TCF and retinoid signaling pathways. Curr Biol 9:1415–1418
26. Epping MT, Wang L, Edel MJ, Carlee L, Hernandez M, Bernards R (2005) The human tumor antigen PRAME is a dominant repressor of retinoic acid receptor signaling. Cell 122:835–847
27. Epping MT, Wang L, Plumb JA, Lieb M, Gronemeyer H, Brown R, Bernards R (2007) A functional genetic screen identifies retinoic acid signaling as a target of histone deacetylase inhibitors. Proc Natl Acad Sci USA 104:17777–17782
28. Fanjul A, Dawson MI, Hobbs PD, Jong L, Cameron JF, Harlev E, Graupner G, Lu XP, Pfahl M (1994) A new class of retinoids with selective inhibition of AP-1 inhibits proliferation. Nature 372:107–111
29. Fernandes I, Bastien Y, Wai T, Nygard K, Lin R, Cormier O, Lee HS, Eng F, Bertos NR, Pelletier N et al (2003) Ligand-dependent nuclear receptor corepressor LCoR functions by histone deacetylase-dependent and -independent mechanisms. Mol Cell 11:139–150

30. Franco PJ, Li G, Wei LN (2003) Interaction of nuclear receptor zinc finger DNA binding domains with histone deacetylase. Mol Cell Endocrinol 206:1–12
31. Fujiki R, Chikanishi T, Hashiba W, Ito H, Takada I, Roeder RG, Kitagawa H, Kato S (2009) GlcNAcylation of a histone methyltransferase in retinoic-acid-induced granulopoiesis. Nature 459:455–459
32. Ghisletti S, Huang W, Ogawa S, Pascual G, Lin ME, Willson TM, Rosenfeld MG, Glass CK (2007) Parallel SUMOylation-dependent pathways mediate gene- and signal-specific transrepression by LXRs and PPARgamma. Mol Cell 25:57–70
33. Gianni M, Bauer A, Garattini E, Chambon P, Rochette-Egly C (2002) Phosphorylation by p38MAPK and recruitment of SUG-1 are required for RA-induced RAR gamma degradation and transactivation. EMBO J 21:3760–3769
34. Glass CK, Rosenfeld MG (2000) The coregulator exchange in transcriptional functions of nuclear receptors. Genes Dev 14:121–141
35. Gurevich I, Aneskievich BJ (2009) Liganded RARalpha and RARgamma interact with but are repressed by TNIP1. Biochem Biophys Res Commun 389:409–414
36. Guttman M, Amit I, Garber M, French C, Lin MF, Feldser D, Huarte M, Zuk O, Carey BW, Cassady JP et al (2009) Chromatin signature reveals over a thousand highly conserved large non-coding RNAs in mammals. Nature 458:223–227
37. Guttman M, Donaghey J, Carey BW, Garber M, Grenier JK, Munson G, Young G, Lucas AB, Ach R, Bruhn L et al (2011) lincRNAs act in the circuitry controlling pluripotency and differentiation. Nature 477:295–300
38. Hah N, Danko CG, Core L, Waterfall JJ, Siepel A, Lis JT, Kraus WL (2011) A rapid, extensive, and transient transcriptional response to estrogen signaling in breast cancer cells. Cell 145:622–634
39. Heintzman ND, Hon GC, Hawkins RD, Kheradpour P, Stark A, Harp LF, Ye Z, Lee LK, Stuart RK, Ching CW et al (2009) Histone modifications at human enhancers reflect global cell-type-specific gene expression. Nature 459:108–112
40. Heintzman ND, Ren B (2009) Finding distal regulatory elements in the human genome. Curr Opin Genet Dev 19:541–549
41. Heintzman ND, Stuart RK, Hon G, Fu Y, Ching CW, Hawkins RD, Barrera LO, Van Calcar S, Qu C, Ching KA et al (2007) Distinct and predictive chromatin signatures of transcriptional promoters and enhancers in the human genome. Nat Genet 39:311–318
42. Hermanson O, Glass CK, Rosenfeld MG (2002) Nuclear receptor coregulators: multiple modes of modification. Trends Endocrinol Metab 13:55–60
43. Hong SH, David G, Wong CW, Dejean A, Privalsky ML (1997) SMRT corepressor interacts with PLZF and with the PML-retinoic acid receptor alpha (RARalpha) and PLZF-RARalpha oncoproteins associated with acute promyelocytic leukemia. Proc Natl Acad Sci USA 94:9028–9033
44. Horlein AJ, Naar AM, Heinzel T, Torchia J, Gloss B, Kurokawa R, Ryan A, Kamei Y, Soderstrom M, Glass CK et al (1995) Ligand-independent repression by the thyroid hormone receptor mediated by a nuclear receptor co-repressor. Nature 377:397–404
45. Hou Z, Peng H, White DE, Negorev DG, Maul GG, Feng Y, Longmore GD, Waxman S, Zelent A, Rauscher FJ 3rd (2010) LIM protein Ajuba functions as a nuclear receptor corepressor and negatively regulates retinoic acid signaling. Proc Natl Acad Sci USA 107:2938–2943
46. Hu Q, Tanasa B, Trabucchi M, Li W, Zhang J, Ohgi KA, Rose DW, Glass CK, Rosenfeld MG (2012) DICER- and AGO3-dependent generation of retinoic acid-induced DR2 Alu RNAs regulates human stem cell proliferation. Nat Struct Mol Biol
47. Hu X, Chen Y, Farooqui M, Thomas MC, Chiang CM, Wei LN (2004) Suppressive effect of receptor-interacting protein 140 on coregulator binding to retinoic acid receptor complexes, histone-modifying enzyme activity, and gene activation. J Biol Chem 279:319–325
48. Hua S, Kittler R, White KP (2009) Genomic antagonism between retinoic acid and estrogen signaling in breast cancer. Cell 137:1259–1271
49. Huq MD, Ha SG, Wei LN (2008) Modulation of retinoic acid receptor alpha activity by lysine methylation in the DNA binding domain. J Proteome Res 7:4538–4545

50. Huq MD, Tsai NP, Khan SA, Wei LN (2007) Lysine trimethylation of retinoic acid receptor-alpha: a novel means to regulate receptor function. Mol Cell Proteomics: MCP 6:677–688
51. Kashyap V, Gudas LJ (2010) Epigenetic regulatory mechanisms distinguish retinoic acid-mediated transcriptional responses in stem cells and fibroblasts. J Biol Chem 285:14534–14548
52. Khetchoumian K, Teletin M, Tisserand J, Mark M, Herquel B, Ignat M, Zucman-Rossi J, Cammas F, Lerouge T, Thibault C et al (2007) Loss of Trim24 (Tif1alpha) gene function confers oncogenic activity to retinoic acid receptor alpha. Nat Genet 39:1500–1506
53. Kim JH, Lee JM, Nam HJ, Choi HJ, Yang JW, Lee JS, Kim MH, Kim SI, Chung CH, Kim KI et al (2007) SUMOylation of pontin chromatin-remodeling complex reveals a signal integration code in prostate cancer cells. Proc Natl Acad Sci USA 104:20793–20798
54. Kim TK, Hemberg M, Gray JM, Costa AM, Bear DM, Wu J, Harmin DA, Laptewicz M, Barbara-Haley K, Kuersten S et al (2010) Widespread transcription at neuronal activity-regulated enhancers. Nature 465:182–187
55. Lafyatis R, Kim SJ, Angel P, Roberts AB, Sporn MB, Karin M, Wilder RL (1990) Interleukin-1 stimulates and all-trans-retinoic acid inhibits collagenase gene expression through its 5′ activator protein-1-binding site. Mol Endocrinol 4:973–980
56. Lai F, Orom UA, Cesaroni M, Beringer M, Taatjes DJ, Blobel GA, Shiekhattar R (2013) Activating RNAs associate with Mediator to enhance chromatin architecture and transcription. Nature 494:497–501
57. Lam MT, Cho H, Lesch HP, Gosselin D, Heinz S, Tanaka-Oishi Y, Benner C, Kaikkonen MU, Kim AS, Kosaka M et al (2013) Rev-Erbs repress macrophage gene expression by inhibiting enhancer-directed transcription. Nature 498:511–515
58. Langlais D, Couture C, Balsalobre A, Drouin J (2012) The Stat3/GR interaction code: predictive value of direct/indirect DNA recruitment for transcription outcome. Mol Cell 47:38–49
59. Laperriere D, Wang TT, White JH, Mader S (2007) Widespread Alu repeat-driven expansion of consensus DR2 retinoic acid response elements during primate evolution. BMC Genom 8:23
60. Le May N, Iltis I, Ame JC, Zhovmer A, Biard D, Egly JM, Schreiber V, Coin F (2012) Poly (ADP-ribose) glycohydrolase regulates retinoic acid receptor-mediated gene expression. Mol Cell 48:785–798
61. Le May N, Fradin D, Iltis I, Bougneres P, Egly JM (2012) XPG and XPF endonucleases trigger chromatin looping and DNA demethylation for accurate expression of activated genes. Mol Cell 47:622–632
62. Le May N, Mota-Fernandes D, Velez-Cruz R, Iltis I, Biard D, Egly JM (2010) NER factors are recruited to active promoters and facilitate chromatin modification for transcription in the absence of exogenous genotoxic attack. Mol Cell 38:54–66
63. Lee HK, Park UH, Kim EJ, Um SJ (2007) MED25 is distinct from TRAP220/MED1 in cooperating with CBP for retinoid receptor activation. EMBO J 26:3545–3557
64. Lee S, Lee DK, Dou Y, Lee J, Lee B, Kwak E, Kong YY, Lee SK, Roeder RG, Lee JW (2006) Coactivator as a target gene specificity determinant for histone H3 lysine 4 methyltransferases. Proc Natl Acad Sci USA 103:15392–15397
65. Lee SW, Cho YS, Na JM, Park UH, Kang M, Kim EJ, Um SJ (2010) ASXL1 represses retinoic acid receptor-mediated transcription through associating with HP1 and LSD1. J Biol Chem 285:18–29
66. Leygue E (2007) Steroid receptor RNA activator (SRA1): unusual bifaceted gene products with suspected relevance to breast cancer. Nucl Receptor Signal 5:e006
67. Li HJ, Haque ZK, Chen A, Mendelsohn M (2007) RIF-1, a novel nuclear receptor corepressor that associates with the nuclear matrix. J Cell Biochem 102:1021–1035
68. Li W, Notani D, Ma Q, Tanasa B, Nunez E, Chen AY, Merkurjev D, Zhang J, Ohgi K, Song X et al (2013) Functional roles of enhancer RNAs for oestrogen-dependent transcriptional activation. Nature 498:516–520
69. Lin F, Xiao D, Kolluri SK, Zhang X (2000) Unique anti-activator protein-1 activity of retinoic acid receptor beta. Cancer Res 60:3271–3280

70. Lindahl T, Wood RD (1999) Quality control by DNA repair. Science 286:1897–1905
71. Liu W, Ma Q, Wong K, Li W, Ohgi K, Zhang J, Aggarwal AK, Rosenfeld MG (2013) Brd4 and JMJD6-associated anti-pause enhancers in regulation of transcriptional pause release. Cell 155:1581–1595
72. Loudig O, Babichuk C, White J, Abu-Abed S, Mueller C, Petkovich M (2000) Cytochrome P450RAI(CYP26) promoter: a distinct composite retinoic acid response element underlies the complex regulation of retinoic acid metabolism. Mol Endocrinol 14:1483–1497
73. Mahony S, Mazzoni EO, McCuine S, Young RA, Wichterle H, Gifford DK (2011) Ligand-dependent dynamics of retinoic acid receptor binding during early neurogenesis. Genome Biol 12:R2
74. Mangelsdorf DJ, Evans RM (1995) The RXR heterodimers and orphan receptors. Cell 83:841–850
75. Mangelsdorf DJ, Thummel C, Beato M, Herrlich P, Schutz G, Umesono K, Blumberg B, Kastner P, Mark M, Chambon P et al (1995) The nuclear receptor superfamily: the second decade. Cell 83:835–839
76. Mangelsdorf DJ, Umesono K, Kliewer SA, Borgmeyer U, Ong ES, Evans RM (1991) A direct repeat in the cellular retinol-binding protein type II gene confers differential regulation by RXR and RAR. Cell 66:555–561
77. Mengus G, May M, Carre L, Chambon P, Davidson I (1997) Human TAF(II)135 potentiates transcriptional activation by the AF-2s of the retinoic acid, vitamin D3, and thyroid hormone receptors in mammalian cells. Genes Dev 11:1381–1395
78. Moon M, Um SJ, Kim EJ (2012) CAC1 negatively regulates RARalpha activity through cooperation with HDAC. Biochem Biophys Res Commun 427:41–46
79. Nagy L, Kao HY, Chakravarti D, Lin RJ, Hassig CA, Ayer DE, Schreiber SL, Evans RM (1997) Nuclear receptor repression mediated by a complex containing SMRT, mSin3A, and histone deacetylase. Cell 89:373–380
80. Nicholson RC, Mader S, Nagpal S, Leid M, Rochette-Egly C, Chambon P (1990) Negative regulation of the rat stromelysin gene promoter by retinoic acid is mediated by an AP1 binding site. EMBO J 9:4443–4454
81. Nouspikel T (2009) DNA repair in mammalian cells: nucleotide excision repair: variations on versatility. Cell Mol Life Sci: CMLS 66:994–1009
82. Onate SA, Tsai SY, Tsai MJ, O'Malley BW (1995) Sequence and characterization of a coactivator for the steroid hormone receptor superfamily. Science 270:1354–1357
83. Pascual G, Fong AL, Ogawa S, Gamliel A, Li AC, Perissi V, Rose DW, Willson TM, Rosenfeld MG, Glass CK (2005) A SUMOylation-dependent pathway mediates transrepression of inflammatory response genes by PPAR-gamma. Nature 437:759–763
84. Pascual G, Sullivan AL, Ogawa S, Gamliel A, Perissi V, Rosenfeld MG, Glass CK (2007) Anti-inflammatory and antidiabetic roles of PPARgamma. Novartis Found Symp 286:183–196; discussion 196–203
85. Passeron T, Valencia JC, Namiki T, Vieira WD, Passeron H, Miyamura Y, Hearing VJ (2009) Upregulation of SOX9 inhibits the growth of human and mouse melanomas and restores their sensitivity to retinoic acid. J Clin Invest 119:954–963
86. Perissi V, Aggarwal A, Glass CK, Rose DW, Rosenfeld MG (2004) A corepressor/coactivator exchange complex required for transcriptional activation by nuclear receptors and other regulated transcription factors. Cell 116:511–526
87. Perissi V, Jepsen K, Glass CK, Rosenfeld MG (2010) Deconstructing repression: evolving models of co-repressor action. Nat Rev Genet 11:109–123
88. Perissi V, Rosenfeld MG (2005) Controlling nuclear receptors: the circular logic of cofactor cycles. Nat Rev Mol Cell Biol 6:542–554
89. Perissi V, Scafoglio C, Zhang J, Ohgi KA, Rose DW, Glass CK, Rosenfeld MG (2008) TBL1 and TBLR1 phosphorylation on regulated gene promoters overcomes dual CtBP and NCoR/SMRT transcriptional repression checkpoints. Mol Cell 29:755–766
90. Polak P, Domany E (2006) Alu elements contain many binding sites for transcription factors and may play a role in regulation of developmental processes. BMC Genom 7:133

91. Qiu J, Shi G, Jia Y, Li J, Wu M, Dong S, Wong J (2010) The X-linked mental retardation gene PHF8 is a histone demethylase involved in neuronal differentiation. Cell Res 20:908–918
92. Resche-Rigon M, Gronemeyer H (1998) Therapeutic potential of selective modulators of nuclear receptor action. Curr Opin Chem Biol 2:501–507
93. Rochette-Egly C, Plassat JL, Taneja R, Chambon P (2000) The AF-1 and AF-2 activating domains of retinoic acid receptor-alpha (RARalpha) and their phosphorylation are differentially involved in parietal endodermal differentiation of F9 cells and retinoid-induced expression of target genes. Mol Endocrinol 14:1398–1410
94. Rosenfeld MG, Lunyak VV, Glass CK (2006) Sensors and signals: a coactivator/corepressor/epigenetic code for integrating signal-dependent programs of transcriptional response. Genes Dev 20:1405–1428
95. Ross-Innes CS, Stark R, Holmes KA, Schmidt D, Spyrou C, Russell R, Massie CE, Vowler SL, Eldridge M, Carroll JS (2010) Cooperative interaction between retinoic acid receptor-alpha and estrogen receptor in breast cancer. Genes Dev 24:171–182
96. Schule R, Rangarajan P, Yang N, Kliewer S, Ransone LJ, Bolado J, Verma IM, Evans RM (1991) Retinoic acid is a negative regulator of AP-1-responsive genes. Proc Natl Acad Sci USA 88:6092–6096
97. Schule R, Umesono K, Mangelsdorf DJ, Bolado J, Pike JW, Evans RM (1990) Jun-Fos and receptors for vitamins A and D recognize a common response element in the human osteocalcin gene. Cell 61:497–504
98. Shao W, Halachmi S, Brown M (2002) ERAP140, a conserved tissue-specific nuclear receptor coactivator. Mol Cell Biol 22:3358–3372
99. Shao W, Rosenauer A, Mann K, Chang CP, Rachez C, Freedman LP, Miller WH Jr (2000) Ligand-inducible interaction of the DRIP/TRAP coactivator complex with retinoid receptors in retinoic acid-sensitive and—resistant acute promyelocytic leukemia cells. Blood 96:2233–2239
100. Surjit M, Ganti KP, Mukherji A, Ye T, Hua G, Metzger D, Li M, Chambon P (2011) Widespread negative response elements mediate direct repression by agonist-liganded glucocorticoid receptor. Cell 145:224–241
101. Si J, Mueller L, Collins SJ (2007) CaMKII regulates retinoic acid receptor transcriptional activity and the differentiation of myeloid leukemia cells. J Clin Invest 117:1412–1421
102. Smith WC, Nakshatri H, Leroy P, Rees J, Chambon P (1991) A retinoic acid response element is present in the mouse cellular retinol binding protein I (mCRBPI) promoter. EMBO J 10:2223–2230
103. Suzukawa K, Colburn NH (2002) AP-1 transrepressing retinoic acid does not deplete coactivators or AP-1 monomers but may target specific Jun or Fos containing dimers. Oncogene 21:2181–2190
104. Tsai MC, Manor O, Wan Y, Mosammaparast N, Wang JK, Lan F, Shi Y, Segal E, Chang HY (2010) Long noncoding RNA as modular scaffold of histone modification complexes. Science 329:689–693
105. Torchia J, Rose DW, Inostroza J, Kamei Y, Westin S, Glass CK, Rosenfeld MG (1997) The transcriptional co-activator p/CIP binds CBP and mediates nuclear-receptor function. Nature 387:677–684
106. Venteclef N, Jakobsson T, Ehrlund A, Damdimopoulos A, Mikkonen L, Ellis E, Nilsson LM, Parini P, Janne OA, Gustafsson JA et al (2010) GPS2-dependent corepressor/SUMO pathways govern anti-inflammatory actions of LRH-1 and LXRbeta in the hepatic acute phase response. Genes Dev 24:381–395
107. Visel A, Blow MJ, Li Z, Zhang T, Akiyama JA, Holt A, Plajzer-Frick I, Shoukry M, Wright C, Chen F et al (2009) ChIP-seq accurately predicts tissue-specific activity of enhancers. Nature 457:854–858
108. Voegel JJ, Heine MJ, Zechel C, Chambon P, Gronemeyer H (1996) TIF2, a 160 kDa transcriptional mediator for the ligand-dependent activation function AF-2 of nuclear receptors. EMBO J 15:3667–3675

109. Wang D, Garcia-Bassets I, Benner C, Li W, Su X, Zhou Y, Qiu J, Liu W, Kaikkonen MU, Ohgi KA et al (2011) Reprogramming transcription by distinct classes of enhancers functionally defined by eRNA. Nature 474:390–394
110. Wang KC, Yang YW, Liu B, Sanyal A, Corces-Zimmerman R, Chen Y, Lajoie BR, Protacio A, Flynn RA, Gupta RA et al (2011) A long noncoding RNA maintains active chromatin to coordinate homeotic gene expression. Nature 472:120–124
111. Wang Z, Zang C, Cui K, Schones DE, Barski A, Peng W, Zhao K (2009) Genome-wide mapping of HATs and HDACs reveals distinct functions in active and inactive genes. Cell 138:1019–1031
112. Yang-Yen HF, Zhang XK, Graupner G, Tzukerman M, Sakamoto B, Karin M, Pfahl M (1991) Antagonism between retinoic acid receptors and AP-1: implications for tumor promotion and inflammation. New Biol 3:1206–1219
113. Yang L, Lin C, Liu W, Zhang J, Ohgi KA, Grinstein JD, Dorrestein PC, Rosenfeld MG (2011) ncRNA- and Pc2 methylation-dependent gene relocation between nuclear structures mediates gene activation programs. Cell 147:773–788
114. Yang Q, Sakurai T, Kakudo K (2002) Retinoid, retinoic acid receptor beta and breast cancer. Breast Cancer Res Treat 76:167–173
115. Zhang XK, Liu Y, Lee MO (1996) Retinoid receptors in human lung cancer and breast cancer. Mutat Res 350:267–277
116. Zhao X, Patton JR, Davis SL, Florence B, Ames SJ, Spanjaard RA (2004) Regulation of nuclear receptor activity by a pseudouridine synthase through posttranscriptional modification of steroid receptor RNA activator. Mol Cell 15:549–558
117. Zhao X, Patton JR, Ghosh SK, Fischel-Ghodsian N, Shen L, Spanjaard RA (2007) Pus3p- and Pus1p-dependent pseudouridylation of steroid receptor RNA activator controls a functional switch that regulates nuclear receptor signaling. Mol Endocrinol 21:686–699
118. Zhou XF, Shen XQ, Shemshedini L (1999) Ligand-activated retinoic acid receptor inhibits AP-1 transactivation by disrupting c-Jun/c-Fos dimerization. Mol Endocrinol 13:276–285

Subject Index

M. A. Asson-Batres and C. Rochette-Egly (eds.), *The Biochemistry of Retinoic Acid Receptors I: Structure, Activation, and Function at the Molecular Level*, Subcellular Biochemistry 70, DOI: 10.1007/978-94-017-9050-5, © Springer Science+Business Media Dordrecht 2014

MIX
Papier aus verantwortungsvollen Quellen
Paper from responsible sources
FSC® C105338

If you have any concerns about our products,
you can contact us on
ProductSafety@springernature.com

In case Publisher is established outside the EU,
the EU authorized representative is:
Springer Nature Customer Service Center GmbH
Europaplatz 3, 69115 Heidelberg, Germany

Printed by Libri Plureos GmbH
in Hamburg, Germany